Ergebnisse der Mathematik
und ihrer Grenzgebiete

Volume 51

3. Folge

A Series of Modern Surveys
in Mathematics

Boris Khesin • Robert Wendt

The Geometry of Infinite-Dimensional Groups

 Springer

Boris Khesin
Robert Wendt
Department of Mathematics
University of Toronto
40 St. George Street
Toronto, ON
Canada M5S 2E4
e-mail: khesin@math.toronto.edu and
        rwendt@math.toronto.edu

ISBN 978-3-540-77262-0   e-ISBN 978-3-540-77263-7

DOI 10.1007/978-3-540-77263-7

Ergebnisse der Mathematik und ihrer Grenzgebiete. 3. Folge / A Series of Modern
Surveys in Mathematics ISSN 0071-1136

Library of Congress Control Number: 2008932850

Mathematics Subject Classification (2000): 22E65, 37K05, 58B25, 53D30

Typesetting: by the author using a Springer TEX macro package
Production: LE-TEX Jelonek, Schmidt & Vöckler GbR, Leipzig
Cover design: WMX Design GmbH, Heidelberg

Printed on acid-free paper

9 8 7 6 5 4 3 2 1

springer.com

To our Teachers:

to Vladimir Igorevich Arnold
and to the memory of Peter Slodowy

Preface

The aim of this monograph is to give an overview of various classes of infinite-dimensional Lie groups and their applications, mostly in Hamiltonian mechanics, fluid dynamics, integrable systems, and complex geometry. We have chosen to present the unifying ideas of the theory by concentrating on specific types and examples of infinite-dimensional Lie groups. Of course, the selection of the topics is largely influenced by the taste of the authors, but we hope that this selection is wide enough to describe various phenomena arising in the geometry of infinite-dimensional Lie groups and to convince the reader that they are appealing objects to study from both purely mathematical and more applied points of view. This book can be thought of as complementary to the existing more algebraic treatments, in particular, those covering the structure and representation theory of infinite-dimensional Lie algebras, as well as to more analytic ones developing calculus on infinite-dimensional manifolds.

This monograph originated from advanced graduate courses and mini-courses on infinite-dimensional groups and gauge theory given by the first author at the University of Toronto, at the CIRM in Marseille, and at the Ecole Polytechnique in Paris in 2001–2004. It is based on various classical and recent results that have shaped this newly emerged part of infinite-dimensional geometry and group theory.

Our intention was to make the book concise, relatively self-contained, and useful in a graduate course. For this reason, throughout the text, we have included a large number of problems, ranging from simple exercises to open questions. At the end of each section we provide bibliographical notes, trying to make the literature guide more comprehensive, in an attempt to bring the interested reader in contact with some of the most recent developments in this exciting subject, the geometry of infinite-dimensional groups. We hope that this book will be useful to both students and researchers in Lie theory, geometry, and Hamiltonian systems.

It is our pleasure to thank all those who helped us with the preparation of this manuscript. We are deeply indebted to our teachers, collaborators, and friends, who influenced our view of the subject: V. Arnold, Ya. Brenier, H. Bursztyn, Ya. Eliashberg, P. Etingof, V. Fock, I. Frenkel, D. Fuchs, A. Kirillov, F. Malikov, G. Misiołek, R. Moraru, N. Nekrasov, V. Ovsienko, C. Roger, A. Rosly, V. Rubtsov, A. Schwarz, G. Segal, M. Semenov-Tian-Shansky, A. Shnirelman, P. Slodowy, S. Tabachnikov, A. Todorov, A. Veselov, F. Wagemann, J. Weitsman, I. Zakharevich, and many others. We are particularly grateful to Alexei Rosly, the joint projects with whom inspired a large part, in particular the "application chapter," of this book, and who made numerous invaluable remarks on the manuscript. We thank the participants of the graduate courses for their stimulating questions and remarks. Our special thanks go to M. Peters and the Springer team for their invariable help and to D. Kramer for careful editing of the text.

We also acknowledge the support of the Max-Planck Institute in Bonn, the Institut des Hautes Etudes Scientifiques in Bures-sur-Yvette, the Clay Mathematics Institute, as well as the NSERC research grants. The work on this book was partially conducted during the period the first author was employed by the Clay Mathematics Institute as a Clay Book Fellow.

Finally, we thank our families (kids included!) for their tireless moral support and encouragement throughout the over-stretched work on the manuscript.

Contents

Introduction

What is a group? Algebraists teach that this is supposedly a set with two operations that satisfy a load of easily-forgettable axioms...

V.I. Arnold *"On teaching mathematics"* [20]

Today one cannot imagine mathematics and physics without Lie groups, which lie at the foundation of so many structures and theories. Many of these groups are of infinite dimension and they arise naturally in problems related to differential and algebraic geometry, knot theory, fluid dynamics, cosmology, and string theory. Such groups often appear as symmetries of various evolution equations, and their applications range from quantum mechanics to meteorology. Although infinite-dimensional Lie groups have been investigated for quite some time, the scope of applicability of a general theory of such groups is still rather limited. The main reason for this is that infinite-dimensional Lie groups exhibit very peculiar features.

Let us look at the relation between a Lie group and its Lie algebra as an example. As is well known, in finite dimensions each Lie group is, at least locally near the identity, completely described by its Lie algebra. This is achieved with the help of the exponential map, which is a local diffeomorphism from the Lie algebra to the Lie group itself. In infinite dimensions, this correspondence is no longer so straightforward. There may exist Lie groups that do not admit an exponential map. Furthermore, even if the exponential map exists for a given group, it may not be a local diffeomorphism. Another pathology in infinite dimensions is the failure of Lie's third theorem, stating that every finite-dimensional Lie algebra is the Lie algebra attached to some finite-dimensional Lie group. In contrast, there exist infinite-dimensional Lie algebras that do not correspond to any Lie group at all.

In order to avoid such pathologies, any version of a general theory of infinite-dimensional Lie groups would have to restrict its attention to certain classes of such groups and study them separately. For example, one might consider the class of Banach Lie groups, i.e., Lie groups that are locally modeled on Banach spaces and behave very much like finite-dimensional Lie groups. For Banach Lie groups the exponential map always exists and is a local diffeomorphism. However, restricting to Banach Lie groups would already exclude the important case of diffeomorphism groups, and so on. This is why the attempts to develop a unified theory of infinite-dimensional differential geometry, and hence, of infinite-dimensional Lie groups, are still far from reaching greater generality.

In the present book, we choose a different approach. Instead of trying to develop a general theory of such groups, we concentrate on various examples of infinite-dimensional Lie groups, which lead to a realm of important applications.

The examples we treat here mainly belong to three general types of infinite-dimensional Lie groups: groups of diffeomorphisms, gauge transformation groups, and groups of pseudodifferential operators. There are numerous interrelations between various groups appearing in this book. For example, the group of diffeomorphisms of a compact manifold acts naturally on the group of currents over this manifold. When this manifold is a circle, this action gives rise to a deep connection between the representation theory of the Virasoro algebra and the Kac–Moody algebras. In the geometric setting of this book, this relation manifests itself in the correspondence between the coadjoint orbits of these groups.

Another strand connecting various groups considered below is the theme of the "ladder" of current groups. We regard the passage from finite-dimensional Lie groups (i.e., "current groups at a point") to loop groups (i.e., current groups on the circle), and then to double loop groups (current groups on the two-dimensional torus) as a "ladder of groups." On the side of dynamical systems this is revealed in the passage from rational to trigonometric and to elliptic Calogero–Moser systems. The passage from ordinary loop groups to double loop groups also serves as the starting point of a "real–complex correspondence" discussed in the chapter on applications of groups. There we study moduli spaces of flat or integrable connections on real and complex surfaces using the geometry of coadjoint orbits of these two types of groups.

Most of main objects studied in the book can be summarized in the table below.

In Chapter II, in a sense, we are moving horizontally, along the first row of this table. We study affine and elliptic groups, their orbits and geometry, as well as the related Calogero–Moser systems. We also describe in this chapter many Lie groups and Lie algebras outside the scope of this table: groups of diffeomorphisms, the Virasoro group, groups of pseudodifferential operators. In the appendices one can find the Krichever–Novikov algebras, $\mathfrak{gl}_\infty$, and other related objects.

In Chapter III we move vertically in this table and mostly focus on the current groups and on their parallel description in topological and holomorphic contexts. While affine and elliptic Lie groups correspond to the base dimension

Base dimension	Real / topological theory	Complex / holomorphic theory
1	affine (or, loop) groups (orbits ~ monodromies over a circle)	elliptic (or, double loop) groups (orbits ~ holomorphic bundles over an elliptic curve)
2	flat connections over a Riemann surface (Poisson structures)	holomorphic bundles over a complex surface (holomorphic Poisson structures)
3	connections over a threefold (Chern–Simons functional, singular homology, classical linking)	partial connections over a complex threefold (holomorphic Chern–Simons functional, polar homology, holomorphic linking)

1, either real or complex, in dimension 2 we describe the spaces of connections on real or complex surfaces, as well as the symplectic and Poisson structures on the corresponding moduli spaces. (In the table the main focus of study is mentioned in the parentheses of the corresponding block.) In dimension 3 the study of the Chern–Simons functional and its holomorphic version leads one to the notions of classical and holomorphic linking, and to the corresponding homology theories. (Although we confined ourselves to three dimensions, one can continue this table to dimension 4 and higher, which brings in the Yang–Mills and many other interesting functionals; see, e.g., [85].)

Note that the objects (groups, connections, etc.) in each row of this table usually dictate the structure of objects in the row above it, although the "interaction of the rows" is different in the real and complex cases. Namely, in the real setting, the lower-dimensional manifolds appear as the boundary of real manifolds of one dimension higher. For the complex case, the low-dimensional complex varieties arise as divisors in higher-dimensional ones; see details in Chapter III.

Overview of the content. Here are several details on the contents of various chapters and sections.

In Chapter I, we recall some notions and facts from Lie theory and symplectic geometry used throughout the book. Starting with the definition of a Lie group, we review the main related concepts of its Lie algebra, the adjoint and coadjoint representations, and introduce central extensions of Lie groups and algebras. We then recall some notions from symplectic geometry, including Arnold's formulation of the Euler equations on a Lie group, which are the equations for the geodesic flow with respect to a one-sided invariant metric on the group. This setting allows one to describe on the same footing many finite- and infinite-dimensional dynamical systems, including the classical Euler equations for both a rigid body and an ideal fluid, the Korteweg–de Vries equation, and the equations of magnetohydrodynamics. Finally, the preliminaries cover the Marsden–Weinstein Hamiltonian reduction, a method often used to describe complicated Hamiltonian systems starting with a simple one on a nonreduced space, by "dividing out" extra symmetries of the system.

Chapter II is the main part of this book, and can be viewed as a walk through the zoo of the various types of infinite-dimensional Lie groups. We tried to describe these groups by presenting their definitions, possible explicit constructions, information on (or, in some cases, even the complete classification of) their coadjoint orbits. We also discuss relations of these groups to various Hamiltonian systems, elaborating, whenever possible, on important constructions related to integrability of such systems. The table of contents is rather self-explanatory.

We start this chapter by introducing the loop group of a compact Lie group, one of the most studied types of infinite-dimensional groups. In Section 1, we construct its universal central extension, the corresponding Lie algebra (called the affine Kac–Moody Lie algebra), and classify the corresponding coadjoint orbits. We also return to discuss the relation of this Lie algebra to the Landau–Lifschitz equation and the Calogero–Moser integrable system in the later sections.

In Section 2 we turn to the group of diffeomorphisms of the circle and its Lie algebra of smooth vector fields. Both the group and the Lie algebra admit universal central extensions, called the Virasoro–Bott group and the Virasoro algebra respectively. It turns out that the coadjoint orbits of the Virasoro–Bott group can be classified in a manner similar to that for the orbits of the loop groups. The Euler equation for a natural right-invariant metric on the Virasoro–Bott group is the famous Korteweg–de Vries (KdV) equation, which describes waves in shallow water. Furthermore, the Euler nature of the KdV helps one to show that this equation is completely integrable.

Section 3 is devoted to various diffeomorphism groups and, in particular, to the group of volume-preserving diffeomorphisms of a compact Riemannian manifold M. The Euler equations on this group are the Euler equations for an ideal incompressible fluid filling M. Enlarging the group of volume-preserving diffeomorphisms by either smooth functions or vector fields on M gives the Euler equations of gas dynamics or of magnetohydrodynamics, respectively. We also mention some results on the Riemannian geometry of diffeomorphism groups and discuss the relation of the latter to the the Marsden–Weinstein symplectic structure on the space of immersed curves in $\mathbb{R}^3$.

Section 4 deals with the group of pseudodifferential symbols (or operators) on the circle. It turns out that this group can be endowed with the structure of a Poisson Lie group, where the corresponding Poisson structures are given by the Adler–Gelfand–Dickey brackets. The dynamical systems naturally corresponding to this group are the Kadomtsev–Petviashvili hierarchy, the higher n-KdV equations, and the nonlinear Schrödinger equation.

Section 5 returns to the loop groups "at the next level": here we deal with their generalizations, elliptic Lie groups and the corresponding Lie algebras. These groups are extensions of the groups of double loops, i.e., the groups of smooth maps from a two-dimensional torus to a finite-dimensional complex

Lie group. The central extension of such a group relies on the choice of complex structure on this torus (i.e., on the choice of the underlying elliptic curve). The coadjoint orbits of the elliptic Lie groups can be classified in terms of holomorphic principal bundles over the elliptic curve.

This section also unifies several classes of the groups considered earlier in the light of an application to the Calogero–Moser systems. It turns out that the integrable types of potentials in these systems (rational, trigonometric, and elliptic ones) can be obtained, respectively, from the finite-dimensional semisimple Lie algebras, the affine algebras, and the elliptic Lie algebras by Hamiltonian reductions.

Chapter III deals with far-reaching applications of the parallelism between the affine and elliptic Lie algebras, which resembles the "real–complex" correspondence. The infinite-dimensional Lie groups we are concerned with here are groups of gauge transformations of principal bundles over real and complex surfaces. We show how the classification of coadjoint orbits of loop groups (respectively, double loop groups) can be used to study the Poisson structure on the moduli space of flat connections (respectively, semistable holomorphic bundles) over a Riemann surface (respectively, a complex surface).

The correspondence between the real and complex cases leads to somewhat surprising analogies between notions in differential topology (such as orientation, boundary, and the Stokes theorem) and those in complex algebraic geometry (a meromorphic differential form, its divisor of poles, and the Cauchy–Stokes formula). These analogies are formalized in the notion of polar homology, and their applications include the construction of a holomorphic linking number for a pair of complex curves in a complex threefold. The definition of the latter is closely related to a holomorphic version of the Chern–Simons functional.

In the appendices we mention several topics serving either as an explanation to some facts used in the main text, or as an indication of further developments. In particular, we include reminders on root systems and some important facts from the theory of compact Lie groups. Other appendices provide brief introductions and guides to the literature on the algebra $\mathfrak{gl}_\infty$, the Krichever–Novikov algebras (generalizing the Virasoro algebra and loop algebras to higher-genus Riemann surfaces), integrable systems on the moduli of flat connections, the Kähler structures on Virasoro orbits, a relation of diffeomorphism groups to optimal mass transport, the Hofer metric on the group of Hamiltonian diffeomorphisms, the Drinfeld–Sokolov reduction, as well as proofs of several statements from the main text.

Numeration system and shortcuts. We have employed a single numeration of definitions, theorems, etc. The Roman numeral in the cross-references addresses to the chapter number, while its absence indicates that the cross-references are within the same chapter.

The different sections in Chapter II can be read to a large degree independently. Furthermore, Chapter III is based on just two sections from Chapter

II: those on the affine groups (Section 1) and on the elliptic Lie groups (Sections 5). The section on polar homology is also rather independent, although motivated by the preceding exposition in Chapter III.

For a first reading we recommend the following "shortcut" through the book: After Chapter I on preliminaries, one can proceed to Sections 1, 2, and 5 of Chapter II and Sections 2 and 3 of Chapter III. The reader more interested in applications to Hamiltonian systems will find them mostly in Sections 2 through 5 of Chapter II, while for applications to moduli spaces of flat connections one may choose to proceed to Chapter III after reading only Sections 1 and 5 of Chapter II.

I

Preliminaries

In this chapter, we collect some key notions and facts from the theory of Lie groups and Hamiltonian systems, as well as set up the notations.

1 Lie Groups and Lie Algebras

This section introduces the notions of a Lie group and the corresponding Lie algebra. Many of the basic facts known for finite-dimensional Lie groups are no longer true for infinite-dimensional ones, and below we illustrate some of the pathologies one can encounter in the infinite-dimensional setting.

1.1 Lie Groups and an Infinite-Dimensional Setting

The most basic definition for us will be that of a (transformation) group.

Definition 1.1 A nonempty collection G of transformations of some set is called a (transformation) *group* if along with every two transformations $g, h \in G$ belonging to the collection, the composition $g \circ h$ and the inverse transformation g^{-1} belong to the same collection G.

It follows directly from this definition that every group contains the identity transformation e. Also, the composition of transformations is an associative operation. These properties, associativity and the existence of the unit and an inverse of each element, are often taken as the definition of an abstract group.[1]

The groups we are concerned with in this book are so-called Lie groups. In addition to being a group, they carry the structure of a smooth manifold such that both the multiplication and inversion respect this structure.

---

[1] Here we employ the point of view of V.I. Arnold, that every group should be viewed as the group of transformations of some set, and the "usual" axiomatic definition of a group only obscures its true meaning (cf. [19], p. 58).

Definition 1.2 A *Lie group* is a smooth manifold G with a group structure such that the multiplication $G \times G \to G$ and the inversion $G \to G$ are smooth maps.

The Lie groups considered throughout this book will usually be infinite-dimensional. So what do we mean by an infinite-dimensional manifold? Roughly speaking, an infinite-dimensional manifold is a manifold modeled on an infinite-dimensional locally convex vector space just as a finite-dimensional manifold is modeled on $\mathbb{R}^n$.

Definition 1.3 Let V, W be *Fréchet spaces*, i.e., complete locally convex Hausdorff metrizable vector spaces, and let U be an open subset of V. A map $f : U \subset V \to W$ is said to be *differentiable* at a point $u \in U$ in a direction $v \in V$ if the limit

$$Df(u; v) = \lim_{t \to 0} \frac{f(u + tv) - f(u)}{t} \tag{1.1}$$

exists. The function is said to be continuously differentiable on U if the limit exists for all $u \in U$ and all $v \in V$, and if the function $Df : U \times V \to W$ is continuous as a function on $U \times V$. In the same way, we can build the second derivative $D^2 f$, which (if it exists) will be a function $D^2 f : U \times V \times V \to W$, and so on. A function $f : U \to W$ is called *smooth* or C^∞ if all its derivatives exist and are continuous.

Definition 1.4 A *Fréchet manifold* is a Hausdorff space with a coordinate atlas taking values in a Fréchet space such that all transition functions are smooth maps.

Remark 1.5 Now one can start defining vector fields, tangent spaces, differential forms, principal bundles, and the like on a Fréchet manifold exactly in the same way as for finite-dimensional manifolds.

For example, for a manifold M, a *tangent vector* at some point $m \in M$ is defined as an equivalence class of smooth parametrized curves $f : \mathbb{R} \to M$ such that $f(0) = m$. The set of all such equivalence classes is the *tangent space* $T_m M$ at m. The union of the tangent spaces $T_m M$ for all $m \in M$ can be given the structure of a Fréchet manifold TM, the tangent bundle of M. Now a smooth vector field on the manifold M is a smooth map $v : M \to TM$, and one defines in a similar vein the directional derivative of a function and the Lie bracket of two vector fields.

Since the dual of a Fréchet space need not be Fréchet, we define differential 1-forms in the Fréchet setting directly, as smooth maps $\alpha : TM \to \mathbb{R}$ such that for any $m \in M$, the restriction $\alpha|_{T_m M} : T_m M \to \mathbb{R}$ is a linear map. Differential forms of higher degree are defined analogously: say, a 2-form on a Fréchet manifold M is a smooth map $\beta : T^{\otimes 2} M \to \mathbb{R}$ whose restriction $\beta|_{T_m^{\otimes 2} M} : T_m^{\otimes 2} M \to \mathbb{R}$ for any $m \in M$ is bilinear and antisymmetric. The differential df of a smooth function $f : M \to \mathbb{R}$ is defined via the directional derivative, and this construction generalizes to smooth n-forms on a Fréchet manifold M to give the *exterior derivative* operator d, which maps n-forms to $(n + 1)$-forms on M; see, for example, [231].

Remark 1.6 More facts on infinite-dimensional manifolds can be found in, e.g., [265, 157]. From now on, whenever we speak of an infinite-dimensional manifold, we implicitly mean a Fréchet manifold (unless we say explicitly otherwise). In particular, our infinite-dimensional Lie groups are *Fréchet Lie groups*.

Instead of Fréchet manifolds, one could consider manifolds modeled on Banach spaces. This would lead to the category of Banach manifolds. The main advantage of Banach manifolds is that strong theorems from finite-dimensional analysis, such as the inverse function theorem, hold in Banach spaces but not necessarily in Fréchet spaces. However, some of the Lie groups we will be considering, such as the diffeomorphism groups, are not Banach manifolds. For this reason we stay within the more general framework of Fréchet manifolds. In fact, for most purposes, it is enough to consider groups modeled on locally convex vector spaces. This is the setting considered by Milnor [265].

1.2 The Lie Algebra of a Lie Group

Definition 1.7 Let G be a Lie group with the identity element $e \in G$. The tangent space to the group G at its identity element is (the vector space of) the *Lie algebra* $\mathfrak{g}$ of this group G. The group multiplication on a Lie group G endows its Lie algebra $\mathfrak{g}$ with the following bilinear operation $[\,,\,] : \mathfrak{g} \times \mathfrak{g} \to \mathfrak{g}$, called the *Lie bracket* on $\mathfrak{g}$.

First note that the Lie algebra $\mathfrak{g}$ can be identified with the set of left-invariant vector fields on the group G. Namely, to a given vector $X \in \mathfrak{g}$ one can associate a vector field $\widetilde{X}$ on G by left translation: $\widetilde{X}(g) = l_{g_*}X$, where $l_g : G \to G$ denotes the multiplication by a group element g from the left, $h \in G \mapsto gh$. Obviously, such a vector field $\widetilde{X}$ is invariant under left translations by elements of G. That is, $l_{g_*}\widetilde{X} = \widetilde{X}$ for all $g \in G$. On the other hand, any left-invariant vector field $\widetilde{X}$ on the group G uniquely defines an element $\widetilde{X}(e) \in \mathfrak{g}$.

The usual Lie bracket (or commutator) $[\widetilde{X}, \widetilde{Y}]$ of two left-invariant vector fields $\widetilde{X}$ and $\widetilde{Y}$ on the group is again a left-invariant vector field on G. Hence we can write $[\widetilde{X}, \widetilde{Y}] = \widetilde{Z}$ for some $Z \in \mathfrak{g}$. We define the *Lie bracket* $[X, Y]$ of two elements X, Y of the Lie algebra $\mathfrak{g}$ of the group G via $[X, Y] := Z$. The Lie bracket gives the space $\mathfrak{g}$ the *structure of a Lie algebra*.

Examples 1.8 Here are several finite-dimensional Lie groups and their Lie algebras:

- $\mathrm{GL}(n, \mathbb{R})$, the set of nondegenerate $n \times n$ matrices, is a Lie group with respect to the matrix product: multiplication and taking the inverse are smooth operations. Its Lie algebra is $\mathfrak{gl}(n, \mathbb{R}) = \mathrm{Mat}(n, \mathbb{R})$, the set of all $n \times n$ matrices.

- $\mathrm{SL}(n, \mathbb{R}) = \{A \in \mathrm{GL}(n, \mathbb{R}) \mid \det A = 1\}$ is a Lie group and a closed subgroup of $\mathrm{GL}(n, \mathbb{R})$. Its Lie algebra is the space of traceless matrices $\mathfrak{sl}(n, \mathbb{R}) = \{A \in \mathfrak{gl}(n, \mathbb{R}) \mid \mathrm{tr}\, A = 0\}$. This follows from the relation

$$\det(I + \epsilon A) = 1 + \epsilon\, \mathrm{tr}\, A + \mathcal{O}(\epsilon^2), \qquad \text{as} \quad \epsilon \to 0,$$

where I is the identity matrix.

- $\mathrm{SO}(n, \mathbb{R})$ is a Lie group of transformations $\{A : \mathbb{R}^n \to \mathbb{R}^n\}$ preserving the Euclidean inner product of vectors (and orientation) in $\mathbb{R}^n$, i.e. $(Au,\ Av) = (u,\ v)$ for all vectors $u, v \in \mathbb{R}^n$. Equivalently, one can define

$$\mathrm{SO}(n, \mathbb{R}) = \{A \in \mathrm{GL}(n, \mathbb{R}) \mid AA^t = I,\ \det A > 0\}.$$

The Lie algebra of $\mathrm{SO}(n)$ is the space of skew-symmetric matrices

$$\mathfrak{so}(n, \mathbb{R}) = \{A \in \mathfrak{gl}(n, \mathbb{R}) \mid A + A^t = 0\},$$

as the relation

$$(I + \epsilon A)(I + \epsilon A^t) = I + \epsilon(A + A^t) + \mathcal{O}(\epsilon^2)$$

shows.

- $\mathrm{Sp}(2n, \mathbb{R})$ is the group of transformations of $\mathbb{R}^{2n}$ preserving the nondegenerate skew-product of vectors.

Exercise 1.9 Give an alternative definition of $\mathrm{Sp}(2n, \mathbb{R})$ with the help of the equation satisfied by the corresponding matrices for the following skew-product of vectors $\langle u, v \rangle := \sum_{j=1}^{n}(u_j v_{j+n} - v_j u_{j+n})$. Find the corresponding Lie algebra.

Exercise 1.10 Show that in all of Examples 1.8, the Lie bracket is given by the usual commutator of matrices: $[A, B] = AB - BA$.

The following examples are the first infinite-dimensional Lie groups we shall encounter.

Example 1.11 Let M be a compact n-dimensional manifold. Consider the set $\mathrm{Diff}(M)$ of diffeomorphisms of M. It is an open subspace of (the Fréchet manifold of) all smooth maps from M to M. One can check that the composition and inversion are smooth maps, so that the set $\mathrm{Diff}(M)$ is a Fréchet

Lie group; see [157].[2] Its Lie algebra is given by $\mathrm{Vect}(M)$, the Lie algebra of smooth vector fields on M.

Given a volume form μ on M, one can define the group of volume-preserving diffeomorphisms

$$SDiff(M) := \{\phi \in Diff(M) \mid \phi^*\mu = \mu\}.$$

It is a Lie group, since $SDiff(M)$ is a closed subgroup of $Diff(M)$. Its Lie algebra $SVect(M) := \{v \in \mathrm{Vect}(M) \mid \mathrm{div}(v) = 0\}$ consists of vector fields on M that are divergence-free with respect to the volume form μ.

Example 1.12 Let M be a finite-dimensional compact manifold and let G be a finite-dimensional Lie group. Set the *group of currents* on M to be $G^M = C^\infty(M, G)$, the group of G-valued functions on M. We can define a multiplication on G^M pointwise, i.e., we set $(\varphi \cdot \psi)(g) = \varphi(g)\psi(g)$ for all $\varphi, \psi \in G^M$. This multiplication gives G^M the structure of a (Fréchet) Lie group, as we discuss below.

Example 1.13 A slight, but important, generalization of the example above is the following: Let G be a finite-dimensional Lie group, and P a principal G-bundle over a manifold M. Denote by $\pi : P \to M$ the natural projection to the base. Define the Lie *group* $\mathrm{Gau}(P)$ *of gauge transformations* (or, simply, the *gauge group*) of P as the group of bundle (i.e., fiberwise) automorphisms: $\mathrm{Gau}(P) = \{\varphi \in \mathrm{Aut}(P) \mid \pi \circ \varphi = \pi\}$. The group multiplication is the natural composition of the bundle automorphisms. (Automorphisms of each fiber of P form a copy of the group G, and all together they define the associated bundle over M with the structure group G. The identity bundle automorphism gives the trivial section of this associated G-bundle, and the gauge transformation group consists of all smooth sections of it; see details in [265].) One can show that this is a Lie group (cf. [157]), and we denote the corresponding Lie algebra by $\mathfrak{gau}(P)$. For a topologically trivial G-bundle P, the group $\mathrm{Gau}(P)$ coincides with the current group G^M.

Exercise 1.14 Describe the Lie brackets for the Lie algebras in the last three examples.

Remark 1.15 For a Lie group G, the Lie bracket on the corresponding Lie algebra $\mathfrak{g}$, which we defined via the usual Lie bracket of left-invariant vector fields on the group, satisfies the following properties:

---

[2] In many analysis questions it is convenient to work with the larger space of diffeomorphisms $Diff^s(M)$ of Sobolev class H^s. For $s > n/2 + 1$ these spaces are smooth Hilbert manifolds. On the other hand, the spaces $Diff^s(M)$ are only topological (but not smooth) groups, since the composition of such diffeomorphisms is not smooth. Indeed, while the right multiplication $r_\phi : \psi \mapsto \psi \circ \phi$ is smooth, the left multiplication $l_\psi : \phi \mapsto \psi \circ \phi$ is only continuous, but not even Lipschitz continuous; see [95].

(*i*) it is antisymmetric in X and Y, i.e., $[X, Y] = -[Y, X]$, and
(*ii*) it satisfies the Jacobi identity:

$$[[X, Y], Z] + [[Z, X], Y] + [[Y, Z], X] = 0.$$

The Jacobi identity can be thought of as an infinitesimal analogue of the associativity of the group multiplication.

1.3 The Exponential Map

Definition 1.16 The *exponential map* from a Lie algebra to the corresponding Lie group $\exp : \mathfrak{g} \to G$ is defined as follows: Let us fix some $X \in \mathfrak{g}$ and let $\widetilde{X}$ denote the corresponding left-invariant vector field. The flow of the field $\widetilde{X}$ is a map $\phi_X : G \times \mathbb{R} \to G$ such that $\frac{d}{dt}\phi_X(g, t) = \widetilde{X}(\phi_X(g, t))$ for all t and $\phi_X(g, 0) = g$. The flow ϕ_X is the solution of an ordinary differential equation, which, if it exists, is unique. In the case that the flow subgroup $\phi_X(e, .)$ exists for all $X \in \mathfrak{g}$, we define the exponential map $\exp : \mathfrak{g} \to G$ via the time-one map $X \mapsto \phi_X(e, 1)$; see Figure 1.1.

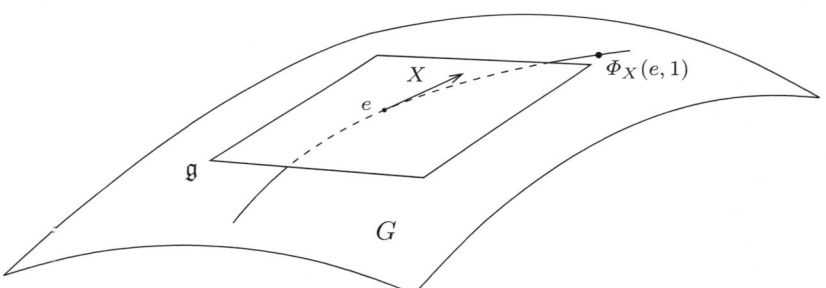

Fig. 1.1. The exponential map on the group G associates to a vector X the time-one map for the trajectory of a left-invariant vector field defined by X at $e \in G$.

Example 1.17 For each of the finite-dimensional Lie groups considered in Example 1.8, the exponential map is given by the usual exponential map for matrices:

$$\exp : A \mapsto \exp(A) = \sum_{n=0}^{\infty} \frac{1}{n!} A^n.$$

Remark 1.18 The definition of the exponential map relies on the existence and uniqueness of solutions of certain first-order differential equations. In general, solutions of differential equations in Fréchet spaces might not be unique.[3] However, the differential equation in the definition of the exponential map is of special type, which secures the solution's uniqueness upon fixing its initial condition. Namely, let $\phi : \mathbb{R} \to G$ be a smooth path in the Lie group G. Its derivative $\phi'(t) := \frac{d}{dt}\phi(t)$ is a tangent vector to the group G at the point $\phi(t)$. Translate this vector back to the identity via left multiplication by $\phi^{-1}(t)$. The corresponding element of the Lie algebra $\mathfrak{g}$ is denoted by $\phi^{-1}(t)\phi'(t)$ and is called the *left logarithmic derivative* of the path ϕ.

Now consider a Lie algebra element $X \in \mathfrak{g}$. By definition of the exponential map, the curve $\phi(t) = \exp(tX)$ satisfies the differential equation $\phi'(t) = \phi(t)X$ with the initial condition $\phi(0) = e$. So for all solutions of this differential equation, the left logarithmic derivative is given by the constant curve $X \in \mathfrak{g}$. Now the uniqueness of the exponential map is implied by the following Exercise.

Exercise 1.19 Show that two smooth paths $\phi, \psi : \mathbb{R} \to G$ have the same left logarithmic derivative for all $t \in \mathbb{R}$ if and only if they are translations of each other by some constant element $g \in G$: $\phi(t) = g\,\psi(t)$ for all $t \in \mathbb{R}$. (Hint: see, e.g., [265].)

Remark 1.20 As far as the existence is concerned, the exponential map exists for all finite-dimensional Lie groups and more generally for Lie groups modeled on Banach spaces, as follows from the general theory of differential equations. However, there may exist infinite-dimensional Lie groups that do not admit an exponential map. Moreover, even in the cases in which the exponential map of an infinite-dimensional group exists, it can exhibit rather peculiar properties; see the examples below.

Example 1.21 For the diffeomorphism group $\mathrm{Diff}(M)$ the exponential map $\exp : \mathrm{Vect}(M) \to \mathrm{Diff}(M)$ has to assign to each vector field on M the time-one map for its flow. However, for a noncompact M this map may not exist: the corresponding vector field may not be complete. Indeed, for example, for the vector field $\xi = x^2 \partial/\partial x$ on the real line $M = \mathbb{R}$, the time-one map of the flow is not defined on the whole of $\mathbb{R}$: the corresponding flow sends some points to infinity for the time less than 1! Fortunately, for compact manifolds M and smooth vector fields, the time-one maps of the corresponding flows, and hence the exponential maps, are well defined.

Note that the group of diffeomorphisms of a noncompact manifold is not complete, and hence it is not a Lie group in our sense. It is an important open problem to find a Lie group that is modeled on a complete space and does not admit an exponential map.

---

[3] For instance, the initial value problem $u(x,0) = f(x)$ for the equation $u_t(x,t) = u_x(x,t)$ with $x \in [0,1]$ has wave-type solutions $u(x,t) = f(x+t)$. For nonzero t such a solution $u(x,t)$ for $x \in [0,1]$ depends on the extension of $f(x)$ to the segment $[-t, 1-t]$. Due to arbitrariness in the choice of a smooth extension of f from $[0,1]$ to $\mathbb{R}$, the solution to this initial value problem is not unique.

Let us return to the current group G^M, where the exponential map exists and can be used to give this group the structure of a Fréchet Lie group. Namely, the space $\mathfrak{g}^M = C^\infty(M, \mathfrak{g})$ endowed with the topology of uniform convergence is a Fréchet space. Moreover, the map $\exp : \mathfrak{g} \to G$ can be used to define a map $\widetilde{\exp} : \mathfrak{g}^M \to G^M$ pointwise. In a sufficiently small neighborhood of $0 \in \mathfrak{z}^M$, the map $\widetilde{\exp}$ is bijective. Thus it can be used to define a local system of open neighborhoods of the identity in G^M. We can use left translation to transfer this system to any point in G^M and thus define a topology on the group G^M. Again using the exponential map, we can define coordinate charts on G^M. This definition implies that multiplication and inversion in G^M are smooth maps. So G^M is an infinite-dimensional Lie groups (see, e.g., [157] for more details).

From the construction of the Lie group structure on G^M, it is clear that its Lie algebra is the current algebra $\mathfrak{g}^M$, and that the exponential map $\mathfrak{g}^M \to G^M$ is the map $\widetilde{\exp}$ described above. Note, however, that $\widetilde{\exp}$ is not, in general, surjective, even if $\exp : \mathfrak{g} \to G$ is surjective. As an example, take the manifold M to be the circle S^1 and G to be the group $SU(2)$.

Exercise 1.22 Show that the map

$$\theta \mapsto \begin{pmatrix} e^{i\theta} & 0 \\ 0 & e^{-i\theta} \end{pmatrix}$$

for $\theta \in S^1 = \mathbb{R}/2\pi\mathbb{Z}$ defines an element in G^{S^1} that does not belong to the image of the exponential map $\widetilde{\exp} : \mathfrak{g}^{S^1} \to G^{S^1}$.

In contrast to the exponential map in the case of the current group G^M, the exponential map $\exp : \mathrm{Vect}(M) \to \mathrm{Diff}(M)$ for the diffeomorphism group of a compact M is not, in general, even locally surjective already for the case of a circle.

Proposition 1.23 (see, e.g., [265, 301, 322]) *The exponential map* $\exp : \mathrm{Vect}(S^1) \to \mathrm{Diff}(S^1)$ *is not locally surjective.*

PROOF. First observe that any nowhere-vanishing vector field on S^1 is conjugate under $\mathrm{Diff}(S^1)$ to a constant vector field. Indeed, if $\xi(\theta) = v(\theta)\frac{\partial}{\partial\theta}$ is such a vector field, we can define a diffeomorphism $\psi : S^1 \to S^1$ via $\psi(\theta) = a \int_0^\theta \frac{dt}{v(t)}$. Here, $a \in \mathbb{R}$ is chosen such that $\psi(2\pi) = 2\pi$. Then $\psi_*(\xi \circ \psi^{-1})$ is a constant vector field on S^1.

From this observation, one can conclude that any diffeomorphism of S^1 that lies in the image of the exponential map and that does not have any fixed points is conjugate to a rigid rotation of S^1. Hence in order to see that the exponential map is not locally surjective, it is enough to construct diffeomorphisms arbitrarily close to the identity that do not have any fixed points and that are not conjugate to a rigid rotation. For this, one can take diffeomorphisms without fixed points, but which have isolated periodic points, i.e., fixed points for a certain nth iteration of this diffeomorphism. Indeed, if such a diffeomorphism ψ belonged to the image of the exponential map, so would its nth power ψ^n. Then the corresponding vector field defining the ψ^n as the time-one map would either have zeros or be nonvanishing everywhere. In the former case, the n-periodic points of ψ must actually be its fixed points, while in the latter case, the diffeomorphism ψ^n, as well as ψ, would be conjugate to a rigid rotation and hence *all* points of ψ would be n-periodic. Both cases give us a contradiction.

Explicitly, a family of such diffeomorphisms can be constructed as follows: Let us identify S^1 with $\mathbb{R}/2\pi\mathbb{Z}$. Then consider the map $\psi_{n,\epsilon} : x \mapsto x + \frac{2\pi}{n} + \epsilon \sin(nx)$. For ϵ small enough, this is indeed a diffeomorphism of S^1. Furthermore, by choosing n large and ϵ small, the diffeomorphisms $\psi_{n,\epsilon}$ can be made arbitrarily close to the identity while having no fixed points. Finally, for $\epsilon \neq 0$, $\psi_{n,\epsilon}$ cannot be conjugate to a rigid rotation. If it were conjugate to a rotation, it would have to be the rotation $\psi_{n,0}$, since $\psi_{n,\epsilon}^n(0) = 0$. But in this case, we would have $\psi_{n,\epsilon}^n = \mathrm{id}$, which is not true for $\epsilon \neq 0$. $\square$

1.4 Abstract Lie Algebras

As we have seen in the last section, the Lie bracket of two left-invariant vector fields $\widetilde{X}$ and $\widetilde{Y}$ on a Lie group G defines a bilinear map $[.\,,.] : \mathfrak{g} \times \mathfrak{g} \to \mathfrak{g}$ of the Lie algebra of G that is antisymmetric in X and Y and satisfies the Jacobi identity (1.2). These properties can be taken as the definition of an abstract Lie algebra:

Definition 1.24 An *(abstract) Lie algebra* is a real or complex vector space $\mathfrak{g}$ together with a bilinear map $[.\,,.] : \mathfrak{g} \times \mathfrak{g} \to \mathfrak{g}$ (the Lie bracket) that is antisymmetric in X and Y and that satisfies the Jacobi identity

$$[[X,Y],Z] + [[Z,X],Y] + [[Y,Z],X] = 0 \,. \tag{1.2}$$

All the Lie algebras we have encountered so far as accompanying the corresponding Lie groups can also be regarded by themselves, i.e., as abstract Lie algebras. A famous theorem of Sophus Lie states that every finite-dimensional (abstract) Lie algebra $\mathfrak{g}$ is the Lie algebra of some Lie group G. In infinite dimensions this is no longer true in general.

Example 1.25 ([205, 207]) To illustrate the failure of Lie's theorem in an infinite-dimensional context, consider the Lie algebra of complex vector fields on the circle $\mathrm{Vect}^{\mathbb{C}}(S^1) = \mathrm{Vect}(S^1) \otimes \mathbb{C}$. Let us show that this Lie algebra cannot be the Lie algebra of any Lie group. First note that $\mathrm{Vect}^{\mathbb{C}}(S^1)$ contains as a subalgebra the Lie algebra $\mathrm{Vect}(S^1)$ of real vector fields on the circle, which is the Lie algebra of the group $\mathrm{Diff}(S^1)$.

Let G_1 denote the group $\mathrm{PSL}(2,\mathbb{R})$ and let G_k denote the k-fold covering of G_1. The group G_2 is isomorphic to $\mathrm{SL}(2,\mathbb{R})$, while for $k > 2$ it is known that the groups G_k have no matrix realization. The group $\mathrm{Diff}(S^1)$ contains each G_k as a subgroup. Namely, G_k is the subgroup corresponding to the Lie subalgebra $\mathfrak{g}_k$ spanned by the vector fields

$$\frac{\partial}{\partial\theta}, \quad \sin(k\theta)\frac{\partial}{\partial\theta}, \quad \cos(k\theta)\frac{\partial}{\partial\theta}\,.$$

(Note that each $\mathfrak{g}_k$ is isomorphic to $\mathfrak{sl}(2,\mathbb{R})$.)

Now suppose that there exists a complexification of the group $\mathrm{Diff}(S^1)$, i.e., a Lie group G corresponding to the complex Lie algebra $\mathrm{Vect}^{\mathbb{C}}(S^1)$. Such a group G would have to contain the complexifications of all the groups G_k. However, for $k > 2$ the groups G_k do not admit complexifications: the only complex groups corresponding to the Lie algebra $\mathfrak{sl}(2,\mathbb{C})$ are $\mathrm{SL}(2,\mathbb{C})$ and $\mathrm{PSL}(2,\mathbb{C})$.

More precisely, if the complex Lie group G existed, the real subgroups G_k would belong to the complex subgroups of G corresponding to complex subalgebras $\mathfrak{g}_k^{\mathbb{C}} \simeq \mathfrak{sl}(2,\mathbb{C})$. But these complex subgroups have to be isomorphic either to $\mathrm{SL}(2,\mathbb{C})$, which contains only $\mathrm{SL}(2,\mathbb{R}) = G_2$, or to $\mathrm{PSL}(2,\mathbb{C})$, which contains only $\mathrm{PSL}(2,\mathbb{R}) = G_1$. Thus the complex group G containing all G_k cannot exist, and hence there is no Lie group for the Lie algebra $\mathrm{Vect}^{\mathbb{C}}(S^1)$.

Lie algebra homomorphisms are defined in the usual way: A map $\rho : \mathfrak{g} \to \mathfrak{h}$ between two Lie algebras is a Lie algebra homomorphism if it satisfies $\rho([X,Y]) = [\rho(X),\rho(Y)]$ for all $X, Y \in \mathfrak{g}$. We will also need another important class of maps between Lie algebras called derivations:

Definition 1.26 A linear map $\delta : \mathfrak{g} \to \mathfrak{g}$ of a Lie algebra $\mathfrak{g}$ to itself is called a *derivation* if it satisfies

$$\delta([X,Y]) = [\delta(X),Y] + [X,\delta(Y)]$$

for all $X, Y \in \mathfrak{g}$.

Exercise 1.27 Define the map $\mathrm{ad}_X : \mathfrak{g} \to \mathfrak{g}$ associated to a fixed vector $X \in \mathfrak{g}$ via

$$\mathrm{ad}_X(Y) = [X,Y]\,.$$

Show that this is a derivation for any choice of X. (Hint: use the Jacobi identity.)

If a derivation of a Lie algebra $\mathfrak{g}$ can be expressed in the form ad_X for some $X \in \mathfrak{g}$, it is called an *inner derivation*; otherwise, it is called an *outer derivation* of $\mathfrak{g}$.

Exercise 1.28 Let δ be a derivation of a Lie algebra $\mathfrak{g}$, and suppose that $\exp(\delta) = \sum_{i=0}^{\infty} \frac{1}{i!} \delta^i$ makes sense (for example, suppose, the map δ is nilpotent). Show that the map $\exp(\delta)$ is an automorphism of the Lie algebra $\mathfrak{g}$.

Definition 1.29 A *subalgebra* of a Lie algebra $\mathfrak{g}$ is a subspace $\mathfrak{h} \subset \mathfrak{g}$ invariant under the Lie bracket in $\mathfrak{g}$. An *ideal* of a Lie algebra $\mathfrak{g}$ is a subalgebra $\mathfrak{h} \subset \mathfrak{g}$ such that $[X, \mathfrak{h}] \subset \mathfrak{h}$ for all $X \in \mathfrak{g}$.

The importance of ideals comes from the fact that if $\mathfrak{h} \subset \mathfrak{g}$ is an ideal, then the quotient space $\mathfrak{g}/\mathfrak{h}$ is again a Lie algebra.

Exercise 1.30 (i) Show that for an ideal $\mathfrak{h} \subset \mathfrak{g}$ the Lie bracket on $\mathfrak{g}$ descends to a Lie bracket on the quotient space $\mathfrak{g}/\mathfrak{h}$.

(ii) Show that if $\rho : \mathfrak{g} \to \widetilde{\mathfrak{g}}$ is a homomorphism of two Lie algebras, then the kernel $\ker \rho$ of ρ is an ideal in $\mathfrak{g}$.

Definition 1.31 A Lie algebra is *simple* (respectively, *semisimple*) if it does not contain nontrivial ideals (respectively, nontrivial abelian ideals).

Any finite-dimensional semisimple Lie algebra is a direct sum of nonabelian simple Lie algebras.

A group analogue of an ideal is the notion of a normal subgroup. A subgroup $H \subset G$ of a group G is called *normal* if $gHg^{-1} \subset H$ for all $g \in G$. Exercise 1.30 translates directly to normal subgroups.

2 Adjoint and Coadjoint Orbits

Writing out a linear operator in a different basis or a vector field in a different coordinate system has a far-reaching generalization as the adjoint representation for any Lie group. In this section we define the adjoint and coadjoint representations and the corresponding orbits for an arbitrary Lie group.

2.1 The Adjoint Representation

A *representation* of a Lie group G on a vector space V is a linear action φ of the group G on V that is smooth in the sense that the map $G \times V \to V$, $(g, v) \mapsto gv$, is smooth. If V is a real vector space, (V, φ) is called a real representation, and if V is complex, it is a complex representation. (Here V is assumed to be a Fréchet space, and, often, a Hilbert space. In the latter case, the representation is said to be unitary if the inner product on V is invariant under the action of G.)

Every Lie group has two distinguished representations: the adjoint and the coadjoint representations. Since they will play a special role in this book, we describe them in more detail.

Any element $g \in G$ defines an automorphism c_g of the group G by conjugation:

$$c_g : h \in G \mapsto ghg^{-1}.$$

The differential of c_g at the identity $e \in G$ maps the Lie algebra of G to itself and thus defines an element $\mathrm{Ad}_g \in \mathrm{Aut}(\mathfrak{g})$, the group of all automorphisms of the Lie algebra $\mathfrak{g}$.

Definition 2.1 The map $\mathrm{Ad} : G \to \mathrm{Aut}(\mathfrak{g})$, $g \mapsto \mathrm{Ad}_g$ defines a representation of the group G on the space $\mathfrak{g}$ and is called the *group adjoint representation*; see Figure 2.1. The orbits of the group G in its Lie algebra $\mathfrak{g}$ are called the *adjoint orbits* of G.

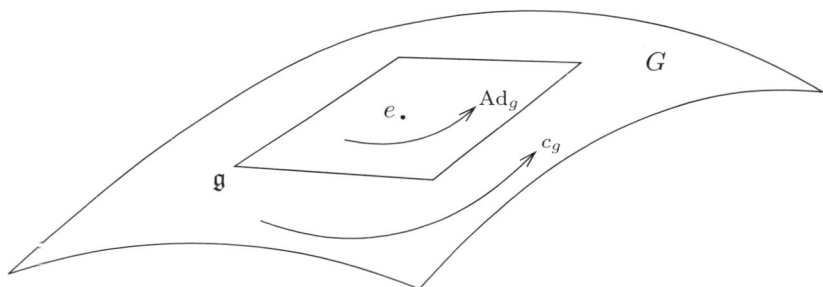

Fig. 2.1. Conjugation c_g on the group G generates the adjoint representation Ad_g on the Lie algebra $\mathfrak{g}$.

The differential of $\mathrm{Ad} : G \to \mathrm{Aut}(\mathfrak{g})$ at the group identity $g = e$ defines a map $\mathrm{ad} : \mathfrak{g} \to \mathrm{End}(\mathfrak{g})$, the *adjoint representation of the Lie algebra* $\mathfrak{g}$.

One can show that the bracket $[\,,\,]$ on the space $\mathfrak{g}$ defined via

$$[X, Y] := \mathrm{ad}_X(Y)$$

coincides with the bracket (or commutator) of the corresponding two left-invariant vector fields on the group G and hence with the Lie bracket on $\mathfrak{g}$ defined in Section 1.2.

Example 2.2

- Let $g \in \mathrm{GL}(n, \mathbb{R})$ and $A \in \mathfrak{gl}(n, \mathbb{R})$. Then $\mathrm{Ad}_g A = gAg^{-1}$. Hence the adjoint orbits are given by sets of similar (i.e., conjugate) matrices in $\mathfrak{gl}(n, \mathbb{R})$. The adjoint representation of $\mathfrak{gl}(n, \mathbb{R})$ is given by $\mathrm{ad}_A(B) = [A, B] = AB - BA$.

- The adjoint orbits of $SO(3, \mathbb{R})$ are spheres centered at the origin of $\mathbb{R}^3 \simeq \mathfrak{so}(3, \mathbb{R})$ and the origin itself.
- The adjoint orbits of $SL(2, \mathbb{R})$ are contained in the sets of similar matrices. By writing $A = \left(\begin{smallmatrix} a & b \\ c & -a \end{smallmatrix} \right) \in \mathfrak{sl}(2, \mathbb{R})$, one sees that the adjoint orbits lie in the level sets of $\Delta = -(a^2 + bc) = \text{const}$: matrices that are conjugate to each other have the same determinant. Note, however, that not all matrices in $\mathfrak{sl}(2, \mathbb{R})$ that have the same determinant are conjugate. For instance, the matrices with determinant $\Delta = 0$ constitute three different orbits: the origin and two other orbits, cones, passing through the matrices $\left(\begin{smallmatrix} 0 & \pm 1 \\ 0 & 0 \end{smallmatrix} \right)$, respectively. For $\Delta \neq 0$ the $SL(2, \mathbb{R})$-orbits are either one-sheet hyperboloids or connected components of the two-sheet hyperboloids $a^2 + bc = \text{const}$, since the group $SL(2, \mathbb{R})$ is connected.
- Let G be the set of orientation-preserving affine transformations of the real line. That is, $G = \{(a, b) \mid a, b \in \mathbb{R}, \ a > 0\}$, and $(a, b) \in G$ acts on $x \in \mathbb{R}$ via $x \mapsto ax + b$. The Lie algebra of G is $\mathbb{R}^2$, and its adjoint orbits are the affine lines

$$\{(\alpha, \beta) \in \mathbb{R}^2 \mid \alpha = \text{const} \neq 0, \ \beta \text{ arbitrary}\},$$

the two rays

$$\{(\alpha, \beta) \in \mathbb{R}^2, \alpha = 0, \beta < 0\} \ \text{ and } \ \{(\alpha, \beta) \in \mathbb{R}^2, \alpha = 0, \beta > 0\},$$

and the origin $\{(0, 0)\}$; see Figure 2.2.
- Let M be a compact manifold. The adjoint orbits of the current group $GL(n, \mathbb{C})^M$ in its Lie algebra $\mathfrak{gl}(n, \mathbb{C})^M$ are given by fixing the (smoothly dependent) Jordan normal form of the current at each point of the manifold M.
- Let M be a compact manifold. The adjoint representation of $\text{Diff}(M)$ on $\text{Vect}(M)$ is given by coordinate changes of the vector field: for a $\phi \in \text{Diff}(M)$ one has $\text{Ad}_\phi : v \mapsto \phi_* v \circ \phi^{-1}$. The adjoint representation of $\text{Vect}(M)$ on itself is given by the negative of the usual Lie bracket of vector fields: $\text{ad}_v w = \frac{\partial v}{\partial x} w(x) - \frac{\partial w}{\partial x} v(x)$ in any local coordinate x.

Exercise 2.3 Verify the latter formula for the action of $\text{Diff}(M)$ on $\text{Vect}(M)$ from the definition of the group adjoint action. (Hint: express the diffeomorphisms corresponding to the vector fields $v(x)$ and $w(x)$ in the form

$$g(t): \quad x \mapsto x + tv(x) + o(t), \quad h(s): \quad x \mapsto x + sw(x) + o(s), \quad t, s \to 0,$$

and find the first several terms of $g(t)h(s)g^{-1}(t)$.)

2.2 The Coadjoint Representation

The dual object to the adjoint representation of a Lie group G on its Lie algebra $\mathfrak{g}$ is called the coadjoint representation of G on $\mathfrak{g}^*$, the dual space to $\mathfrak{g}$.

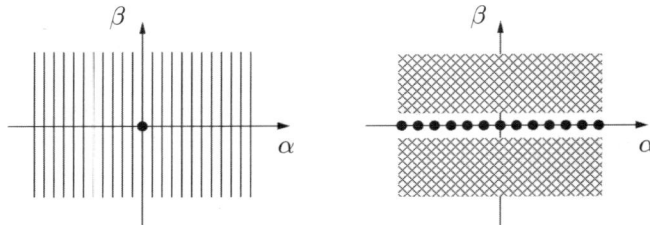

Fig. 2.2. Adjoint and coadjoint orbits of the group of affine transformations on the line.

Definition 2.4 The *coadjoint representation* Ad^* of the group G on the space $\mathfrak{g}^*$ is the dual of the adjoint representation. Let $\langle \, , \, \rangle$ denote the pairing between $\mathfrak{g}$ and its dual $\mathfrak{g}^*$. Then the *coadjoint action of the group G* on the dual space $\mathfrak{g}^*$ is given by the operators $\mathrm{Ad}_g^* : \mathfrak{g}^* \to \mathfrak{g}^*$ for any $g \in G$ that are defined by the relation

$$\langle \mathrm{Ad}_g^*(\xi), X \rangle := \langle \xi, \mathrm{Ad}_{g^{-1}}(X) \rangle \tag{2.3}$$

for all ξ in $\mathfrak{g}^*$ and $X \in \mathfrak{g}$. The orbits of the group G under this action on $\mathfrak{g}^*$ are called the *coadjoint orbits* of G.

The differential $\mathrm{ad}^* : \mathfrak{g} \to \mathrm{End}(\mathfrak{g}^*)$ of the group representation $\mathrm{Ad}^* : G \to \mathrm{Aut}(\mathfrak{g}^*)$ at the group identity $e \in G$ is called the *coadjoint representation of the Lie algebra* $\mathfrak{g}$. Explicitly, at a given vector $Z \in \mathfrak{g}$ it is defined by the relation

$$\langle \mathrm{ad}_Z^*(\xi), X \rangle = -\langle \xi, \mathrm{ad}_Z(X) \rangle.$$

Remark 2.5 The dual space of a Fréchet space is not necessarily again a Fréchet space. In this case, instead of considering the full dual space to an infinite-dimensional Lie algebra $\mathfrak{g}$, we will usually confine ourselves to considering only appropriate "smooth duals," the functionals from a certain G-invariant Fréchet subspace $\mathfrak{g}_s^* \subset \mathfrak{g}^*$. Natural smooth duals will be different according to the type of the infinite-dimensional groups considered, but they all have a (weak) nondegenerate pairing with the corresponding Lie algebra $\mathfrak{g}$ in the following sense: for every nonzero element $X \in \mathfrak{g}$, there exists some element $\xi \in \mathfrak{g}_s^*$ such that $\langle \xi, X \rangle \neq 0$, and the other way around. This ensures that the coadjoint action is uniquely fixed by equation (2.3). The pair $(\mathfrak{g}_s^*, \mathrm{Ad}^* |_{\mathfrak{g}_s^*})$ is called the regular (or smooth) part of the coadjoint representation of G, and, abusing notations, we will usually skip the index s.

Example 2.6

- In the first three cases of Example 2.2, there exists a G-invariant inner product on $\mathfrak{g}$ that induces an isomorphism between $\mathfrak{g}$ and $\mathfrak{g}^*$ respecting the group actions. Hence the adjoint and coadjoint representations of the groups G are isomorphic, and the coadjoint orbits coincide with the adjoint ones.

- The group of affine transformations of the real line in Example 2.2 has two 2-dimensional coadjoint orbits, the upper and lower half-planes in $\mathbb{R}^2$, and a set of zero-dimensional orbits, namely, the points $(\alpha, 0)$ for each $\alpha \in \mathbb{R}$ (see Figure 2.2).
- For a compact manifold M with some fixed volume form $d\mathrm{Vol}$, we can define a nondegenerate G^M-invariant product on the current Lie algebra $\mathfrak{gl}(n, \mathbb{R})^M$ by setting

$$\langle X, Y \rangle = \int_M \mathrm{tr}(X(x) \cdot Y(x)) \, d\mathrm{Vol}(x)$$

for $X, Y \in \mathfrak{gl}(n, \mathbb{R})^M$. This inner product can be used to identify the current algebra $\mathfrak{g}^M$ with a subspace in its dual $\mathfrak{g}_s^* \subset \mathfrak{g}^*$. The space $\mathfrak{g}_s^*$ is called the smooth (or regular) part of $\mathfrak{g}^*$. Thanks to the nondegenerate pairing, the smooth part of the coadjoint representation of G^M is isomorphic to the adjoint representation.

Note that each of the finite-dimensional coadjoint orbits above is even-dimensional. This is a consequence of the general fact that coadjoint orbits are symplectic manifolds, which we discuss later.

Remark 2.7 In what follows we pay particular attention to the structure and description of coadjoint orbits of infinite-dimensional Lie groups. We are interested in coadjoint orbits mostly because they appear as natural phase spaces of dynamical systems. Another reason to study coadjoint orbits comes from the orbit method. This is a general principle due to A. Kirillov, which asserts that the information on the set of unitary representations of a Lie group G is contained in the group coadjoint orbits. This method has become a powerful tool in the study of Lie groups and it has been worked out in detail for large classes of finite-dimensional Lie groups such as nilpotent and compact Lie groups (see [206]). In infinite dimensions, the correspondence between coadjoint orbits and unitary representations has been fully understood only for certain types of groups, e.g., for affine Lie groups (cf. [132, 322, 385]), although there are some indications that it works for other classes as well.

3 Central Extensions

In this section we collect several basic facts about central extensions of Lie groups and Lie algebras. One can think of a central extension of a Lie group G as a new bigger Lie group $\widetilde{G}$ fibered over the initial group G in such a way that the fiber over the identity $e \in G$ lies in the center of $\widetilde{G}$.

Central extensions of Lie groups appear naturally in representation theory and quantum mechanics when one lifts a group projective representation to an ordinary one: one often needs to pass to a central extension of the group to be able to do this. For us the main advantage of these extensions is that for many infinite-dimensional groups their central extensions have simpler and "more regular" structure of the coadjoint orbits, as well as more interesting dynamical systems related to them.

3.1 Lie Algebra Central Extensions

Definition 3.1 A *central extension of a Lie algebra* $\mathfrak{g}$ by a vector space $\mathfrak{n}$ is a Lie algebra $\widetilde{\mathfrak{g}}$ whose underlying vector space $\widetilde{\mathfrak{g}} = \mathfrak{g} \oplus \mathfrak{n}$ is equipped with the following Lie bracket:

$$[(X, u), (Y, v)]^{\sim} = ([X, Y], \omega(X, Y))$$

for some continuous bilinear map $\omega : \mathfrak{g} \times \mathfrak{g} \to \mathfrak{n}$. (Note that ω depends only on X and Y, but not on u and v, which means that the extension is *central*: the space $\mathfrak{n}$ belongs to the center of the new Lie algebra, i.e., it commutes with all of $\widetilde{\mathfrak{g}}$: $[(0, u), (Y, v)] = 0$ for all $Y \in \mathfrak{g}$ and $u, v \in \mathfrak{n}$.) The skew symmetry and the Jacobi identity for the new Lie bracket on $\widetilde{\mathfrak{g}}$ are equivalent to the following conditions on the map ω. Such a map $\omega : \mathfrak{g} \times \mathfrak{g} \to \mathfrak{n}$ has to be a 2-*cocycle on the Lie algebra* $\mathfrak{g}$, i.e., ω has to be bilinear and antisymmetric, and it has to satisfy the *cocycle identity*

$$\omega([X, Y], Z) + \omega([Z, X], Y) + \omega([Y, Z], X) = 0$$

for any triple of elements $X, Y, Z \in \mathfrak{g}$. (Here and below we always require Lie algebra cocycles to be continuous maps.)

A 2-cocycle ω on $\mathfrak{g}$ with values in $\mathfrak{n}$ is called a 2-*coboundary* if there exists a linear map $\alpha : \mathfrak{g} \to \mathfrak{n}$ such that $\omega(X, Y) = \alpha([X, Y])$ for all $X, Y \in \mathfrak{g}$. One can easily see that the central extension defined by such a 2-coboundary becomes the trivial extension by the zero cocycle after the change of coordinates $(X, u) \to (X, u - \alpha(X))$.

Hence in describing different central extensions we are interested only in the 2-cocycles modulo 2-coboundaries, i.e., in the *second cohomology* $H^2(\mathfrak{g}; \mathfrak{n})$ of the Lie algebra $\mathfrak{g}$ with values in $\mathfrak{n}$: $H^2(\mathfrak{g}; \mathfrak{n}) = \mathcal{Z}(\mathfrak{g}; \mathfrak{n})/\mathcal{B}(\mathfrak{g}; \mathfrak{n})$, where $\mathcal{Z}(\mathfrak{g}; \mathfrak{n})$ is the vector space of all 2-cocycles on $\mathfrak{g}$ with values in $\mathfrak{n}$, and $\mathcal{B}(\mathfrak{g}; \mathfrak{n})$ is the subspace of 2-coboundaries.

Remark 3.2 A central extension of a Lie algebra $\mathfrak{g}$ by an abelian Lie algebra $\mathfrak{n}$ can be defined by the exact sequence

$$\{0\} \longrightarrow \mathfrak{n} \longrightarrow \widetilde{\mathfrak{g}} \longrightarrow \mathfrak{g} \longrightarrow \{0\}$$

of Lie algebras such that $\mathfrak{n}$ lies in the center of $\widetilde{\mathfrak{g}}$. A morphism of two central extensions is a pair (ν, μ) of Lie algebra homomorphisms $\nu : \mathfrak{n} \to \mathfrak{n}'$ and $\mu : \widetilde{\mathfrak{g}} \to \widetilde{\mathfrak{g}}'$ such that the following diagram is commutative:

$$
\begin{array}{ccccccccc}
0 & \longrightarrow & \mathfrak{n} & \longrightarrow & \widetilde{\mathfrak{g}} & \xrightarrow{\pi} & \mathfrak{g} & \longrightarrow & 0 \\
 & & \downarrow{\nu} & & \downarrow{\mu} & & \downarrow{\mathrm{id}} & & \\
0 & \longrightarrow & \mathfrak{n}' & \longrightarrow & \widetilde{\mathfrak{g}}' & \xrightarrow{\pi'} & \mathfrak{g} & \longrightarrow & 0.
\end{array}
\tag{3.4}
$$

Two extensions are said to be equivalent if the map μ is an isomorphism and $\nu = \mathrm{id}$.

Exercise 3.3 Prove the following equivalence:

Proposition 3.4 *There is a one-to-one correspondence between the equivalence classes of central extensions of $\mathfrak{g}$ by $\mathfrak{n}$ and the elements of $H^2(\mathfrak{g}; \mathfrak{n})$.*

Example 3.5 Consider the abelian Lie algebra $\mathfrak{g} = \mathbb{R}^2$, and let $\omega \in \Lambda^2(\mathbb{R}^2)$ be an arbitrary skew-symmetric bilinear form on $\mathbb{R}^2$. Then ω defines a 2-cocycle on $\mathbb{R}^2$ with values in $\mathbb{R}$ (in this case, the cocycle condition is trivial, since $\mathfrak{g}$ is abelian). The resulting central extension is $\widetilde{\mathfrak{g}} = \mathbb{R}^2 \oplus \mathbb{R}$ with Lie bracket $[(v_1, h_1), (v_2, h_2)] = (0, \omega(v_1, v_2))$. Moreover, since $\mathfrak{g}$ is abelian, $\mathcal{B}(\mathfrak{g}; \mathbb{R}) = \{0\}$ whence $H^2(\mathfrak{g}; \mathbb{R}) = \Lambda^2(\mathbb{R}^2) \cong \mathbb{R}$. Note that all $\omega \neq 0 \in \Lambda^2(\mathbb{R}^2)$ lead to isomorphic Lie algebras. The algebra $\widetilde{\mathfrak{g}}$ with a nonzero ω, i.e., a representative of this isomorphism class, is called the three-dimensional *Heisenberg algebra*.

By taking a nondegenerate skew-symmetric form ω in $\mathbb{R}^{2n}$, we can define in the same way the $(2n + 1)$-dimensional Heisenberg algebra.

An infinite-dimensional analogue of the Heisenberg algebra is as follows. Consider the space $\mathfrak{g} = \{f \in C^\infty(S^1) \mid \int_{S^1} f \, d\theta = 0\}$ of smooth functions on the circle with zero mean and regard it as an abelian Lie algebra. Define the 2-cocycle by $\omega(f, g) = \int_{S^1} f'g \, d\theta$. (One can view this algebra and the corresponding cocycle as the "limit" $n \to \infty$ of the example above by considering the functions in Fourier components.)

Exercise 3.6 Check the skew-symmetry and the cocycle identity for $\omega(f, g)$.

Definition 3.7 A central extension $\widetilde{\mathfrak{g}}$ of $\mathfrak{g}$ is called *universal* if for any other central extension $\widetilde{\mathfrak{g}}'$, there is a unique morphism $\widetilde{\mathfrak{g}} \to \widetilde{\mathfrak{g}}'$ of the central extensions. If it exists, the universal central extension of a Lie algebra $\mathfrak{g}$ is unique up to isomorphism.

Remark 3.8 A sufficient condition for a Lie algebra $\mathfrak{g}$ to have a universal central extension is that $\mathfrak{g}$ be perfect, i.e., that it coincide with its own derived algebra: $\mathfrak{g} = [\mathfrak{g}, \mathfrak{g}]$ (see, e.g., [276]). Any finite-dimensional semisimple Lie algebra is perfect. The universal central extension of a semisimple Lie algebra $\mathfrak{g}$ coincides with $\mathfrak{g}$ itself: such algebras do not admit nontrivial central extensions.

No abelian Lie algebra is perfect. Nevertheless, abelian Lie algebras can still have universal central extensions: for instance, the three-dimensional Heisenberg algebra is the universal central extension of the abelian algebra $\mathbb{R}^2$.

Example 3.9 Let M be a finite-dimensional manifold. One can show that the Lie algebra $\text{Vect}(M)$ of vector fields on M is perfect. The universal central extension of $\text{Vect}(M)$ for the case $M = S^1$ is called the Virasoro algebra, and we describe it in detail in Section 2 of Chapter II.

Example 3.10 For a simple Lie algebra $\mathfrak{g}$ and any n-dimensional compact manifold M, the current Lie algebra $\mathfrak{g}^M$ is perfect. (More generally, for any perfect finite-dimensional Lie algebra $\mathfrak{g}$ the Lie algebra $\mathfrak{g}^M$ is perfect.) Its universal central extension $\widetilde{\mathfrak{g}}^M$ can be constructed as follows. Let $\langle \, , \, \rangle$ be a nondegenerate symmetric invariant bilinear form on $\mathfrak{g}$, where the invariance means that $\langle [A, B], C \rangle = \langle A, [B, C] \rangle$ for all A, B, $C \in \mathfrak{g}$. Denote by $\Omega^1(M)$ the set of 1-forms on M and let $d\Omega^0(M)$ be the subset of exact 1-forms. Now we define the 2-cocycle ω on $\mathfrak{g}^M$ with values in $\Omega^1(M)/d\Omega^0(M)$ via

$$\omega(X, Y) := \langle X, dY \rangle,$$

where $X, Y \in \mathfrak{g}^M$. The antisymmetry of ω is immediate, while the cocycle identity follows from the Jacobi identity in $\mathfrak{g}^M$ and the invariance of the bilinear form. So ω defines a central extension of $\mathfrak{g}^M$. For a proof of universality of this central extension see, e.g., [322, 247].

In the case of $M = S^1$ the corresponding space $\Omega^1(S^1)/d\Omega^0(S^1)$ is one-dimensional. The current algebra on S^1 is called the loop algebra associated to $\mathfrak{g}$, and it has the universal central extension by the $\mathbb{R}$- (or $\mathbb{C}$)-valued 2-cocycle

$$\omega(X, Y) := \int_{S^1} \langle X, dY \rangle.$$

We discuss loop algebras and their generalizations in detail in Sections 1 and 5 of Chapter II.

3.2 Central Extensions of Lie Groups

Central extensions of Lie groups can be defined similarly to those of Lie algebras. However, unlike the case of Lie algebras, not all group extensions can be described explicitly by cocycles. This is why we start with the alternative definition of the extensions via exact sequences.

Definition 3.11 A *central extension* $\widetilde{G}$ of a Lie group G by an abelian Lie group H is an exact sequence of Lie groups

$$\{e\} \to H \to \widetilde{G} \to G \to \{e\}$$

such that the image of H lies in the center of $\widetilde{G}$. (Here $\{e\}$ is the trivial group containing only the identity element.) Morphisms and equivalence of two central extensions are defined analogously to the case of Lie algebras.

If the central extension $\widetilde{G}$ is topologically a direct product of G and H, $\widetilde{G} = G \times H$ (or, equivalently, if there is a smooth section in the principal H-bundle $\widetilde{G} \to G$),[4] one can define the multiplication in $\widetilde{G}$ as follows:

$$(g_1, h_1) \cdot (g_2, h_2) = (g_1 g_2, \gamma(g_1, g_2) h_1 h_2)$$

for a smooth map $\gamma : G \times G \to H$, which is similar to the case of Lie algebra central extensions. The associativity of this multiplication corresponds to the so-called *group cocycle identity* on the map γ.

Definition 3.12 Let G and H be Lie groups and suppose H is abelian. A smooth map $\gamma : G \times G \to H$ that satisfies

$$\gamma(g_1 g_2, g_3)\gamma(g_1, g_2) = \gamma(g_1, g_2 g_3)\gamma(g_2, g_3)$$

is called a smooth *group 2-cocycle* on G with values in H.

A smooth 2-cocycle on G with values in H is called a 2-*coboundary* if there exists a smooth map $\lambda : G \to H$ such that $\gamma(g_1, g_2) = \lambda(g_1)\lambda(g_2)\lambda(g_1 g_2)^{-1}$. As before, the group 2-coboundaries correspond to the trivial group extensions, after a possible change of coordinates (more precisely, of the trivializing section for $\widetilde{G} \to G$). Similarly, two group 2-cocycles define isomorphic extensions if they differ by a 2-coboundary. This explains the following fact.

Proposition 3.13 *There is a one-to-one correspondence between the set of central extensions of G by H that admit a smooth section and the elements in the second cohomology group $H^2(G, H) := Z(G, H)/B(G, H)$. Here $Z(G, H)$ and $B(G, H)$ denote respectively the sets of smooth 2-cocycles and 2-coboundaries on G, with the natural abelian group structure.*

However, in contrast to the case of Lie algebras, there exist central extensions of Lie groups that do not admit a smooth section, and hence cannot be defined by smooth 2-cocycles. We will encounter examples for such groups in Chapter II.

A central extension of a Lie group G always defines a central extension of the corresponding Lie algebra. The converse need not be true: the existence of a Lie group for a given Lie algebra is not guaranteed in infinite dimensions. Instead, one says that a central extension $\widetilde{\mathfrak{g}}$ of a Lie algebra $\mathfrak{g}$ lifts to the group level if there exists a central extension $\widetilde{G}$ of the group G whose Lie algebra is given by $\widetilde{\mathfrak{g}}$. If the group central extension $\widetilde{G}$ by H is defined by a group 2-cocycle γ, one can recover the Lie algebra 2-cocycle defining the corresponding central extension $\widetilde{\mathfrak{g}}$ of the Lie algebra $\mathfrak{g}$ directly from the group cocycle γ by appropriate differentiation.

---

[4] We always require central extensions of Lie groups to have smooth local sections, in order to secure the existence of a continuous linear section for the corresponding Lie algebra extensions.

Proposition 3.14 *Let H be an abelian Lie group with a Lie algebra $\mathfrak{h}$, and let γ be an H-valued 2-cocycle on G defining a central extension $\widetilde{G}$. Then the $\mathfrak{h}$-valued 2-cocycle ω defining the corresponding central extension $\widetilde{\mathfrak{g}}$ of the Lie algebra $\mathfrak{g}$ is given by*

$$\omega(X, Y) = \frac{d^2}{dt\,ds}\Big|_{t=0,s=0}\gamma(g_t, h_s) - \frac{d^2}{dt\,ds}\Big|_{t=0,s=0}\gamma(h_s, g_t),$$

where g_t is a smooth curve in G such that $\frac{d}{dt}\big|_{t=0}g_t = X$, and h_s is a smooth curve in G such that $\frac{d}{ds}\big|_{s=0}h_s = Y$.

Exercise 3.15 Prove the above proposition.

Example 3.16 Let G be $\mathbb{R}^2 = \{(a, b)\}$ with the natural abelian group structure. The three-dimensional *Heisenberg group* $\widetilde{G}$ can be defined as the following matrix group:

$$\widetilde{G} = \left\{ \begin{pmatrix} 1 & a & c \\ 0 & 1 & b \\ 0 & 0 & 1 \end{pmatrix} \mid a, b, c \in \mathbb{R} \right\},$$

and is is a central extension of the group G. One verifies directly that the central extension is defined via the $\mathbb{R}$-valued group 2-cocycle γ given by $\gamma((a, b), (a', b')) = ab'$. Using Proposition 3.14, we see that the infinitesimal form of the cocycle γ is given by

$$\omega((A, B), (A', B')) = AB' - A'B,$$

so that the Lie algebra of $\widetilde{G}$ is the three-dimensional Heisenberg algebra discussed in Example 3.5.

4 The Euler Equations for Lie Groups

The Euler equations form a class of dynamical systems closely related to Lie groups and to the geometry of their coadjoint orbits. To describe them we start with generalities on Poisson structures and Hamiltonian systems, before bridging them to Lie groups. Although the manifolds considered in this section are finite-dimensional, we will see later in the book that most of the notions and formulas discussed here are applicable in the infinite-dimensional context (where the dual $\mathfrak{g}^*$ of a Lie algebra $\mathfrak{g}$ stands for its smooth dual).

4.1 Poisson Structures on Manifolds

Definition 4.1 A *Poisson structure* on a manifold M is a bilinear operation on functions

$$\{\,,\,\}: \quad C^\infty(M) \times C^\infty(M) \to C^\infty(M)$$

satisfying the following properties:
(i) antisymmetry:
$$\{f, g\} = -\{g, f\},$$
(ii) the Jacobi identity:
$$\{f, \{g, h\}\} + \{g, \{h, f\}\} + \{h, \{f, g\}\} = 0, \quad \text{and}$$
(iii) the Leibniz identity:
$$\{f, gh\} = \{f, g\}h + \{f, h\}g$$
for any functions $f, g, h \in C^\infty(M)$.

The first two properties mean that a Poisson structure defines a Lie algebra structure $\{\ ,\ \}$ on the space $C^\infty(M)$ of smooth functions on M, while the third property implies that $\{f, \cdot\} : C^\infty(M) \to C^\infty(M)$ is a derivation for any function $f \in C^\infty(M)$. Since each derivation on the space of functions is the Lie derivative along an appropriate vector field, the Poisson structure can be thought of as a map from functions to the corresponding vector fields on the manifold:

Definition 4.2 Let $H : M \to \mathbb{R}$ be any smooth function on a Poisson manifold M. Such a function H defines a vector field ξ_H on M by $L_{\xi_H} g = \{H, g\}$ for any test function $g \in C^\infty(M)$. The vector field ξ_H is called the *Hamiltonian field* corresponding to the *Hamiltonian function* H with respect to the Poisson bracket $\{\ ,\ \}$.

We call a function $F : M \to \mathbb{R}$ a *Casimir function* on a Poisson manifold M if it generates the zero Hamiltonian field, i.e., if the Poisson bracket of the function F with any other function vanishes everywhere on M.

Remark 4.3 Let M be a manifold with a Poisson structure $\{\ ,\ \}$, and we fix some point $m \in M$. All Hamiltonian vector fields on M evaluated at the point $m \in M$ span a subspace of the tangent space $T_m M$. Thus, the Poisson structure defines a distribution of such subspaces on the manifold M (i.e., a subbundle of the tangent bundle TM) by varying the point m.[5] Note that the dimension of this distribution can differ from one point to another. This distribution is integrable, according to the Frobenius theorem, since the commutator of two Hamiltonian vector fields is again Hamiltonian. Therefore, it gives rise to a (possibly singular) *foliation* of the Poisson manifold M [384].

---

[5] Alternatively, a Poisson structure can be defined by specifying a *bivector field* Π on the manifold M, i.e., a section of $TM^{\wedge 2}$:

$$\{f, g\} = \Pi(df, dg).$$

Such a bivector field defines a distribution on M as the images of the map $\Pi : T^*M \to TM$.

The leaves of this foliation are called *symplectic leaves*. (In short, two points belong to the same symplectic leaf if they can be joined by a path whose velocity at any point is a Hamiltonian vector.)

Definition 4.4 A pair (N, ω) consisting of a manifold N and a 2-form ω on N is called a *symplectic manifold* if ω is closed $(d\omega = 0)$ and nondegenerate. (In the case of infinite dimensions, the form ω is required to be nondegenerate in the sense that for each point $p \in N$ and any nonzero vector $X \in T_p N$, there exists another vector $Y \in T_p M$ such that $\omega_p(X, Y) \neq 0$.) The 2-form ω is called the *symplectic form* on the manifold N.

The reason for the name "symplectic leaves" in Remark 4.3 is that one can define a symplectic 2-form ω on each leaf. It suffices to fix its values on Hamiltonian vector fields ξ_f and ξ_g at any point:

$$\omega(\xi_f, \xi_g) := \{f, g\},$$

since the tangent space of each leaf is generated by Hamiltonian fields.

Exercise 4.5 Show that the 2-form ω defined on the leaf through a point $m \in M$ is closed and nondegenerate.

Note that Casimir functions, by definition, are constant on the leaves of the above foliation, and the codimension of generic symplectic leaves on a Poisson manifold M is equal to the number of (locally functionally) independent Casimir functions on M.

Example 4.6 For the Poisson structure

$$\{f, g\} := \frac{\partial f}{\partial x}\frac{\partial g}{\partial y} - \frac{\partial f}{\partial y}\frac{\partial g}{\partial x}$$

in $\mathbb{R}^3 = \{(x, y, z)\}$, its symplectic leaves are the planes $z = \text{const}$. The coordinate function z, or any function $F = F(z)$, is a Casimir function for this Poisson manifold.

Remark 4.7 Locally, a Poisson manifold near any point p splits into the product of a symplectic space and a Poisson manifold whose rank at p is zero [384]. The symplectic space is a neighborhood of the symplectic leaf passing through p, while the Poisson manifold of zero rank represents the transverse Poisson structure at the point p.

Below we will see that this splitting works in many (but not all!) infinite-dimensional examples: Poisson structures can have infinite-dimensional symplectic leaves and finite-dimensional Poisson transversals.

4.2 Hamiltonian Equations on the Dual of a Lie Algebra

Let G be a Lie group (finite- or infinite-dimensional) with Lie algebra $\mathfrak{g}$, and let $\mathfrak{g}^*$ denote (the smooth part of) its dual.

Definition 4.8 The natural *Lie–Poisson* (or *Kirillov–Kostant* Poisson) structure $\{\ ,\ \}_{LP}$ on the dual Lie algebra $\mathfrak{g}^*$,

$$\{\ ,\ \}_{LP} : C^\infty(\mathfrak{g}^*) \times C^\infty(\mathfrak{g}^*) \to C^\infty(\mathfrak{g}^*)\,,$$

is defined via

$$\{f, g\}_{LP}(m) := \langle [df_m,\ dg_m],\ m \rangle$$

for any $m \in \mathfrak{g}^*$ and any two smooth functions f, g on $\mathfrak{g}^*$; see Figure 4.1. (Here df_m is the differential of the smooth function f taken at the point m, understood as an element of the space $\mathfrak{g}$ itself, and $\langle\ ,\ \rangle$ is the natural pairing between the dual spaces $\mathfrak{g}$ and $\mathfrak{g}^*$.)

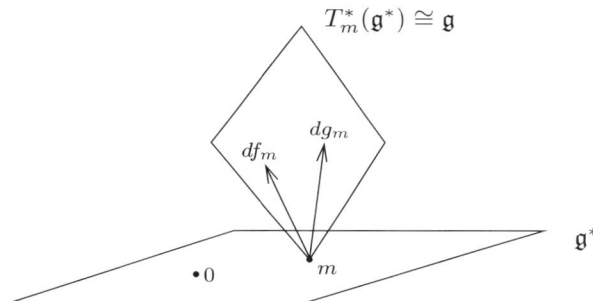

Fig. 4.1. Defining the Lie–Poisson structure: $df_m,\ dg_m \in \mathfrak{g}$, while $m \in \mathfrak{g}^*$.

Proposition 4.9 *The Hamiltonian equation corresponding to a function H and the natural Lie–Poisson structure $\{\ ,\ \}_{LP}$ on $\mathfrak{g}^*$ is given by*

$$\frac{d}{dt} m(t) = -\operatorname{ad}^*_{dH_{m(t)}} m(t).$$

This equation is called the Euler–Poisson equation *on $\mathfrak{g}^*$.*

PROOF.   Let $f \in C^\infty(\mathfrak{g}^*)$ be an arbitrary function. Then

$$\begin{aligned}
L_{\xi_H} f(m) &= \{H, f\}(m) = \langle [dH_m, df_m], m \rangle \\
&= \langle \operatorname{ad}_{dH_m}(df_m), m \rangle = -\langle df_m, \operatorname{ad}^*_{dH_m}(m) \rangle.
\end{aligned}$$

Since the Lie derivative of a function f along a vector field is the evaluation of the function's differential df on this vector field, this implies that $\xi_H(m) = -\operatorname{ad}^*_{dH_m}(m)$, which is the assertion.   □

Corollary 4.10 *The symplectic leaves of* $\{\ ,\ \}_{LP}$ *on* $\mathfrak{g}^*$ *are the coadjoint orbits of* G. *In particular, all (finite-dimensional) coadjoint orbits have even dimension.*

PROOF.   Denote by $\mathcal{O}_m$ the coadjoint orbit through a point $m \in \mathfrak{g}^*$ in the dual space. Let $H \in C^\infty(\mathfrak{g}^*)$ be a function on the dual. For any vector $v \in \mathfrak{g}$ of the Lie algebra, one can represent it as $v = dH_m$ by taking an appropriate function H. Therefore, one can obtain as Hamiltonian vectors $\operatorname{ad}^*_{dH_m}(m)$ at the point m all vectors in the image of $\operatorname{ad}^*_\mathfrak{g}(m)$, i.e., all vectors in the tangent space to the orbit $T_m\mathcal{O}_m := T_m(\operatorname{Ad}^*_G(m))$. By definition, all Hamiltonian vectors span the tangent space to any symplectic leaf at each point m, which proves that coadjoint orbits are exactly the symplectic leaves of the Lie–Poisson bracket.   □

Corollary 4.11 *Let* $A : \mathfrak{g} \to \mathfrak{g}^*$ *be an invertible self-adjoint operator.*[6] *For the quadratic Hamiltonian function* $H : \mathfrak{g}^* \to \mathbb{R}$ *defined by* $H(m) := \frac{1}{2}\langle m, A^{-1}m \rangle$ *the corresponding Hamiltonian equation is*

$$\frac{d}{dt}m(t) = -\operatorname{ad}^*_{A^{-1}m(t)}m(t)\,. \tag{4.5}$$

Indeed, $dH_m(m) = A^{-1}m$ for any $m \in \mathfrak{g}^*$.

Definition 4.12 An invertible self-adjoint operator $A : \mathfrak{g} \to \mathfrak{g}^*$ defining the quadratic Hamiltonian H is called an *inertia operator on* $\mathfrak{g}$.

4.3 A Riemannian Approach to the Euler Equations

It turns out that the Euler–Poisson equations with quadratic Hamiltonians have a beautiful Riemannian reformulation.

V. Arnold suggested in [12] the following general setup for the Euler equation describing a geodesic flow on an arbitrary Lie group. Consider a (possibly infinite-dimensional) Lie group G, which can be thought of as the configuration space of some physical system. (Examples from [12, 18]: SO(3) for a rigid

---

[6] Note that one can define the "self-adjointness property" for an operator from a space to its dual, similarly to a self-adjoint operator acting on a given space with respect to a fixed pairing.

body or the group $SDiff(M)$ of volume-preserving diffeomorphisms for an ideal fluid filling a domain M.) The tangent space at the identity of the Lie group G is the corresponding Lie algebra $\mathfrak{g}$. Fix some (positive definite) quadratic form, the energy, on $\mathfrak{g}$. We consider left (or right) translations of this quadratic form to the tangent space at any point of the group (the "translational symmetry" of the energy). In this way, the energy defines a left- (respectively, right-) invariant Riemannian metric on the group G. The geodesic flow on G with respect to this energy metric represents extremals of the least-action principle, i.e., possible motions of our physical system.[7] To describe a geodesic on the Lie group G with an initial velocity $v(0)$, we transport its velocity vector at any moment t to the identity of the group using the left (respectively, right) translation. This way we obtain the evolution law for $v(t)$ on the Lie algebra $\mathfrak{g}$.

To fix the notation, let $(\ ,\ )$ be some left-invariant metric on the group G. The geodesic flow with respect to this metric is a dynamical system on the tangent bundle TG of the group G. We can pull back this system to the Lie algebra $\mathfrak{g}$ of the group G by left translation. That is, if $g(t)$ is a geodesic in the group G with tangent vector $g'(t)$, then the pullback $v(t) = l^*_{g(t)^{-1}} g'(t)$ is an element of the Lie algebra $\mathfrak{g}$. (In the case of a right-invariant metric, we set $v(t) = r^*_{g(t)^{-1}} g'(t)$.) Hence, the geodesic equations for $g(t)$ give us a dynamical system

$$\frac{d}{dt} v(t) = B(v(t)) \qquad (4.6)$$

on the Lie algebra $\mathfrak{g}$ of the group G, where $B : \mathfrak{g} \to \mathfrak{g}$ is a (nonlinear) operator.

Definition 4.13 The dynamical system (4.6) on the Lie algebra $\mathfrak{g}$ describing the evolution of the velocity vector of a geodesic in a left-invariant metric on the Lie group G is called the *Euler (or Euler–Arnold) equation* corresponding to this metric on G.

It turns out that the Euler equation for a Lie group G can be viewed as a Hamiltonian equation on the dual of the Lie algebra $\mathfrak{g}$ in the following way. Observe that the metric $(\ ,\ )_e$ at the identity $e \in G$ defines a nondegenerate bilinear form on the Lie algebra $\mathfrak{g}$, and therefore, it also determines an inertia operator $A : \mathfrak{g} \to \mathfrak{g}^*$ such that $(v, w)_e = \langle A(v), w \rangle$ for all $v, w \in \mathfrak{g}$. This identification $A : \mathfrak{g} \to \mathfrak{g}^*$ allows one to rewrite the Euler equation on the dual space $\mathfrak{g}^*$; see Figure 4.2. Now, setting $m = A(v)$, one can relate the geodesic equation (4.6) on the Lie algebra $\mathfrak{g}$ to the Hamiltonian equation on the dual $\mathfrak{g}^*$ with respect to the Hamiltonian function $H(m) = \frac{1}{2}\langle m, A^{-1}m \rangle$:

---

[7] Usually, the finite-dimensional examples below are related to the left invariance, while the infinite-dimensional ones with the right invariance of the metric. In particular, for a rigid body one has to consider left translations on SO(3), while for fluids, one must consider the right ones on $SDiff(M)$.

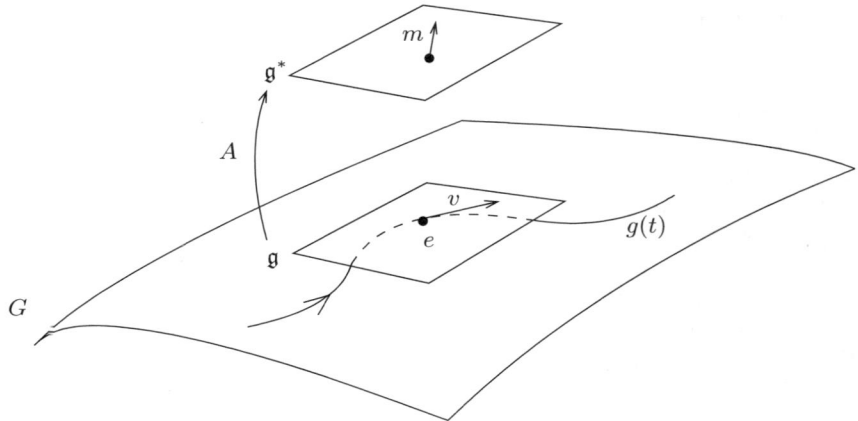

Fig. 4.2. The vector v in the Lie algebra $\mathfrak{g}$ traces the evolution of the velocity vector of a geodesic $g(t)$ on the group. The inertia operator A sends v to a vector m in the dual space $\mathfrak{g}^*$.

Theorem 4.14 (Arnold [12, 13, 18]) *For the left-invariant metric on a group generated by an inertia operator* $A : \mathfrak{g} \to \mathfrak{g}^*$, *the Euler (or the geodesic) equation (4.6) assumes the form*

$$\frac{d}{dt}m(t) = -\operatorname{ad}^*_{A^{-1}m(t)} m(t)$$

on the dual space $\mathfrak{g}^*$.[8]

We postpone the proof of this theorem until the end of this section.

Remark 4.15 The underlying reason for the Riemannian reformulation is the fact that any geodesic problem in *Riemannian* geometry can be described in terms of *symplectic* geometry. Geodesics on M are extremals of a quadratic Lagrangian on TM (coming from the metric on M). They can also be described by the Hamiltonian flow on T^*M for the quadratic Hamiltonian function obtained from the Lagrangian via the Legendre transform.

If the manifold is a group G with a left-invariant metric, then there exists the group action on the tangent bundle TG, as well as on the cotangent bundle T^*G. The left translations on the group trivialize the cotangent bundle $T^*G \cong G \times \mathfrak{g}^*$ and identify any cotangent space of G with $\mathfrak{g}^*$. By taking the quotient with respect to the group action, from the (symplectic) cotangent bundle T^*G we obtain the dual Lie algebra $\mathfrak{g}^* = T^*G|_e$ equipped with the negative

---

[8] Note that these signs are different from the conventions in the book [24], since we have used a different definition of Ad^* here (see equation (2.3)) in order to have the group coadjoint representation, rather than the antirepresentation.

of the Lie–Poisson structure (cf. Section 5 below, on symplectic reduction). The Hamiltonian function on T^*G is dual to the Riemannian metric (viewed as a form on TG), and its restriction to $\mathfrak{g}^*$ is the quadratic form $H(m) = \frac{1}{2}\langle m, A^{-1}m\rangle$, where $m \in \mathfrak{g}^*$.

The geodesics of a left-invariant metric on G correspond to the Hamiltonian function $H(m)$ with respect to the standard Lie–Poisson structure.

Remark 4.16 Instead of using a left-invariant metric on G, we could have used a right-invariant one. This changes the signs in the Euler equation, so that one obtains

$$\frac{d}{dt}m = ad^*_{A^{-1}m}(m).$$

Now the geodesics in a right-invariant metric correspond to the Hamiltonian $-H(m)$.

Example 4.17 Let us consider the group SO(3). The Lie algebra $\mathfrak{so}(3)$ of SO(3) can be identified with $\mathbb{R}^3$ such that the Lie bracket on $\mathfrak{g}$ is the cross product on $\mathbb{R}^3$: $[u, v] = v \times u$. Let A be a symmetric nondegenerate 3×3 matrix, which we view as an inertia operator for a left-invariant metric on SO(3). Then by Arnold's theorem, the Euler equation on $\mathfrak{so}(3)^*$ is given by

$$\frac{d}{dt}m = m \times A^{-1}m.$$

For $A = \mathrm{diag}(I_1, I_2, I_3)$ one obtains the classical Euler equations for a rigid body in $\mathbb{R}^3$:

$$\frac{d}{dt}m_i = (I_k^{-1} - I_j^{-1})m_j m_k$$

for (i, j, k) being a cyclic permutation of $(1, 2, 3)$. Similarly, for $G = \mathrm{SO}(n)$, one obtains the Euler equation for a higher-dimensional rigid body (see Remark 4.28 below).

Example 4.18 Many other conservative dynamical systems in mathematical physics also describe geodesic flows on appropriate Lie groups. In Table 4.1 we list several examples of such systems to emphasize the range of applications of this approach. The choice of a group G (column 1) and an energy metric E (column 2) defines the corresponding Euler equations (column 3).

We discuss many of these examples later in the book. There are plenty of other interesting systems that fit into this framework, such as, e.g., the super-KdV equation or gas dynamics. This list is by no means complete, and we refer to [24, 252] for more details.

Group	Metric	Equation
SO(3)	$\langle \omega, A\omega \rangle$	Euler top
SO(3) $\ltimes \mathbb{R}^3$	quadratic forms	Kirchhoff equation for a body in a fluid
SO(n)	Manakov's metrics	n-dimensional top
Diff(S^1)	L^2	Hopf (or, inviscid Burgers) equation
Virasoro	L^2	KdV equation
Virasoro	H^1	Camassa–Holm equation
Virasoro	$\dot{H}^1$	Hunter–Saxton (or Dym) equation
SDiff(M)	L^2	Euler ideal fluid
SDiff(M)	H^1	averaged Euler flow
SDiff(M) $\ltimes$ SVect(M)	$L^2 + L^2$	Magnetohydrodynamics
Maps(S^1, SO(3))	H^{-1}	Heisenberg magnetic chain

Table 4.1: Euler equations related to various Lie groups.

Now we return to the proof of Arnold's theorem.

PROOF OF THEOREM 4.14.   Consider the energy function (or Lagrangian) $L : TG \to \mathbb{R}$ defined by the left-invariant metric (,) on the group G:

$$L(g, v) = \frac{1}{2}(v, v)_g \, ,$$

where (,)$_g$ is the metric at the point $g \in G$. Then by definition, a geodesic path $g(t)$ on G satisfies the variational principle

$$\delta \int L(g(t), g'(t))dt = 0 \qquad (4.7)$$

with fixed endpoints. (Here and later, δ denotes the variational derivative, and the prime $'$ stands for the time derivative d/dt.)

To simplify the notation, we write $g^{-1}(t)g'(t)$ for $l^*_{g^{-1}(t)}g'(t)$. (If G is a matrix group, this notation agrees with the usual meaning of the expression $g^{-1}(t)g'(t)$ as a matrix product.) Since the metric (,) on the group G is left-invariant, we can write

$$(g'(t), g'(t))_{g(t)} = (g^{-1}(t)g'(t), g^{-1}(t)g'(t))_e \, .$$

Then we can calculate

$$\delta \int \frac{1}{2}(g^{-1}g', g^{-1}g')_e dt = \int (\delta(g^{-1}g'), g^{-1}g')_e dt \, . \qquad (4.8)$$

Note that we have

$$\delta(g^{-1}g') = g^{-1}\delta g' - g^{-1}\delta g g^{-1} g' = (g^{-1}\delta g)' + [g^{-1}g', g^{-1}\delta g] \, ,$$

since

$$(g^{-1}\delta g)' = g^{-1}\delta g' - g^{-1}g'g^{-1}\delta g\,.$$

Thus, the left-hand-side in equation (4.8) becomes

$$\int (\delta(g^{-1}g'), g^{-1}g')_e dt = \int ((g^{-1}\delta g)', g^{-1}g')_e dt + \int ([g^{-1}g', g^{-1}\delta g], g^{-1}g')_e dt$$

$$= -\int (g^{-1}\delta g, (g^{-1}g')')_e dt + \int ([g^{-1}g', g^{-1}\delta g], g^{-1}g')_e dt\,,$$

where we have used integration by parts in the last step. (Since we confined ourselves to variations of the path g with fixed endpoints, we do not pick up any boundary terms in the integration by parts.)

Now set $v(t) := g^{-1}(t)g'(t)$, and let $A : \mathfrak{g} \to \mathfrak{g}^*$ be the inertia operator defined by the metric $(\ ,\ )_e$: $(u, w)_e = \langle u, Aw \rangle$. Then the right-hand side in the latter equation becomes

$$-\int (g^{-1}\delta g, (g^{-1}g')')_e dt + \int ([g^{-1}g', g^{-1}\delta g], g^{-1}g')_e dt$$

$$= -\int \langle g^{-1}\delta g, (Av)' \rangle dt + \int \langle \mathrm{ad}_v(g^{-1}\delta g), Av \rangle dt$$

$$= -\int \langle g^{-1}\delta g, (Av)' \rangle dt - \int \langle g^{-1}\delta g, \mathrm{ad}_v^*(Av) \rangle dt = 0\,.$$

This implies

$$(Av)' = -\mathrm{ad}_v^*(Av)\,.$$

Rewriting this equation in terms of $m = Av$ finishes the proof of Theorem 4.14. $\qquad\square$

4.4 Poisson Pairs and Bi-Hamiltonian Structures

A *first integral* (or a conservation law) for a vector field ξ on a manifold M is a function on M invariant under the flow of this field. In this section we will show that if the vector field ξ is a Hamiltonian vector field with respect to two different Poisson structures on the manifold M that are compatible in a certain sense, there is a way of constructing first integrals for such a field.

Definition 4.19 Two Poisson structures $\{\ ,\ \}_0$ and $\{\ ,\ \}_1$ on a manifold M are said to be *compatible* (or *form a Poisson pair*) if for every $\lambda \in \mathbb{R}$ the linear combination $\{\ ,\ \}_0 + \lambda \{\ ,\ \}_1$ is again a Poisson bracket on M.

A dynamical system $\frac{d}{dt}m = \xi(m)$ on M is called *bi-Hamiltonian* if the vector field ξ is Hamiltonian with respect to both structures $\{\ ,\ \}_0$ and $\{\ ,\ \}_1$.

Our main example of a manifold that admits a Poisson pair is the dual space $\mathfrak{g}^*$ of a Lie algebra $\mathfrak{g}$. One Poisson structure on the space $\mathfrak{g}^*$ is given by the usual Lie–Poisson bracket $\{\ ,\ \}_{LP}$. We can define a second Poisson structure on $\mathfrak{g}^*$ by "freezing" the Lie–Poisson bracket at any point $m_0 \in \mathfrak{g}^*$:

Definition 4.20 The constant Poisson bracket on $\mathfrak{g}^*$ associated to a point $m_0 \in \mathfrak{z}^*$ is the bracket $\{ \ , \ \}_0$ defined on two smooth functions f, g on $\mathfrak{g}^*$ by

$$\{f, g\}_0(m) := \langle [df_m, dg_m], m_0 \rangle .$$

The Poisson bracket $\{ \ , \ \}_0$ depends on the *freezing point* $m_0 \in \mathfrak{g}^*$. Note that at the point m_0 itself the two Poisson brackets $\{ \ , \ \}_{LP}$ and $\{ \ , \ \}_0$ coincide. While the symplectic leaves of the Lie–Poisson bracket $\{ \ , \ \}_{LP}$ are the coadjoint orbits $\mathcal{O}_m$ of the Lie group G, the symplectic leaves of the constant bracket $\{ \ , \ \}_0$ are given by all translations of the tangent space $T_{m_0}\mathcal{O}_{m_0}$ to the coadjoint orbit $\mathcal{O}_{m_0}$ through the point m_0 (see Figure 4.3).

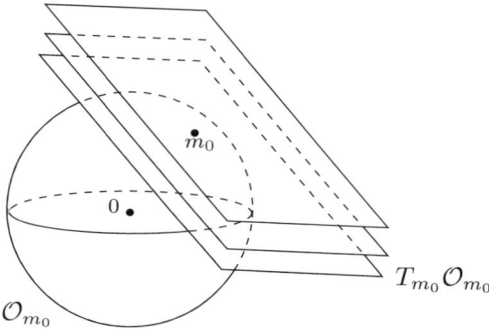

Fig. 4.3. Coadjoint orbit $\mathcal{O}_{m_0}$ through m_0 and leaves of the Poisson bracket frozen at m_0.

Lemma 4.21 *The Poisson brackets $\{ \ , \ \}_{LP}$ and $\{ \ , \ \}_0$ are compatible for every "freezing point" $m_0 \in \mathfrak{g}^*$.*

PROOF. We have to check that $\{ \ , \ \}_\lambda = \{ \ , \ \}_{LP} + \lambda \{ \ , \ \}_0$ is a Poisson bracket on $\mathfrak{g}^*$ for all $\lambda \in \mathbb{R}$. The latter is true since $\{ \ , \ \}_\lambda$ is simply the bracket $\{ \ , \ \}_{LP}$ shifted by $-\lambda m_0$. $\qquad\square$

Remark 4.22 Explicitly, the Hamiltonian equation on $\mathfrak{g}^*$ with the Hamiltonian function F and computed with respect to the constant Poisson structure frozen at a point $m_0 \in \mathfrak{g}^*$ has the following form:

$$\frac{dm}{dt} = -\mathrm{ad}^*_{dF_m} m_0 , \tag{4.9}$$

as a modification of Proposition 4.9 shows.

Going back to the general situation, let $\{\ ,\ \}_0$ and $\{\ ,\ \}_1$ be a Poisson pair on a manifold M. In this case one can generate a bi-Hamiltonian dynamical system by producing a sequence of Hamiltonians in involution, according to the following *Lenard–Magri scheme* [246, 321]. Consider the Poisson bracket $\{\ ,\ \}_\lambda = \{\ ,\ \}_0 + \lambda\{\ ,\ \}_1$ for any λ. Let h_λ be a *Casimir function* on M for this bracket, i.e., a function on the manifold M that is parametrized by λ and satisfies $\{h_\lambda, f\}_\lambda = 0$ for all smooth functions $f \in C^\infty(M)$ and $\lambda \in \mathbb{R}$. Furthermore, suppose that the function h_λ can be expanded into a power series in λ, i.e., that we can write

$$h_\lambda = \sum_{i=0}^{\infty} \lambda^i h_i, \tag{4.10}$$

where each coefficient h_i is a smooth function on M. Any function h_i defines a Hamiltonian vector field ξ_i on M with respect to the Poisson bracket $\{\ ,\ \}_1$ by setting $\{h_i, f\}_1 = L_{\xi_i} f$ for all $f \in C^\infty(M)$.

Theorem 4.23 *The functions h_i, $i = 0, 1, \dots$ are Hamiltonians of a hierarchy of bi-Hamiltonian systems. In other words, each function h_i generates the Hamiltonian vector field ξ_i on M with respect to the Poisson bracket $\{\ ,\ \}_1$, which is also Hamiltonian for the other bracket $\{\ ,\ \}_0$ with the Hamiltonian function $-h_{i+1}$:*

$$\{h_i, f\}_1 = L_{\xi_i} f = -\{h_{i+1}, f\}_0$$

for any f. Other functions h_j, $j \neq i$, are first integrals of the corresponding dynamical systems ξ_i.

In other words, the functions h_i, $i = 0, 1, \dots$ are in involution with respect to each of the two Poisson brackets $\{\ ,\ \}_0$ and $\{\ ,\ \}_1$:

$$\{h_i, h_j\}_k = 0$$

for all $i \neq j$ and for $k = 0, 1$.

PROOF.   Since h_λ is a Casimir function for the Poisson bracket $\{\ ,\ \}_\lambda$, we have $\{h_\lambda, f\}_\lambda = 0$ for all smooth functions f on M. Substituting for h_λ its power series expansion (4.10), we get

$$0 = \{h_\lambda, f\}_\lambda = \left\{ \sum_{i=0}^{\infty} \lambda^i h_i, f \right\}_\lambda = \left\{ \sum_{i=0}^{\infty} \lambda^i h_i, f \right\}_0 + \lambda \left\{ \sum_{i=0}^{\infty} \lambda^i h_i, f \right\}_1.$$

Collecting the coefficients at the powers of λ we find that $\{h_0, f\}_0 = 0$ and

$$\{h_i, f\}_0 = -\{h_{i-1}, f\}_1.$$

The first identity expresses the fact that h_0 is a Casimir function for the bracket $\{\ ,\ \}_0$. The next one says that the Hamiltonian field for h_1 with respect to $\{\ ,\ \}_0$ coincides with the Hamiltonian field for $-h_0$ and the bracket $\{\ ,\ \}_1$, and so on.

To see that every function h_i is a first integral for the equation generated by h_j with respect to each bracket, we have to show that $\{h_i, h_j\}_k = 0$ for $i \neq j$ and $k = 0, 1$. Indeed, for instance, for $i < j$ and $k = 1$ we have

$$\{h_i, h_j\}_1 = -\{h_i, h_{j+1}\}_0 = \{h_{i-1}, h_{j+1}\}_1 = \cdots = -\{h_0, h_{i+j+1}\}_0 = 0,$$

since h_0 is a Casimir for the bracket $\{.,.\}_0$, i.e., in involution with any function, and in particular, with h_{i+j+1}. $\square$

Remark 4.24 The fact that $\{h_i, h_j\}_k = 0$ for $k = 0, 1$ means that the functions h_j are first integrals of the Hamiltonian vector fields ξ_i. So if the functions h_j are independent, Theorem 4.23 provides us with an infinite list of first integrals for each of the fields ξ_i. In this case one says that the h_i are the Hamiltonians of a hierarchy of bi-Hamiltonian systems. We will treat the KdV equation as a bi-Hamiltonian system from this viewpoint in Section II.2.4.

Exercise 4.25 Suppose that a manifold M admits two compatible Poisson structures $\{\ ,\ \}_0$ and $\{\ ,\ \}_1$. Show that if symplectic leaves of $\{\ ,\ \}_\lambda = \{\ ,\ \}_0 + \lambda\{\ ,\ \}_1$ are of codimension greater than 1, and if there are several independent Casimirs $h_\lambda^{(1)}, h_\lambda^{(2)}, \ldots$, then all the coefficients of their expansions in λ are in mutual involution with respect to both brackets, e.g., $\{h_i^{(1)}, h_j^{(2)}\}_k = 0$.

4.5 Integrable Systems and the Liouville–Arnold Theorem

The more first integrals a dynamical system has, the less chaotically it behaves. For a Hamiltonian system the notion of complete integrability corresponds to the "least chaotic" and "most ordered" structure of its trajectories.

Definition 4.26 A Hamiltonian system on a symplectic $2n$-dimensional manifold M is called *(completely) integrable* if it has n integrals in involution that are functionally independent almost everywhere on M. The Hamiltonian function is one of the above first integrals. (Alternatively, one can avoid specifying which of them is a Hamiltonian and describe an *integrable system* as a set of n functions $f_1, \ldots, f_n$ that are functionally independent almost everywhere and commute pairwise,

$$\{f_i, f_j\} = 0 \quad \text{for all } 1 \leq i, j \leq n,$$

with respect to the natural Poisson bracket defined by the symplectic structure on M.)

Example 4.27 Every Hamiltonian system with one degree of freedom is completely integrable, since it always possesses one first integral, the Hamiltonian function itself. This purely dimensional argument implies, for example, that the Euler equation of a three-dimensional rigid body is a completely integrable Hamiltonian system on the coadjoint orbits of SO(3).

Indeed, the configuration space of the Euler top, a three-dimensional rigid body with a fixed point, is the set of all rotations of the Euclidean space, i.e., the Lie group SO(3). The motion of the body is described by the Euler equation on the body angular momentum m in the corresponding phase space, $\mathfrak{so}(3)^*$; see Example 4.17. Note that the conservation of the total momentum $|m|^2$ corresponds to the restriction of the angular momentum evolution to a particular coadjoint orbit, a two-dimensional sphere centered at the origin of $\mathfrak{so}(3)^* \cong \mathbb{R}^3$. Hence the Euler equation for the rigid body is a Hamiltonian system on a two-dimensional symplectic sphere, while the Hamiltonian function is given by the kinetic energy of the body.

Remark 4.28 A more complicated example is a rotation of an n-dimensional rigid body, where the dimensional consideration is not sufficient. Free motions of a body with a fixed point at its mass center are described by the geodesic flow on the group SO(n) of all rotations of Euclidean space $\mathbb{R}^n$. The group SO(n) can be regarded as the configuration space of this system. The left-invariant metric on SO(n) is defined by the quadratic form $-\operatorname{tr}(\omega D \omega)$, where $\omega \in \mathfrak{so}(n)$ is the body's angular velocity and $D = \operatorname{diag}(d_1, \ldots, d_n)$ defines the inertia ellipsoid. The corresponding inertia operator $A : \mathfrak{so}(n) \to \mathfrak{so}(n)^*$ has a very special form: $A(\omega) = D\omega + \omega D$.

Now the evolution of the angular momentum is in the space $\mathfrak{so}(n)^*$, the phase space of the n-dimensional top. The dimension of generic coadjoint orbits in $\mathfrak{so}(n)^*$ is equal to the integer part of $(n-1)^2/2$. Therefore the energy invariance alone is insufficient to guarantee the integrability of the Euler equation for an n-dimensional rigid body. The existence of sufficiently many first integrals and complete integrability in the general n-dimensional case were established by Manakov in [249]. In this paper the argument translation (or freezing) method was discovered and applied to find first integrals in this problem.

We note that the above inertia operators (or equivalently, the corresponding left-invariant metrics) form a variety of dimension n in the $n(n-1)/2$-dimensional space of equivalence classes of symmetric matrices on the Lie algebra $\mathfrak{so}(n)$. For $n > 3$ such quadratic forms are indeed very special in the space of all quadratic forms on this space. The geodesic flow on the group SO(n) equipped with an arbitrary left-invariant Riemannian metric is, in general, nonintegrable.

Other examples of integrable systems include, for instance, the geodesics on an ellipsoid [281] and the Calogero–Moser systems [280, 66, 67].

The following Liouville–Arnold theorem explains how the sufficient number of first integrals simplifies the Hamiltonian system. Consider a common level set of the first integrals

$$M_c = \{m \in M \mid f_i(m) = c_i, \ i = 1, \ldots, n\}.$$

Theorem 4.29 (Liouville–Arnold [11, 18]) *For a compact manifold M, connected components of noncritical common level sets M_c of the n first integrals are n-dimensional tori, while the Hamiltonian system defines a (quasi-) periodic motion on each of them. In a neighborhood of such a component in M there are coordinates $(\varphi_1, \ldots, \varphi_n, I_1, \ldots, I_n)$, where φ_i are angular coordinates along the tori and I_i are first integrals, such that the dynamical system assumes the form $\dot{\varphi}_i = \Omega_i(I_1, \ldots, I_n)$ and the symplectic form is $\omega = \sum_{i=1}^{n} dI_i \wedge d\varphi_i$.*

The coordinates φ_i and I_i are called the *angle* and *action coordinates*, respectively. For the case of a noncompact M, one has a natural $\mathbb{R}^n$-action on the levels M_c, coming from the commuting Hamiltonian vector fields corresponding to the Hamiltonian functions $f_i, i = 1, \ldots, n$.

Note that the symplectic form ω vanishes identically on any level set M_c, so that each regular level set is a Lagrangian submanifold of the symplectic manifold M. (By definition, a *Lagrangian submanifold* $L \subset M$ of a symplectic manifold M is an isotropic submanifold of maximal dimension: for a $2n$-dimensional M, a Lagrangian submanifold L is n-dimensional and satisfies $\omega|_L \equiv 0$.)

Remark 4.30 While in finite dimensions there are many definitions of complete integrability of a Hamiltonian system and they are all more or less equivalent, this question is more subtle in infinite dimensions. One can start defining such systems based on the existence of action-angle coordinates, or on bi-Hamiltonian structures, or on the existence of an infinite number of "sufficiently independent" first integrals, or, even by requiring an explicit solvability. These definitions lead, generally speaking, to inequivalent notions, and precise relations between these definitions in infinite dimensions are yet to be better understood.

There are, however, examples of infinite-dimensional systems in which most, if not all, of these definitions work. This is the case, for example, for the celebrated Korteweg–de Vries equation. Other systems for which several approaches are also known are the Kadomtsev–Petviashvili equation, the Camassa–Holm equation, and many others, some of which we will encounter later in the book.

5 Symplectic Reduction

The Noether theorem in classical mechanics states that a Lagrangian system with extra symmetries has an invariant of motion. Hence in describing such a system one can reduce the dimensionality of the problem by "sacrificing this invariance." The notion of symplectic reduction can be thought of as a Hamiltonian analogue of the latter: If a symplectic manifold admits an appropriate group action, then this action can be "factored out." The quotient is a new symplectic manifold of lower dimension.

This construction can be used in both ways. On the one hand, one can reduce the dimensionality of certain systems that admit extra symmetries. On the other hand, certain complicated physical systems can be better understood by realizing them as the result of symplectic reduction from much simpler systems in higher dimensions.

5.1 Hamiltonian Group Actions

Consider a finite-dimensional symplectic manifold (M, ω), i.e., a manifold M equipped with a nondegenerate closed 2-form ω. Let G be a connected Lie group with Lie algebra $\mathfrak{g}$ and suppose that the exponential map exists. If the group G acts smoothly on M, each element X of the Lie algebra $\mathfrak{g}$ defines a vector field ξ_X on the manifold M as an infinitesimal action of the group:

$$\xi_X(m) := \frac{d}{dt}|_{t=0} \exp(tX)m\,.$$

The *action* of the group G on the manifold M is called *symplectic* if it leaves the symplectic form ω invariant, i.e., if $g^*\omega = \omega$ for all $g \in G$.

Exercise 5.1 Show that for the symplectic group action, the vector field ξ_X for any $X \in \mathfrak{g}$ is symplectic, i.e., the 1-form $\iota_{\xi_X}\omega$ is closed. (Hint: use the *Cartan homotopy formula* on differential forms, $L_\xi = \iota_\xi d + d\iota_\xi$, where L_ξ means the Lie derivative along ξ and the operators ι_ξ and d stand for the inner and outer derivatives of forms.)

The closedness of the 1-form means that it is locally exact, and hence the field ξ_X is locally Hamiltonian: in a neighborhood of each point of the manifold M, there exists a function H_X such that $\iota_{\xi_X}\omega = dH_X$. In general, this field is not necessarily defined by a univalued Hamiltonian function on the whole of M. Even if we suppose that such a Hamiltonian function exists, it is defined only up to an additive constant.

Definition 5.2 The *action* of a Lie group G on M is called *Hamiltonian* if for every $X \in \mathfrak{g}$ there exists a globally defined Hamiltonian function H_X that can be chosen in such a way that the map $\mathfrak{g} \to C^\infty(M)$, associating to X the corresponding Hamiltonian H_X, is a Lie algebra homomorphism of the Lie algebra $\mathfrak{g}$ to the Poisson algebra of functions on M:

$$H_{[X,Y]} = \{H_X, H_Y\}\,.$$

Exercise 5.3 Prove that for a Hamiltonian G-action on M the Lie algebra isomorphism is *equivariant*, i.e.,

$$H_{\mathrm{Ad}_g X}(m) = H_X(g(m))$$

for all $g \in G$, $X \in \mathfrak{g}$, and $m \in M$.

Definition 5.4 Assume that the action of a group G on M is Hamiltonian. Then the *moment map* is the map $\Phi : M \to \mathfrak{g}^*$ defined by

$$H_X(m) = \langle \Phi(m), X \rangle ,$$

where $\langle \, , \, \rangle$ denotes the pairing between $\mathfrak{g}$ and $\mathfrak{g}^*$.

In other words, given a vector X from the Lie algebra $\mathfrak{g}$, the moment map sends points of the manifold to the values of the Hamiltonian function H_X at those points.

Summarizing the above definitions, a symplectic G-action on a symplectic manifold M is called *Hamiltonian* if there exists a G-equivariant smooth map $\Phi : M \to \mathfrak{g}^*$ (the *moment map*) such that for all $X \in \mathfrak{g}$, we have $d\langle \Phi, X \rangle = \iota_{\xi_X} \omega$. Any vector field ξ_X on M that comes from an element $X \in \mathfrak{g}$ for such a group action has the Hamiltonian function $H_X = \langle \Phi, X \rangle$.

Exercise 5.5 Consider $M = \mathbb{R}^2$ with the standard symplectic form $\omega = dp \wedge dq$ and the group $U(1)$ acting on $\mathbb{R}^2$ by rotations. Show that this action is Hamiltonian with the moment map $\Phi(p, q) = \frac{1}{2}(p^2 + q^2)$.

Exercise 5.6 Consider the coadjoint action of a Lie group G on the dual of its Lie algebra. Show that this action restricted to any coadjoint orbit $\mathcal{O} \subset \mathfrak{g}^*$ is Hamiltonian with the moment map being the inclusion $\iota : \mathcal{O} \hookrightarrow \mathfrak{g}^*$.

Exercise 5.7 Generalize the definition of Hamiltonian group actions to Poisson manifolds and show that the coadjoint action of a Lie group G on the dual $\mathfrak{g}^*$ of its Lie algebra is Hamiltonian with the moment map given by the identity map $\mathrm{id} : \mathfrak{g}^* \to \mathfrak{g}^*$.

5.2 Symplectic Quotients

Let (M, ω) be a symplectic manifold with a Hamiltonian action of the group G. The equivariance of the moment map $\Phi : M \to \mathfrak{g}^*$ implies that the inverse image $\Phi^{-1}(\lambda)$ of a point $\lambda \in \mathfrak{g}^*$ is a union of G_λ-orbits, where $G_\lambda := \{g \in G \mid \mathrm{Ad}_g^*(\lambda) = \lambda\}$ is the stabilizer of λ. The symplectic reduction theorem below states that if λ is a regular value of the moment map, and if the set $\Phi^{-1}(\lambda)/G_\lambda$ of G_λ-orbits in $\Phi^{-1}(\lambda)$ is a manifold, then it acquires a natural symplectic structure from the one on M.

Exercise 5.8 Suppose that G is a finite-dimensional Lie group acting on the symplectic manifold M in a Hamiltonian way with a moment map Φ. For $m \in M$, let $G.m$ denote the G-orbit through m and let G_m denote the stabilizer of m with the Lie algebra $\mathfrak{g}_m$. Show that the kernel of the differential $d\Phi_m$ of the moment map Φ at any point $m \in M$ is given by

$$\ker(d\Phi_m) = (T_m(G.m))^\omega := \{\xi \in T_m(M) \mid \omega(\xi, \chi) = 0 \text{ for all } \chi \in T_m(G.m)\}.$$

(Here $(T_m(G.m))^\omega$ is the symplectic orthogonal complement to the tangent space $T_m(G.m)$ in $T_m(M)$.)

Show that the image of the differential $d\Phi_m$ is

$$\operatorname{im}(d\Phi_m) = \operatorname{ann}(\mathfrak{g}_m) := \{\lambda \in \mathfrak{g}^* \mid \lambda(X) = 0 \text{ for all } X \in \mathfrak{g}_m\}.$$

Conclude that an element $\lambda \in \mathfrak{g}^*$ is a regular value of the moment map, i.e., $d\Phi_m$ is surjective for all $m \in \Phi^{-1}(\lambda)$, if and only if for all $m \in \Phi^{-1}(\lambda)$ the stabilizer G_m is discrete.

The restriction of the symplectic form ω to the level set $\Phi^{-1}(\lambda)$ of the moment map is not necessarily symplectic, since it might acquire a kernel.

Exercise 5.9 Show that the foliation of $\Phi^{-1}(\lambda)$ by the kernels of ω is the foliation into (connected components of) G_λ-orbits. (Hint: for a regular value λ the preimage $\Phi^{-1}(\lambda)$ is a smooth submanifold of M, and the exercise above gives

$$\begin{aligned}
\ker \omega|_{\Phi^{-1}(\lambda)} &= T_m\Phi^{-1}(\lambda) \cap (T_m\Phi^{-1}(\lambda))^\omega \\
&= T_m\Phi^{-1}(\lambda) \cap (\ker d\Phi_m)^\omega \\
&= T_m\Phi^{-1}(\lambda) \cap T_m(G.m) = T_m(G_\lambda.m).)
\end{aligned}$$

Hence, if the quotient space of the level $\Phi^{-1}(\lambda)$ over the G_λ-action is reasonably nice, the 2-form ω descends to a symplectic form on this quotient; see Figure 5.1. This is made precise in the following reduction theorem.

Theorem 5.10 (Marsden–Weinstein [254], Meyer [261]) *Suppose that λ is a regular value of the moment map and suppose that $\Phi^{-1}(\lambda)/G_\lambda$ is a manifold (this condition is satisfied if, for example, G_λ is compact and acts freely on $\Phi^{-1}(\lambda)$). Then there exists a unique symplectic structure ω_λ on the reduced space $\Phi^{-1}(\lambda)/G_\lambda$ such that*

$$\iota^*\omega = \pi^*\omega_\lambda.$$

(Here, ι denotes the embedding $\Phi^{-1}(\lambda) \hookrightarrow M$ and π stands for the projection $\Phi^{-1}(\lambda) \to \Phi^{-1}(\lambda)/G_\lambda$.)

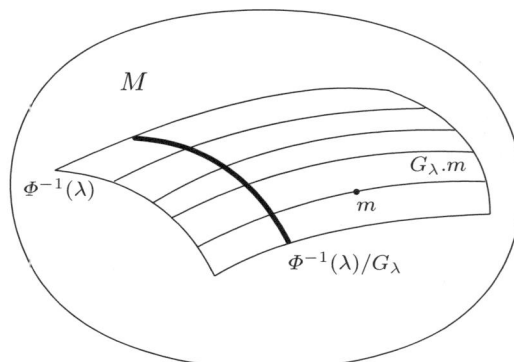

Fig. 5.1. The orbits of the stabilizer group G_λ in the preimage $\Phi^{-1}(\lambda)$.

The resulting manifold $\Phi^{-1}(\lambda)/G_\lambda$ of the above *symplectic reduction* (also known as *Hamiltonian* or *Marsden–Weinstein reduction*) is called the *symplectic quotient*.

Finally, if $H : M \to \mathbb{R}$ is a Hamiltonian function invariant under the G-action, it descends to a function H_λ on the quotient space $\Phi^{-1}(\lambda)/G_\lambda$. Furthermore, if two G-invariant functions F and H on M Poisson commute with respect to the Poisson structure on M defined by the symplectic form ω, the corresponding functions on the quotient $\Phi^{-1}(\lambda)/G_\lambda$ still Poisson commute with respect to the quotient Poisson structure.

Example 5.11 Consider the manifold $M = \mathbb{C}^{n+1}$ with its standard symplectic structure $\omega = \frac{i}{2} \sum dz_i \wedge d\bar{z}_i$. The group $U(1) = \mathbb{R}/\mathbb{Z}$ acts on $\mathbb{C}^{n+1}$ by rotation: $z \mapsto e^{2\pi i t} z$. The moment map for this action is given by $\Phi(z) = \pi\|z\|^2$. The reduced space $\Phi^{-1}(1)/U(1)$ is the complex projective space $\mathbb{CP}^n$ with the symplectic form being (a multiple of) the Fubini–Study form.

Let $f_i : \mathbb{C}^{n+1} \to \mathbb{R}$ denote the function $f_i(z) = \|z_i\|^2$. The functions f_i are invariant under the $U(1)$-action on $\mathbb{C}^{n+1}$, and the corresponding Hamiltonian functions on the symplectic quotient generate the rotations in the $\mathbb{C}$-hyperplanes $\{z_i = \text{const}\}$ in $\mathbb{CP}^n$.

6 Bibliographical Notes

There are many books on Lie groups and Lie algebras covering the material of this chapter ([345, 383, 93, 60] to name a few). The treatment of various aspects of infinite-dimensional Lie groups can be found in [12, 157, 265], as well as in the thorough monographs [301, 322]. The calculus on infinite-dimensional manifolds is developed in [157, 222]. For more details on central extensions (Section 3) we recommend [322, 276], while for the Euler equations on groups (Section 4) see [18, 24, 252].

The Lie–Poisson structure on the dual space of a Lie algebra was already known to Sophus Lie. The symplectic structure on the coadjoint orbits of a Lie group goes back to Kirillov, Kostant, and Souriau. The local description of a Poisson manifold, including its symplectic realizations and the transversal Poisson structure, was given in [384]. A generalization of Poisson manifolds are Jacobi manifolds, functions on which form a Lie algebra (but not necessarily a Poisson algebra). Jacobi manifolds are also called local Lie algebras, and their local structure was described in [199].

There is vast literature on integrable systems; see e.g., [18, 31, 90, 165, 280, 299, 330], and in particular, on integrability of the Euler equations on dual Lie algebras [47, 48, 114, 300, 311]. The Euler equations for an n-dimensional rigid body were considered back in the nineteenth century by Cayley, Frahm, and others; cf., e.g., [128, 266]. Schottky proved the integrability of these equations in the four-dimensional case [339], by giving an explicit theta-function solution for SO(4). The idea of freezing the Lie–Poisson bracket, which resolved the integrability issue in the general n-dimensional case, was proposed by Manakov in [249].

Among other mechanisms of integrability, not discussed in this book, we would like to mention the Adler–Kostant–Symes theorem [3, 213, 362, 328], the R-matrix method, see, e.g., [348, 330]; integrability of geodesic flows on quadrics and related systems, see [281, 23]; as well as discrete analogues of the Euler equation on Lie groups [282, 373, 45].

For the description of the symplectic reduction we followed [154] and [259]. The construction of symplectic reduction does not generalize directly to the case of infinite-dimensional manifolds and infinite-dimensional Lie groups. In spite of the lack of a sufficiently broad general theory, there are many concrete constructions in various infinite-dimensional situations. In particular, Hamiltonian actions of loop groups and, more generally, Hamiltonian actions of gauge transformation groups on Riemann surfaces, which we are going to deal with further in this book, have been considered, for example, in the papers [28, 260].

II

Infinite-Dimensional Lie Groups: Their Geometry, Orbits, and Dynamical Systems

1 Loop Groups and Affine Lie Algebras

For any finite-dimensional Lie group G one can define the corresponding loop group LG as the set of smooth maps from a circle to the group G endowed with pointwise multiplication.

We are interested in the coadjoint representation of the group LG or, rather, of a central extension $\widehat{LG}$ of the group LG. While the loop group itself turns out to be "too simple" to give rise to a rich theory, its central extension, called the affine (Kac–Moody) group corresponding to G, possesses a beautiful geometry and is related to many other fields in mathematics and mathematical physics. It turns out that the coadjoint orbits of the affine groups have finite codimension and are closely related to their finite-dimensional counterparts. This makes such groups attractive for representation theory: a complete classification of their coadjoint orbits indicates existence of rich representation theory for them, according to Kirillov's orbit method.

Here we describe the geometric features of loop groups and their extensions. We also comment on more general current groups G^M of smooth maps from a manifold M to a finite-dimensional Lie group G and consider certain classes of such groups in subsequent sections. Throughout this section, G denotes a finite-dimensional connected and simply connected Lie group, and $\mathfrak{g}$ stands for its Lie algebra.

1.1 The Central Extension of the Loop Lie algebra

Definition 1.1 The *loop algebra* $L\mathfrak{g}$ is the Lie algebra of smooth maps from the circle S^1 to a finite-dimensional Lie algebra $\mathfrak{g}$ with the pointwise Lie bracket.

In this section we are mostly interested in a nontrivial central extension of the loop algebra $L\mathfrak{g}$, which is essentially unique if the finite-dimensional Lie algebra $\mathfrak{g}$ is simple. First we recall the definition of an invariant bilinear form on a Lie algebra $\mathfrak{g}$.

Definition 1.2 A bilinear form $\langle\,,\,\rangle : \mathfrak{g} \times \mathfrak{g} \to \mathbb{R}$ is *invariant* if $\langle A, [B, C]\rangle = \langle [A, B\,], C\rangle$ for all $A, B, C \in \mathfrak{g}$.

For instance, for matrix Lie algebras $\mathfrak{g}$ the pairing $\langle A, B\rangle := \mathrm{tr}(AB)$ is invariant.

Any invariant bilinear form on a Lie algebra $\mathfrak{g}$ gives rise to a Lie algebra 2-cocycle and hence to a central extension of the corresponding loop algebra $L\mathfrak{g}$ in the following way:

Definition / Proposition 1.3 *Let $\langle\,,\,\rangle$ be an invariant bilinear form on $\mathfrak{g}$. The map $\omega : L\mathfrak{g} \times L\mathfrak{g} \to \mathbb{R}$ defined by*

$$\omega(X, Y) := \frac{1}{2\pi}\int_{S^1}\langle X, dY\rangle = \frac{1}{2\pi}\int_0^{2\pi}\langle X(\theta), \frac{d}{d\theta}Y(\theta)\rangle d\theta$$

is a 2-cocycle on the loop algebra $L\mathfrak{g}$. The corresponding one-dimensional central extension of $L\mathfrak{g}$, given by the commutator

$$[(X(\theta), \alpha), (Y(\theta), \beta)] = ([X, Y](\theta), \omega(X, Y)),$$

is denoted by $\widehat{L\mathfrak{g}}$ and is called the affine Lie algebra *corresponding to the algebra $\mathfrak{g}$ and the form $\langle\,,\,\rangle$.*

PROOF.   The antisymmetry of the map ω follows from integration by parts. The cocycle identity for ω is verified by a direct calculation also using integration by parts and the invariance of the inner product on $\mathfrak{g}$.   □

Exercise 1.4 Verify the cocycle identity for ω.

Remark 1.5 In the algebraic literature, one is usually interested in the algebra $L\mathfrak{g}_{\mathrm{pol}}$ of *polynomial loops* in $\mathfrak{g}$ (i.e., trigonometric polynomials with values in $\mathfrak{g}$). Obviously, the construction of the cocycle ω on $L\mathfrak{g}$ restricts to $L\mathfrak{g}_{\mathrm{pol}}$, and we shall denote the corresponding central extension by $\widehat{L\mathfrak{g}}_{\mathrm{pol}}$. If $\mathfrak{g}$ is a simple Lie algebra, there exists a *unique* (up to a scalar factor) invariant bilinear form on $\mathfrak{g}$, and the corresponding affine Lie algebra is called the *affine Kac–Moody algebra* corresponding to $\mathfrak{g}$.

The (complexified) algebra $L\mathfrak{g}_{\mathrm{pol}}$ can be thought of as the space of $\mathfrak{g}$-valued Laurent polynomials

$$L\mathfrak{g}_{\mathrm{pol}} := \mathfrak{g} \otimes \mathbb{C}[z, z^{-1}],$$

while in the cocycle formula the integral over the circle is replaced by taking the residue at $z = 0$:

$$\omega(X, Y) := \text{res}\,|_{z=0}\Big\langle X(z), dY(z)\Big\rangle.$$

The equivalence of polynomial loops with the latter description is delivered by the change of variable $\theta \to z = e^{i\theta}$, which sends Fourier polynomials in θ to Laurent polynomials in z and identifies $d/d\theta$ with $iz \cdot d/dz$.

It turns out that if the Lie algebra $\mathfrak{g}$ is semisimple, the affine Lie algebras corresponding to $\mathfrak{g}$ for different invariant forms exhaust all nontrivial central extensions of the loop algebra $L\mathfrak{g}$:

Proposition 1.6 *For the Lie algebra $\mathfrak{g}$ of a semisimple compact Lie group G, any continuous 2-cocycle ω on the loop algebra $L\mathfrak{g}$ is cohomologous to a cocycle of the form*

$$\frac{1}{2\pi}\int_0^{2\pi}\Big\langle X(\theta), \frac{d}{d\theta}Y(\theta)\Big\rangle d\theta$$

for some bilinear invariant form $\langle\ ,\ \rangle$ on $\mathfrak{g}$.

PROOF. (See [322]): We first note that it suffices to consider cocycles ω invariant under conjugation by constant loops $g \in G$. Indeed, if ω is not invariant under conjugation, the cocycle

$$\widetilde{\omega} = \int_G g^*\omega\, d\text{Vol}$$

is invariant and belongs to the same cohomology class as ω. (Here $d\text{Vol}$ denotes the normalized Haar measure on G, i.e., the left-invariant volume form such that $\int_G d\text{Vol} = 1$.)

Let us consider the complexification $L\mathfrak{g}_{\mathbb{C}} = L\mathfrak{g} \otimes \mathbb{C}$ of the loop algebra $L\mathfrak{g}$. An element $X \in L\mathfrak{g}_{\mathbb{C}}$ can be expanded into a Fourier series $X = \sum X_r z^r$, where z denotes the function $e^{i\theta}$, and the X_r are elements of the finite-dimensional Lie algebra $\mathfrak{g}$. Any 2-cocycle ω on $L\mathfrak{g}$ can be extended to a bilinear map $L\mathfrak{g}_{\mathbb{C}} \times L\mathfrak{g}_{\mathbb{C}} \to \mathbb{C}$. By continuity, this cocycle ω is completely determined by its values on Laurent polynomials in z (i.e., on Fourier polynomials in θ). Let us define maps $\omega_{p,q} : \mathfrak{g}_{\mathbb{C}} \times \mathfrak{g}_{\mathbb{C}} \to \mathbb{C}$ by

$$\omega_{p,q}(A, B) := \omega(Az^p, Bz^q).$$

Each of these maps is G-invariant and hence symmetric. (This follows from the general fact that for a semisimple Lie algebra $\mathfrak{g}$, any G-invariant bilinear map $\mathfrak{g}_{\mathbb{C}} \times \mathfrak{g}_{\mathbb{C}} \to \mathbb{C}$ has to be symmetric. The latter is an easy consequence of the Schur lemma.) Using the antisymmetry of the cocycle ω and the symmetry for $\omega_{p,q}$ we obtain

$$\omega_{p,q}(A, B) = -\omega_{q,p}(B, A) = -\omega_{q,p}(A, B),$$

i.e., $\omega_{p,q} = -\omega_{q,p}$. Moreover, the cocycle identity for ω implies

$$\omega([Az^p, Bz^q, Cz^r) + \omega([Cz^r, Az^p], Bz^q) + \omega([Bz^q, Cz^r], Az^p) = 0.$$

Rewriting this in terms of the maps $\omega_{p,q}$, we get

$$\omega_{p+q,r}([A, B], C) + \omega_{r+p,q}([C, A], B) + \omega_{q+r,p}([B, C], A) = 0.$$

Using the G-invariance and symmetry of $\omega_{p,q}$ we can rewrite the second term on the left-hand side of the equation above as

$$\omega_{r+p,q}([C, A], B) = \omega_{r+p,q}([A, B], C)$$

and similarly for the third term. Finally, using the fact that $\mathfrak{g} = [\mathfrak{g}, \mathfrak{g}]$ for semisimple $\mathfrak{g}$, we obtain

$$\omega_{p+q,r} + \omega_{r+p,q} + \omega_{q+r,p} = 0. \tag{1.1}$$

From now on, we have only to exploit the latter relation. Setting $p = q = 0$ in this relation yields

$$\omega_{0,r} = 0.$$

On the other hand, setting $r = -p - q$ gives

$$\omega_{p+q,-p-q} + \omega_{-p,p} + \omega_{-q,q} = 0 \, ;$$

hence by induction we get

$$\omega_{n,-n} = n\omega_{1,-1} \, .$$

Finally, putting $r = n - p - q$ in (1.1), we obtain

$$\omega_{p+q,n-p-q} + \omega_{n-q,q} + \omega_{n-p,p} = 0 \, ,$$

which implies $\omega_{n-k,k} = k\omega_{n-1,1}$ for all n and k. Hence

$$0 = k\omega_{0,n} = k\omega_{n-n,n} = nk\omega_{n-1,1} = n\omega_{n-k,k} \, ,$$

from which we deduce that $\omega_{p,q} = 0$ for all p, q with $p + q \neq 0$.

Returning to the cocycle ω, we obtain for $X, Y \in L\mathfrak{g}$,

$$w(X, Y) = \omega \left(\sum_p X_p z^p, \sum_q Y_q z^q \right) = \sum_{p,q} \omega_{p,q}(X_p, Y_q)$$

$$= \sum_p p\,\omega_{1,-1}(X_p, Y_{-p}) = \frac{i}{2\pi} \int_0^{2\pi} \sum_p \omega_{1,-1} \left(X_p e^{ip\theta}, \frac{d}{d\theta} Y_{-p} e^{-ip\theta} \right) d\theta$$

$$= \frac{i}{2\pi} \int_0^{2\pi} \omega_{1,-1}(X(\theta), Y'(\theta)) d\theta \, ,$$

which is of the required form, since $\omega_{1,-1}$ is an invariant bilinear form on the Lie algebra $\mathfrak{g}$ (recall that here $z = e^{i\theta}$). $\qquad\square$

Remark 1.7 Let $\mathfrak{g}$ be a simple finite-dimensional Lie algebra and M a compact oriented connected manifold. Consider the *current algebra* $\mathfrak{g}^M = C^\infty(M, \mathfrak{g})$, the Lie algebra of smooth maps from M to $\mathfrak{g}$ with the pointwise bracket. Denote by $\langle\,,\,\rangle$ the nondegenerate bilinear form on $\mathfrak{g}$, which is unique up to a multiple. One can show that the universal central extension of $\mathfrak{g}^M$ is given by the 2-cocycle

$$\omega(X, Y) = \langle X, dY\rangle$$

with values in the quotient space $\Omega^1(M)/d\Omega^0(M)$ of all 1-forms in M modulo all exact 1-forms; see [322].

Note that if the dimension of the manifold M is at least 2, then the space $\Omega^1(M)/d\Omega^0(M)$ is already infinite-dimensional. However, if M is one-dimensional, i.e., a circle $M = S^1$, any 1-form on M is closed. Hence $\Omega^1(M)/d\Omega^0(M)$ is exactly the first de Rham cohomology group of M, which is one-dimensional. Thus in the latter case, the universal central extension of $\mathfrak{g}^M = L\mathfrak{g}$ is exactly the central extension from Proposition 1.6.

To visualize the $\Omega^1(M)/d\Omega^0(M)$-valued central extension of the current algebra $\mathfrak{g}^M$ for higher-dimensional M, we fix an element in the dual space to $\Omega^1(M)/d\Omega^0(M)$ and associate to it an $\mathbb{R}$-valued central extension. First note that the (smooth) dual to the quotient space $V = \Omega^1(M)/d\Omega^0(M)$ is the space $V^* = Z^{n-1}(M)$ of all closed $(n-1)$-forms on M. Indeed, there is a natural pairing between any 1-form $u \in \Omega^1(M)$ and an $(n-1)$-form γ on the compact manifold: $\langle u, \gamma\rangle := \int_M u \wedge \gamma$. For a closed form $\gamma \in Z^{n-1}(M)$ this pairing gives zero if and only if the form u is exact: $u = df$, i.e., if $u \in d\Omega^0(M)$.

Now the real-valued 2-cocycle ω_γ associated to the form $\gamma \in Z^{n-1}(M)$ is

$$\omega_\gamma(X, Y) = \int_M \langle X, dY\rangle \wedge \gamma.$$

Note that if the closed $(n-1)$-form γ is not smooth, but a singular δ-type form (i.e., a de Rham current) supported on a closed curve $\Gamma \subset M$, then the cocycle $\omega_\gamma(X, Y)$ degenerates exactly to the 2-cocycle of the affine algebra situated on the curve Γ: $\int_M \langle X, dY\rangle \wedge \gamma = \int_\Gamma \langle X, dY\rangle$. In a sense, the 2-cocycle ω_γ on the current algebra $\mathfrak{g}^M$ for a smooth $(n-1)$-form γ on M is a diffused version of the 2-cocycle for the affine algebra on the curve Γ.

Example 1.8 Current Lie algebras and their extensions have the following natural generalization. Consider the gauge transformation Lie algebra $\mathfrak{gau}(P)$ associated with a principal G-bundle $P \to M$ on a compact manifold M, where G is a compact simple Lie group. Fix some connection d_A in the bundle P. Then $\omega_A(X, Y) := \langle X, d_A Y\rangle$ defines a Lie algebra 2-cocycle (with values in $\Omega^1(M)/d\Omega^0(M)$) on the gauge algebra $\mathfrak{gau}(P)$ of the bundle P. The corresponding central extension does not depend on the connection d_A.

1.2 Coadjoint Orbits of Affine Lie Groups

In the last section, we constructed the affine Lie algebra $\widehat{L\mathfrak{g}}$ corresponding to a finite-dimensional Lie algebra $\mathfrak{g}$. Now we are interested in the adjoint and coadjoint representations of the corresponding *affine Lie group* $\widehat{LG}$. This group itself will be explicitly constructed in the next section.

Note that in order to understand the adjoint and coadjoint orbits of the group $\widehat{LG}$ we do not have to know what the extended group looks like, but we need just its nonextended version! (Indeed, the would-be group $\widehat{LG}$ should be a *central extension* of the loop group LG for the Lie group G. On the other hand, one can immediately see that for any Lie group, its center acts trivially in the adjoint representation of the group. In particular, the center of the centrally extended group $\widehat{LG}$ acts trivially, and in order to study the adjoint action of this large group, we have only to understand the action of the nonextended group LG on the Lie algebra $\widehat{L\mathfrak{g}}$. The same consideration is valid for the coadjoint action.) Here is the explicit description of the group adjoint action.

Proposition 1.9 *In the adjoint representation of the affine group* $\widehat{LG}$, *an element* $g \in LG$ *acts on* $(X, c) \in \widehat{L\mathfrak{g}} = L\mathfrak{g} \oplus \mathbb{R}$ *via*

$$\mathrm{Ad}_g(X, c) = \left(\mathrm{Ad}_g(X), c - \frac{1}{2\pi} \int_0^{2\pi} \left\langle g^{-1}(\theta) \frac{d}{d\theta} g(\theta), X(\theta) \right\rangle d\theta \right).$$

PROOF.   First, one has to check that the map above defines a *group representation* of the loop group LG on $\widehat{L\mathfrak{g}} = L\mathfrak{g} \oplus \mathbb{R}$, i.e., $\mathrm{Ad}_{gh}(X, c) = \mathrm{Ad}_g(\mathrm{Ad}_h(X, c))$ for all $g, h \in LG$ and $(X, c) \in \widehat{L\mathfrak{g}} = L\mathfrak{g} \oplus \mathbb{R}$. The loops g and h act on the first factor by the pointwise adjoint action, so there is nothing to check. For the second factor we calculate

$$c - \frac{1}{2\pi} \int \langle (gh)^{-1}(gh)', X \rangle d\theta = c - \frac{1}{2\pi} \int \langle g^{-1}g', \mathrm{Ad}_h(X) \rangle d\theta$$
$$- \frac{1}{2\pi} \int \langle h^{-1}h', X \rangle d\theta,$$

as required. (Here and below the prime $'$ stands for $d/d\theta$.)

In order to see that the action defined above is indeed the *adjoint* representation of the group $\widehat{LG}$, we have to check that the corresponding infinitesimal action coincides with the adjoint action of the centrally extended loop algebra $\widehat{L\mathfrak{g}}$ on itself. To this end, let g_s be a path in LG with a tangent vector $Y \in L\mathfrak{g}$, i.e., g_s is a family of loops in G depending smoothly on a parameter $s \in \mathbb{R}$ such that $g_0 = e$, the constant loop, and $\frac{d}{ds}|_{s=0} g_s = Y$. Then we have

$$\frac{d}{ds}\bigg|_{s=0} \frac{1}{2\pi} \int_0^{2\pi} \langle g_s^{-1}(\theta) g_s'(\theta), X(\theta) \rangle d\theta = \frac{1}{2\pi} \int_0^{2\pi} \langle Y'(\theta), X(\theta) \rangle d\theta.$$

So we obtain

$$
\frac{d}{ds}\Big|_{s=0} \left(\mathrm{Ad}_{g_s}(X), c - \frac{1}{2\pi} \int_0^{2\pi} \langle g_s^{-1}(\theta)g_s'(\theta), X(\theta) \rangle d\theta \right)
$$
$$
= \left([Y, X], -\frac{1}{2\pi} \int_0^{2\pi} \langle Y'(\theta), X(\theta) \rangle d\theta \right) = ([Y, X], \omega(Y, X)) =: \mathrm{ad}_Y(X, c),
$$

where we have used integration by parts in the last step. This implies that we differentiated the group action $\mathrm{Ad}_{g_s}(X, c)$ with respect to s. □

Let us turn to the coadjoint representation of the affine group $\widehat{LG}$ on its dual Lie algebra $\widehat{L\mathfrak{g}}^*$. The smooth dual $\widehat{L\mathfrak{g}}_s^*$ for this algebra $\widehat{L\mathfrak{g}}$ can be thought of as the space of the following pairs:

$$
\widehat{L\mathfrak{g}}_s^* = \{(A, a) \mid A \in L\mathfrak{g}, \ a \in \mathbb{R}\} .
$$

It possesses a nondegenerate pairing with $\widehat{L\mathfrak{g}}$ via

$$
\langle (X, c), (A, a) \rangle = \frac{1}{2\pi} \int_0^{2\pi} \langle X(\theta), A(\theta) \rangle d\theta + ca .
$$

Abusing notation, we are going to drop the index s in the sequel, although in addition to $\widehat{L\mathfrak{g}}_s^*$, the whole dual space $\widehat{L\mathfrak{g}}^*$ includes also "singular functionals."[9] As in the case of the adjoint representation, to describe the coadjoint representation of the group $\widehat{LG}$ it is enough to consider the action of the loop group LG on $\widehat{L\mathfrak{g}}^*$, since the former factors through the map $\widehat{LG} \to LG$ (i.e., the center of $\widehat{LG}$ acts trivially in the coadjoint representation as well).

Proposition 1.10 *In the coadjoint representation of* $\widehat{LG}$ *on* $\widehat{L\mathfrak{g}}_s^*$*, an element* $g \in LG$ *acts via*

$$
\mathrm{Ad}_g^*(A, a) = \left(\mathrm{Ad}_g(A) + a \left(\frac{d}{d\theta} g \right) g^{-1}, a \right) .
$$

Proof.   Recall that the coadjoint action of LG on $\widehat{L\mathfrak{g}}_s^*$ is defined via

$$
\langle \mathrm{Ad}_g^*(A, a), (X, c) \rangle = \langle (A, a), \mathrm{Ad}_{g^{-1}}(X, c) \rangle .
$$

Now the required statement follows from the above formula for Ad_g by a direct calculation. □

---

[9] Formally, elements of the dual space $\widehat{L\mathfrak{g}}^*$ are pairs (A, a), where $A(\theta) \, d\theta$ is a $\mathfrak{g}$-valued de Rham current, not necessarily represented by a smooth 1-form; cf. the discussion of currents in Section II.3.5.

Corollary 1.11 *In the coadjoint representation of the algebra $\widehat{L\mathfrak{g}}$ on its dual $\widehat{L\mathfrak{g}}^*$, an element $X \in L\mathfrak{g}$ acts via*

$$\mathrm{ad}_X^*(A, a) = ([X, A] + aX', 0).$$

Exercise 1.12 (*i*) Verify this formula for $\mathrm{ad}_X^*(A, a)$ by differentiating the group coadjoint action above, as in the proof of Proposition 1.9.

(*ii*) Verify the same formula directly from the definition of the Lie bracket in the affine algebra.

(*iii*) Check that the formula for $\mathrm{Ad}_g^*(A, a)$ indeed defines a representation of the corresponding group, i.e., that $\mathrm{Ad}_g^* \mathrm{Ad}_h^* = \mathrm{Ad}_{gh}^*$.

This exercise gives an alternative way of verifying the formula for the loop group coadjoint action on $\widehat{L\mathfrak{g}}^*$ without calculating the corresponding group adjoint action.

Remark 1.13 Let $\mathfrak{g}$ be the Lie algebra $\mathfrak{gl}(n, \mathbb{R})$. The above proposition and corollary show that the dual space $\widehat{L\mathfrak{g}}^* = \{(A, a)\}$ can be identified with the space of matrix-valued first-order linear differential operators on the circle:

$$\widehat{L\mathfrak{g}}^* = \left\{ -a\frac{d}{d\theta} + A \mid A \in L\mathfrak{g}, \ a \in \mathbb{R} \right\}.$$

This identification is natural in the sense that the group coadjoint action Ad_g^* on the dual space $\widehat{L\mathfrak{g}}^*$ coincides with the gauge action on differential operators:

$$g: \ -a\frac{d}{d\theta} + A \mapsto -a\frac{d}{d\theta} + gAg^{-1} + a\left(\frac{d}{d\theta}g\right)g^{-1}.$$

(Note that in terms of the connection $\nabla = -a\frac{d}{d\theta} + A$, the gauge action of a loop g can be written is the usual conjugation: $\nabla \mapsto g \circ \nabla \circ g^{-1}$.) Accordingly, the algebra coadjoint action ad_X^* on $\widehat{L\mathfrak{g}}^*$ can be written in the form of a commutator:

$$\left[X, -a\frac{d}{d\theta} + A\right] = a\frac{dX}{d\theta} + [X, A] = \mathrm{ad}_X^*(A, a).$$

(Here we used the Leibniz rule for the operators of multiplication by X and differentiation in θ: $\frac{d}{d\theta} \circ X - X \circ \frac{d}{d\theta} = X'$.) The *cocentral direction* $a\frac{d}{d\theta}$ in $\widehat{L\mathfrak{g}}^*$ is dual to the central direction in the affine Lie algebra $\widehat{L\mathfrak{g}}$.

A similar consideration holds for any Lie algebra $\mathfrak{g}$, where instead of matrix differential operators one has the identification of the dual space with the space of connections in the (topologically trivial) G-bundle over S^1.

Remark 1.14 Sometimes it is convenient to combine the affine algebra with its dual by adding the "cocentral direction" $a\frac{d}{d\theta}$ to $\widehat{L\mathfrak{g}}$. Namely, the map

$$X \mapsto \frac{d}{d\theta}X$$

is a derivation of the Lie algebra $L\mathfrak{g}$. It extends to a derivation of the centrally extended algebra $\widehat{L\mathfrak{g}}$ by the trivial action on the center. Thus one can define the semidirect product $\widetilde{L\mathfrak{g}} = \widehat{L\mathfrak{g}} \rtimes \mathbb{R}d$, which is also sometimes called the *affine Lie algebra* or *doubly extended loop algebra corresponding* to $\mathfrak{g}$. The advantage of the doubly extended loop algebra $\widetilde{L\mathfrak{g}}$ is that unlike $\widehat{L\mathfrak{g}}$, it admits a nondegenerate invariant bilinear form given by

$$\left\langle \left(X + a\frac{d}{d\theta}, c \right), \left(Y + b\frac{d}{d\theta}, d \right) \right\rangle = \frac{1}{2\pi} \int_0^{2\pi} \langle X(\theta), Y(\theta) \rangle d\theta + ad + bc.$$

This invariant bilinear form allows one to identify the adjoint and coadjoint actions on the doubly extended affine Lie algebra $\widetilde{L\mathfrak{g}}$.

Now we would like to study the orbits in the coadjoint representation of the group $\widehat{LG}$. Since for an element $(A, a) \in \widehat{L\mathfrak{g}}^*$, the value a is invariant under the action of LG, we can study the orbits in a fixed hyperplane $a = \text{const}$. To simplify the notation, let us fix $a = 1$. Other hyperplanes with $a \neq 0$ differ from this one by an overall rescaling. We will comment on the case $a = 0$ later (see Remark 1.20).

Theorem 1.15 ([329]) *Let G be a compact, connected, and simply connected Lie group. There is a one-to-one correspondence between the set of LG-orbits in the $(a = 1)$-hyperplane in $\widehat{L\mathfrak{g}}^*$ and the set of conjugacy classes in the group G. Moreover, the stabilizer of an element $(A, 1) \in \widehat{L\mathfrak{g}}^*$ is isomorphic to the centralizer of the corresponding conjugacy class in G.*

Corollary 1.16 *Every coadjoint orbit in the $(a = 1)$-hyperplane in $\widehat{L\mathfrak{g}}^*$ is isomorphic to LG/H for some subgroup $H \subset G$. In particular, the codimension of such orbits in the hyperplane is always finite and not greater than $\dim G$.*

PROOF OF THEOREM 1.15.   Let us associate to each $(A, 1) \in \widehat{L\mathfrak{g}}_s^*$ the differential equation

$$\frac{d}{d\theta}\psi - A\psi = 0, \tag{1.2}$$

and let ψ be a solution of this differential equation. Since A assumes values in the Lie algebra $\mathfrak{g}$, any solution $\psi(\theta)$ of equation (1.2) with $\psi(0) \in G$ stays in the group G and hence defines a path $\psi : \mathbb{R} \to G$. (We can think of G as a matrix group to invoke the intuition from the theory of differential equations, although all the steps can be translated to purely group-theoretical language; see, e.g., [322].)

First we note that the gauge action on differential operators becomes rather simple in terms of the action on the solutions. Namely, suppose that the solution ψ of equation $\psi' = A\psi$ is multiplied on the left by a function $g : \mathbb{R} \to G$. Then the G-valued function $\phi(\theta) = g(\theta)\psi(\theta)$ is a solution to the new differential equation $\phi' = \tilde{A}\phi$, where

$$\tilde{A} := gAg^{-1} + g'g^{-1},$$

i.e., $\tilde{A}$ is gauge equivalent to A. Thus, the gauge action by g on the differential operator $-d/d\theta + A$ corresponds to the multiplication of solutions by g on the left.

Now recall that A is a loop in $\mathfrak{g}$, i.e., $A(\theta)$ is periodic in θ. This implies that the map $\psi(\theta + 2\pi)$ is also a solution of equation (1.2). Note that all solutions of the linear differential equation $\psi' = A\psi$ have the form $\psi(\theta) \cdot h$, where h is a constant matrix $h \in G$. In particular, the ratio of two solutions

$$M_\psi := \psi(\theta + 2\pi)^{-1}\psi(\theta),$$

the *monodromy* of ψ, is a constant matrix from the same group: $M_\psi \in G$. The monodromy matrices M_ψ corresponding to different solutions or to different initial conditions $\psi(0)$ can differ by conjugation. In this way, we assign to each element $(A, 1)$ the corresponding conjugacy class of the monodromy M_ψ in the group G.

One can see that this construction assigns the same conjugacy class of the group G to each element in the coadjoint orbit of $\widehat{LG}$ containing $(A, 1)$. Indeed, different elements of the same coadjoint orbit correspond to gauge-equivalent differential operators. Hence their solutions differ by multiplication: $\phi(\theta) = g(\theta)\psi(\theta)$ with a periodic $g(\theta) = g(\theta + 2\pi)$. Then

$$M_\phi = \phi(\theta + 2\pi)^{-1}\phi(\theta) = \psi(\theta + 2\pi)^{-1}g(\theta + 2\pi)^{-1}g(\theta)\psi(\theta)$$
$$= \psi(\theta + 2\pi)^{-1}\psi(\theta) = M_\psi,$$

i.e., the monodromy matrix (defined modulo conjugation in G) is the same. Therefore, the conjugacy class of G corresponding to $\mathrm{Ad}^*_g(A, 1)$ is the same as the conjugacy class corresponding to $(A, 1)$.

Conversely, two differential operators with the same (or conjugate) monodromy matrix belong to the same coadjoint orbit in $\widehat{L\mathfrak{g}}^*$. Indeed, choose the solutions $\phi(\theta)$ and $\psi(\theta)$ so that their monodromy matrices would be the same. Then define $g(\theta) := \phi(\theta)\psi(\theta)^{-1}$. Due to the equality of monodromies, $g(\theta)$ is periodic: $g(\theta) = g(\theta + 2\pi)$. Thus we have defined an element of the loop group $g \in LG$ that sends one of the solutions to the other $g : \psi \mapsto g\psi = \phi$, and hence makes the corresponding differential operators gauge-equivalent.

Finally, we have to check that the map from the LG-orbits in the ($a = 1$)-hyperplane to the conjugacy classes in G is surjective. This is a consequence of the surjectivity of the exponential map of the finite-dimensional compact connected Lie group G (see Appendix A.2). Indeed, fix some $g \in G$, and let $H \in \mathfrak{g}$ be an element of the inverse image of g under the exponential map. Then we can view $(H, 1)$ as an element of the $(a = 1)$-hyperplane in $\widehat{L\mathfrak{g}}^*$, and the conjugacy class of G corresponding to the coadjoint orbit through $(H, 1)$ contains g.   □

Remark 1.17 The above proof shows, in particular, that every coadjoint orbit of $\widehat{LG}$ in the $(a = 1)$-hyperplane contains a constant element.

It also shows that the stabilizer of an element $(A, 1) \in \widehat{L\mathfrak{g}}^*$ in the coadjoint representation of LG is conjugated to the stabilizer of the corresponding conjugacy class in G. The stabilizer of $(0, 1) \in \widehat{L\mathfrak{g}}^*$ is the whole group G, so that the coadjoint orbit through $(0, 1)$ is isomorphic to LG/G.

To give a more precise description of the stabilizers, let $T \subset G$ be a maximal torus in the compact group G. Then every element of G is conjugate to some element of the torus T. (For $G = \mathrm{SU}(n)$, this is the classical fact that every unitary matrix is diagonalizable.) Furthermore, two elements of the maximal torus T are conjugate in the group G if and only if they are conjugate under the group $W = N_G(T)/T$, where $N_G(T)$ is the normalizer of T in G. The group W is a finite group, the so-called *Weyl group* of G. It follows from the discussion above that the set of conjugacy classes in G, and hence the set of coadjoint orbits in the $(a = 1)$-hyperplane in $\widehat{L\mathfrak{g}}^*$, can be identified with the quotient T/W of the maximal torus over the Weyl group. This set, in turn, is a convex polytope, the *fundamental alcove* of G (see Appendix A.2). The codimension of a conjugacy class in G is determined by its position in this polytope. For example, the stabilizer of every conjugacy class in the interior of the fundamental alcove is the maximal torus T itself. Hence the corresponding coadjoint orbits of LG are isomorphic to LG/T.

Corollary 1.18 *The conjugacy class of the monodromy is the only invariant of the coadjoint action of the affine group. The group invariants of G define Casimir functions on the dual space $\widehat{L\mathfrak{g}}^*$ with respect to the coadjoint action of the group $\widehat{LG}$.*

Indeed, the one-to-one correspondence in Theorem 1.15 between the coadjoint LG-orbits and conjugacy classes in G is furnished by monodromies of the corresponding differential operators.

Example 1.19 Consider the affine algebra for $\mathfrak{g} = \mathfrak{sl}(2, \mathbb{R})$ and the hyperplane $a = 1$ in its dual. The corresponding monodromies belong to the group $G = \mathrm{SL}(2, \mathbb{R})$, and the trace of the monodromy matrices is a Casimir function on the dual space to the affine algebra $\widehat{L\mathfrak{g}}^*$.

The codimension of the orbits is equal to 1 or 3. The only orbits of codimension 3 are those corresponding to the monodromy $M = \mathrm{id}$ or $-\mathrm{id}$. If M is a Jordan 2×2 block or M has distinct eigenvalues, the codimension is equal to 1.[1)] We are going to study the adjacency of the orbits in this case in more detail when we discuss the Virasoro coadjoint orbits in the next section.

Remark 1.20 It is curious to compare the orbit classification for *affine* (i.e., extended loop) groups with that for the *nonextended* loop groups LG. The smooth dual $L\mathfrak{g}^*$ of the corresponding loop algebra is isomorphic to $L\mathfrak{g}$. The corresponding LG-action on this dual is given by the pointwise conjugation action of LG. (In the absence of the cocycle, the actions at different points of the circle are not related!) By definition, smooth currents $A(\theta)$ and $B(\theta)$ belong to the same coadjoint orbit if and only if $A(\theta)$ and $B(\theta)$ are conjugate in $\mathfrak{g}$ for each $\theta \in S^1$ and the conjugating map is a smooth loop in G. (For a simply connected group G there are no topological obstructions for the existence of such a loop for smooth and pointwise conjugate A and B, since the stabilizer of any element in G is connected.) Hence the classification of coadjoint orbits for the loop group reduces to the classification of *families of matrices* up to conjugation; cf. [15, 17]. Every function $f : S^1 \to \mathbb{R}$ that depends only on the eigenvalues of the family $A(\theta)$ is invariant under the coadjoint action of the group LG, i.e. it defines a Casimir for $L\mathfrak{g}^*$. This shows that every coadjoint orbit of the nonextended loop group has infinite codimension! This differs drastically from the extended case, i.e., orbits of the affine groups $\widehat{LG}$.

Note that one can view the corresponding dual space $L\mathfrak{g}^*$ of the nonextended loop algebra as the ($a = 0$)-hyperplane in the extended dual, $\widehat{L\mathfrak{g}}^*$. The corresponding affine group action in this hyperplane is exactly the loop group action, i.e., it is given by the pointwise conjugation action of LG. Thus, the affine coadjoint orbits for $a = 0$ are all of *infinite codimension*, while the orbits of the same group $\widehat{LG}$ in other hyperplanes $a \neq 0$ are all of *finite codimension*.

1.3 Construction of the Central Extension of the Loop Group

As we learned in Section 1.1, for a finite-dimensional semisimple Lie algebra $\mathfrak{g}$ every central extension of the loop algebra $L\mathfrak{g}$ is given by means of the cocycle

$$\omega(X, Y) = \frac{1}{2\pi} \int_0^{2\pi} \langle X(\theta), Y'(\theta) \rangle d\theta , \tag{1.3}$$

where $\langle \, , \, \rangle : \mathfrak{g} \times \mathfrak{g} \to \mathbb{R}$ is some invariant bilinear form on $\mathfrak{g}$.

---

[10] Note that the group $\mathrm{SL}(2, \mathbb{R})$ is not simply connected, and hence the loop group of $\mathrm{SL}(2, \mathbb{R})$ is not connected. Different differential operators from $\widehat{L\mathfrak{g}}^*$ with the same monodromy can be conjugated by an $\mathrm{SL}(2, \mathbb{R})$-loop that does not belong to the identity component of LG. Nevertheless, the same correspondence of the coadjoint LG-orbits and G-conjugacy classes still holds.

Suppose that the Lie algebra $\mathfrak{g}$ is simple. Then up to a multiple, there exists a unique invariant bilinear form on $\mathfrak{g}$, the *Killing form*. Throughout this section, we shall normalize the Killing form in such a way that the long roots in $\mathfrak{g}$ have the square length 2 (see Appendix A.1 for some facts about root systems). Abusing notation slightly, we will often write $\mathrm{tr}(XY) = \langle X, Y \rangle$ for the normalized Killing form.

The choice of the bilinear form $\langle \, , \, \rangle$ fixes the 2-cocycle ω and hence the central extension $\widehat{L\mathfrak{g}}$ of the loop algebra $L\mathfrak{g}$. The goal of this section is to show that this central extension lifts to the group level. In other words, we are looking for a central extension

$$\{e\} \to S^1 \to \widehat{LG} \to LG \to \{e\}$$

of the loop group LG whose Lie algebra is given by $\widehat{L\mathfrak{g}}$. If such a central extension $\widehat{LG}$ of LG exists, it defines an S^1-bundle $\widehat{LG} \to LG$. The problem in constructing the central extension $\widehat{LG}$ comes from the fact that this S^1-bundle $\widehat{LG} \to LG$ turns out to be *topologically nontrivial*. This is why it is impossible to write down a continuous group 2-cocycle that defines the corresponding group extension. Indeed, otherwise this would imply the existence of a global section and hence the triviality of the bundle.

Theorem 1.21 ([322]) *Let G be a simple compact simply connected Lie group and let ω be the 2-cocycle on $L\mathfrak{g}$ defined by formula (1.3), where the bilinear form $\langle \, , \, \rangle$ on $\mathfrak{g}$ is the (normalized) Killing form. Then the Lie algebra central extension of $L\mathfrak{g}$ defined by the cocycle ω lifts to a group central extension $\widehat{LG}$ of LG.*

This centrally extended group $\widehat{LG}$ is called the *affine Lie group*. The rest of this section is devoted to the proof of this theorem by explicitly constructing the extended loop group as a certain quotient group (cf. [262]). In fact, the construction below shows that the central extension of the loop algebra $L\mathfrak{g}$ defined by a multiple $k\omega$ of the cocycle ω lifts to a central extension of the loop group LG whenever k is an integer.

Step 1. Before we start constructing the extended loop group $\widehat{LG}$, let us give an alternative construction of the centrally extended loop algebra $\widehat{L\mathfrak{g}}$.

Let D be a closed two-dimensional disk whose boundary is the circle $S^1 = \partial D$. The loop algebra $L\mathfrak{g}$ can be regarded as a quotient $L\mathfrak{g} = \mathfrak{g}^D / \mathfrak{g}^D_{S^1}$, where $\mathfrak{g}^D := C^\infty(D, \mathfrak{g})$ is the Lie algebra of smooth maps from the disk D to $\mathfrak{g}$, and

$$\mathfrak{g}^D_{S^1} = \{X \in C^\infty(D, \mathfrak{g}) \mid X|_{S^1} = 0\}$$

denotes the subalgebra of maps that vanish on $\partial D = S^1$, the boundary of D.

Define a bilinear map $\omega^D : \mathfrak{g}^D \times \mathfrak{g}^D \to \mathbb{R}$ by

$$\omega^D(X, Y) = \frac{1}{2\pi} \int_D \mathrm{tr}(dX \wedge dY).$$

Here dX and dY are $\mathfrak{g}$-valued 1-forms on the disk, and in order to obtain the 2-form on D we consider their wedge product as 1-forms on the disk and the Killing pairing of their values in the target space, in $\mathfrak{g}$. (It helps, however, to think of G as a matrix group, so that the standard matrix calculus is applicable. Then dX and dY should be viewed as matrix-valued 1-forms on the disk D, while $dX \wedge dY$ is the usual matrix product combined with the wedge product. So $\operatorname{tr}(dX \wedge dY)$ is a 2-form on D, which can be integrated over the disk D.) One can easily see that ω^D is a 2-cocycle on the current Lie algebra $\mathfrak{g}^D$. Hence it defines a central extension $\widehat{\mathfrak{g}^D}$ of the current Lie algebra $\mathfrak{g}^D$. Furthermore, by the Stokes formula, we have

$$\frac{1}{2\pi} \int_D \operatorname{tr}(dX \wedge dY) = \frac{1}{2\pi} \int_D d\operatorname{tr}(XdY) = \frac{1}{2\pi} \int_{\partial D = S^1} \operatorname{tr}(XdY).$$

This shows that the 2-cocycle ω^D depends only on the boundary values of X and Y (and it coincides with the cocycle of the affine algebra on the boundary $\partial D = S^1$). In particular, it vanishes on the subalgebra $\mathfrak{g}^D_{S^1} \subset \mathfrak{g}^D$, so that $\mathfrak{g}^D_{S^1}$ is an ideal in the Lie algebra $\widehat{\mathfrak{g}}^D$. This implies that the quotient $\widehat{\mathfrak{g}}^D / \mathfrak{g}^D_{S^1}$ with respect to this ideal is isomorphic to the affine algebra $\widehat{L\mathfrak{g}}$.

  Step 2. Our goal is to construct the group extension $\widehat{LG}$ as a quotient, similar to the construction of the Lie algebra extension $\widehat{L\mathfrak{g}}$ above. First note that one can easily present the *nonextended* group LG as a quotient: $LG = G^D / G^D_{S^1}$, where $G^D = C^\infty(D, G)$ is the current group on the disk and

$$G^D_{S^1} = \{g \in C^\infty(D, G) \mid g|_{S^1} = e \in G\}$$

is the subgroup of currents based on the boundary $\partial D = S^1$.

  Now we are looking for a central extension $\widehat{G}^D$ of the Lie group G^D whose Lie algebra is given by $\widehat{\mathfrak{g}}^D$. After that, we are going to embed $G^D_{S^1}$ as a normal subgroup in $\widehat{G}^D$ such that the Lie algebra of $G^D_{S^1}$ will be isomorphic to $\mathfrak{g}^D_{S^1}$. In this way, we construct the Lie group $\widehat{LG} = \widehat{G}^D / G^D_{S^1}$, which is a central extension of LG and whose Lie algebra is $\widehat{L\mathfrak{g}} = \widehat{\mathfrak{g}}^D / \mathfrak{g}^D_{S^1}$.

Lemma 1.22 *There exists a one-dimensional central extension $\widehat{G}^D$ of the group G^D by $S^1 \approx \mathrm{U}(1)$ that is topologically trivial $(\widehat{G}^D = G^D \times \mathrm{U}(1))$ and whose Lie algebra is $\widehat{\mathfrak{g}}^D$.*

PROOF.   Let $g \in G^D$ be a map from the disk D to G. Then $g^{-1}dg$ and $(dg)g^{-1}$ are $\mathfrak{g}$-valued 1-forms on D. Let us define a map

$$\gamma : G^D \times G^D \to \mathbb{R}$$

by the formula

$$\gamma(g_1, g_2) = \frac{1}{4\pi} \int_D \operatorname{tr}(g_1^{-1}dg_1 \wedge dg_2 g_2^{-1}).$$

Exercise 1.23 Verify the group 2-cocycle identity

$$\gamma(g_1g_2, g_3) + \gamma(g_1, g_2) = \gamma(g_1, g_2g_3) + \gamma(g_2, g_3),$$

which shows that γ is a 2-cocycle on G^D with values in the additive group $\mathbb{R}$. (Hint: it follows from

$$(g_1g_2)^{-1} d(g_1g_2) = g_2^{-1}g_1^{-1} dg_1 g_2 + g_2^{-1} dg_2$$

and

$$d(g_1g_2)(g_1g_2)^{-1} = dg_1 g_1^{-1} + g_1 dg_2 g_2^{-1} g_1^{-1}.)$$

By exponentiating the cocycle γ we obtain the U(1)-valued map $e^{i\gamma}$, which is a group 2-cocycle on the current group G^D with values in the multiplicative group U(1). It defines the desired central extension $\widehat{G}^D$ of G^D by U(1) with the multiplication in $\widehat{G}^D$ given by

$$(g_1, a_1)(g_2, a_2) = (g_1g_2(x,y), a_1 a_2 e^{i\gamma(g_1, g_2)}).$$

Finally, we have to check that the Lie algebra of $\widehat{G}^D$ is given by $\widehat{\mathfrak{g}}^D$. To do this, we calculate the infinitesimal version of the cocycle $e^{i\gamma}$: Let g_s and h_t be two smooth curves in G^D such that $\frac{d}{dt}|_{t=0} g_t = X$ and $\frac{d}{ds}|_{s=0} h_s = Y$. According to Proposition 3.14 of Chapter I, the infinitesimal version of the group 2-cocycle $e^{i\gamma}$ is given by

$$\frac{d^2}{dt\,ds}\Big|_{t=0, s=0} \exp(i\gamma(g_t, h_s)) - \frac{d^2}{dt\,ds}\Big|_{t=0, s=0} \exp(i\gamma(h_s, g_t))$$

$$= -\frac{1}{2\pi} \int_D \mathrm{tr}(dX \wedge dY) = -\omega^D(X, Y).$$

$\square$

Step 3. Our final goal is to embed $G_{S^1}^D$ as a normal subgroup into $\widehat{G}^D$. Let B be a three-dimensional ball bounded by the two-sphere S^2. In turn, this sphere is represented as the union of two disks D and D' glued together along the common boundary. Given a G-valued map $g \in G_{S^1}^D$ on the disk D that is based on $S^1 = \partial D$, one can extend it trivially to the other disk D': $g|_{D'} \equiv e \in G$, and hence to the whole of S^2. Now choose an arbitrary extension $\widetilde{g}$ of the group current g from S^2 to the three-dimensional ball B (see Figure 1.1). This is possible due to the following topological fact.

Fact 1 (see, e.g., [60]): *Let G be a finite-dimensional Lie group. Then $\pi_2(G) = \{e\}$, that is any continuous map $f : S^2 \to G$ can be extended to a continuous map $\hat{f} : B \to G$.*

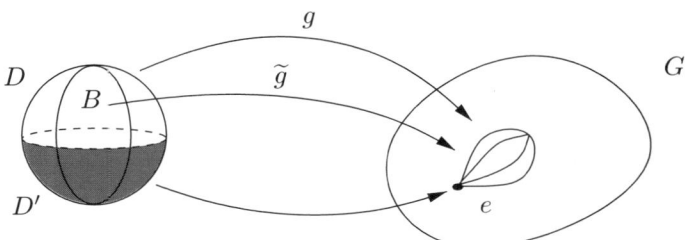

Fig. 1.1. Extension of the map $g : D \to G$ to a map $\tilde{g} : B \to G$ such that $\widetilde{g}(D') = e$.

This allows us to define the number

$$\lambda(\tilde{g}) = \frac{1}{12\pi} \int_B \mathrm{tr}(\tilde{g}^{-1}d\tilde{g})^{\wedge 3} .$$

The origin of the above expression for $\lambda(\tilde{g})$ is as follows.

Fact 2 (see, e.g.. [60]): *Let G be a compact simple and simply connected Lie group Then the third cohomology group $H^3(G, \mathbb{Z})$, with values in $\mathbb{Z}$, is generated by a single element, a closed differential left-invariant 3-form η on G. For a smooth map $\tilde{g} : B \to G$ from a three-dimensional manifold B to the group G, the pullback of the form η for the map $\tilde{g}$ is given by the following 3-form on B:*

$$\tilde{g}^*\eta = \frac{1}{24\pi^2} \, \mathrm{tr}(\tilde{g}^{-1}d\tilde{g})^{\wedge 3} .$$

Note that this form η can be defined as the left-invariant differential 3-form on the group G whose value at the identity of the group, evaluated on three tangent vectors $A, B, C \in T_e G = \mathfrak{g}$, is given by $\frac{1}{8\pi^2}\langle A, [B, C]\rangle$.

Remark 1.24 Both of the above facts follow from the Hopf theorem on compact simple Lie groups; see Appendix A.2 and, in particular, Proposition A.2.16 for the discussion, references, and more details on compact Lie groups.

Summarizing the above, we see that the image $g(S^2)$ is a spheroid in the group G, while the extension $\tilde{g}$ defines a topological 3-dimensional ball $\tilde{g}(B)$ filling this spheroid. The value $\lambda(\tilde{g})$ is the (normalized) integral of the form η over this topological ball. Although the number $\lambda(\tilde{g})$ depends on the extension $\tilde{g}$ from the sphere S^2 to the ball B, the following lemma shows that this dependence is easily controlled.

Lemma 1.25 *Consider extensions $\tilde{g}, \tilde{h} : B \to G$ of two group currents $g, h : S^2 \to G$ from the sphere S^2 to the ball B. Then*

$$\lambda(\tilde{g}\tilde{h}) = \lambda(\tilde{g}) + \lambda(\tilde{h}) + \gamma(g, h). \tag{1.4}$$

Moreover, for any current $g : S^2 \to G$ *the number* $e^{i\lambda(\tilde{g})}$ *does not depend on the choice of the extension* $\tilde{g}$ *of the current* g *from* S^2 *to* B.

PROOF.   Let $\tilde{g}$ and $\tilde{h}$ be extensions of g and h to the ball B. Then

$$\lambda(\tilde{g}\tilde{h}) = \frac{1}{12\pi} \int_B \mathrm{tr} \left((\tilde{g}\tilde{h})^{-1} d(\tilde{g}\tilde{h}) \right)^{\wedge 3} = \frac{1}{12\pi} \int_B \mathrm{tr} \left(\tilde{h}^{-1}\tilde{g}^{-1} d\tilde{g}\tilde{h} + \tilde{h}^{-1} d\tilde{h} \right)^{\wedge 3}$$

$$= \lambda(\tilde{g}) + \lambda(\tilde{h}) + \frac{1}{4\pi} \int_B \mathrm{tr} \left(\tilde{g}^{-1} d\tilde{g} \wedge \tilde{g}^{-1} d\tilde{g} \wedge d\tilde{h}\tilde{h}^{-1} \right.$$

$$\left. + \tilde{g}^{-1} d\tilde{g} \wedge d\tilde{h}\tilde{h}^{-1} \wedge d\tilde{h}\tilde{h}^{-1} \right)$$

$$= \lambda(\tilde{g}) + \lambda(\tilde{h}) + \frac{1}{4\pi} \int_B d \left(\mathrm{tr} \left(\tilde{g}^{-1} d\tilde{g} \wedge d\tilde{h}\tilde{h}^{-1} \right) \right)$$

$$= \lambda(\tilde{g}) + \lambda(\tilde{h}) + \gamma(g, h),$$

where we have used the Stokes formula in the last step.

Next, we show that the value $\lambda(\tilde{g})$ (modulo 2π) depends only on the current g on the sphere, but not on its extension $\tilde{g}$ to the ball. Indeed, any two extensions $\tilde{g}$ and $\tilde{g}'$ of the group current g from S^2 to B differ by a map $\tilde{h} : B \to G$ such that $\tilde{g} = \tilde{g}'\tilde{h}$ on the ball B, while on the boundary $\tilde{h}|_{\partial B} \equiv e$.

This means that $\tilde{h}$ defines a map from the ball B to the group G, which sends the whole boundary sphere $S^2 = \partial B$ into the unit element $e \in G$. In other words, the image $\tilde{h}(B)$ is a closed 3-cycle (a topological three-sphere) in G. Then by definition of the integral 3-form η, its integral $\int \eta$ over any closed 3-cycle in G, and in particular, over the image $\tilde{h}(B)$, is an integer. This gives that

$$\lambda(\tilde{h}) = 2\pi \int_B \tilde{h}^* \eta = 2\pi \int_{\tilde{h}(B)} \eta \in 2\pi \mathbb{Z}.$$

Finally, since $\tilde{h}|_{\partial B} \equiv e$, we have $\gamma(g, \tilde{h}|_{\partial B}) = 0$ by the definition of γ. Hence by the first part of the lemma, we have the following additivity: $\lambda(\tilde{g}\tilde{h}) = \lambda(\tilde{g}) + \lambda(\tilde{h})$. The latter means that $\lambda(\tilde{g}') = \lambda(\tilde{g})$ (modulo 2π), i.e., the value $e^{i\lambda(\tilde{g})}$ indeed does not depend on the choice of extension $\tilde{g}$, but it depends on g only. $\square$

In particular, we can use the notation $e^{i\lambda(g)}$ with g instead of $\tilde{g}$.

Now we can define the embedding $\psi : G_{S^1}^D \to \widehat{G}^D$ of the group of based currents $G_{S^1}^D$ into the centrally extended group currents $\widehat{G}^D$ via

$$g \mapsto (g, e^{i\lambda(g)}).$$

Lemma 1.26 *The map* $\psi : G_{S^1}^D \to \widehat{G}^D$ *defines an embedding of* $G_{S^1}^G$ *as a normal subgroup in* $\widehat{G}^D$.

PROOF.   The preceding lemma implies that the map ψ is well defined and the injectivity of ψ is obvious. Furthermore, the established relation (1.4) between λ and γ exactly means that $\psi(G^D_{S^1})$ is indeed a subgroup of $\widehat{G}^D$. The normality of $G^D_{S^1}$ follows from the identity

$$\lambda(\tilde{g}\tilde{h}\tilde{g}^{-1}) = \lambda(\tilde{h}) + \gamma(gh, g^{-1}) + \gamma(h, g^{-1}),$$

which can be deduced directly from relation (1.4).   □

Now the construction of the required group extension is completed with the following observation.

Lemma 1.27 *Under the embedding $G^D_{S^1} \hookrightarrow \widehat{G}^D$, the Lie algebra of $G^D_{S^1}$ gets identified with the Lie subalgebra $\mathfrak{g}^D_{S^1} \subset \widehat{\mathfrak{g}}^D$.*

PROOF.   The form $\mathrm{tr}(\tilde{g}^{-1}d\tilde{g})^{\wedge 3}$ is a homogeneous polynomial of degree 3 in the derivatives of $\tilde{g}$. Hence all its derivatives at $\tilde{g} \equiv e$ vanish (see Figure 1.2). Therefore the tangent space to this subgroup $G^D_{S^1}$ in the group $\widehat{G}^D$ coincides with the "horizontal" embedding $\mathfrak{g}^D_{S^1} \hookrightarrow \widehat{\mathfrak{g}}^D$, $X \mapsto (X, 0)$.   □

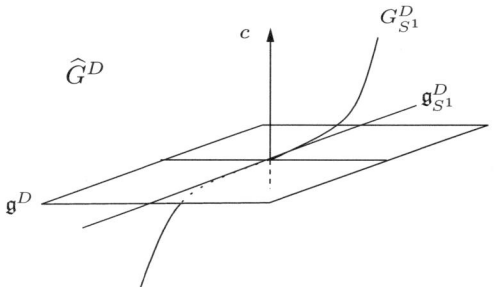

Fig. 1.2. $G^D_{S^1}$ as a subset of $\widehat{G}^D$ in a neighborhood of the identity of the group $\widehat{G}^D$. Here c denotes the central direction of the Lie algebra $\widehat{\mathfrak{g}}^D$.

Remark 1.28 Let us briefly remark on other central extensions of the loop group LG. Each multiple $k\omega^D$ of the cocycle ω^D defines a central extension of the Lie algebra $\mathfrak{g}^D$ and hence gives rise to an extension of the Lie algebra $L\mathfrak{g}$. Obviously, each of these extensions of $\mathfrak{g}^D$ lifts to a central extension $(\widehat{G}^D)^k$ of the group G^D. The problem, however, is that there exists a topological obstruction to the existence of an embedding of the group $G^D_{S^1}$ into the corresponding extension of G^D. Indeed, in order to embed $G^D_{S^1}$ into $(\widehat{G}^D)^k$, we have to use the map $g \mapsto (g, e^{ik\lambda(\tilde{g})})$. However, Lemma 1.25 shows that the number $e^{ik\lambda(\tilde{g})}$ depends on the choice of $\tilde{g}$ unless k is an integer. Hence, we can lift the central extension of the loop algebra $L\mathfrak{g}$ with the cocycle $k\omega$ to the group level whenever $k \in \mathbb{Z}$. It can be shown [322] that these exhaust all possible central extensions of the loop group LG. Moreover, the group extension $\widehat{LG}$ with $k = 1$ considered above is singled out among other extensions of the loop group $\widehat{LG}$ by the fact that it is *simply connected*.

1.4 Bibliographical Notes

The standard reference for the material of this section is the by now classical monograph by Pressley and Segal [322]. The book [178] describes the theory of Kac–Moody Lie algebras, while the corresponding groups are treated from an algebraic point of view in [225].

The central extension of the loop algebra goes back to Kac [177] and Moody [275]. The classification of coadjoint orbits of the affine groups was given by Reyman and Semenov-Tian-Shansky [329], Segal [342], and Frenkel [132].

Normal forms of *families of matrices* up to conjugation are described in [15, 17]. They are related to the description of the coadjoint orbits of nonextended loop groups in the $GL(n)$ case. Such families also provide versal deformations of the orbits of $GL(n)$-affine groups in terms of deformations of the corresponding monodromy operators.

For a finite-dimensional group G that admits an outer automorphism σ of finite order one can define the *twisted loop group* $LG_\sigma = \{g : \mathbb{R} \to G \mid g(\theta + 2\pi) = \sigma(g(\theta))$ for all $\theta \in \mathbb{R}\}$. Central extensions of the twisted loop groups are constructed in a similar fashion as for the untwisted ones. The corresponding Lie algebras are called the *twisted affine Lie algebras*. The classification of the coadjoint orbits of the latter algebras is similar to that of the untwisted ones, except that one has to use conjugacy classes of nonconnected compact Lie groups [385].

Irreducible highest-weight representations of affine Lie algebras in both the twisted and nontwisted cases naturally correspond to, respectively, twisted and standard conjugacy classes in the corresponding compact Lie group [270]. In [132] it is shown how the orbit classification for affine groups gives an analogue of Kirillov's character formula for compact Lie groups in the loop group setting and relates it to the Wiener integration on the space of continuous paths in the compact group.

An important class of algebras, related to the affine ones, is given by the Krichever–Novikov-type algebras [219, 220, 336, 353], which we discuss in Appendix A.3. Loop algebras along with their central extensions can be regarded as subalgebras of the $\mathfrak{gl}_\infty$ algebra (see Appendix A.9), which relates them to infinite-dimensional Grassmannians and provides a universal treatment of the corresponding soliton equations [75, 176, 372].

The quotient construction of the central extension of the loop group, as described in Section 1.3, has appeared in the physics literature on the Wess–Zumino–Witten (or Wess–Zumino–Novikov–Witten) model [263]. Similar constructions of this group cocycle, its variations, and generalizations can be found in the papers by Faddeev and Shatashvili [109], Polyakov and Wiegmann [319], Witten [387], Mickelsson [262], Pressley and Segal [322], Losev, Moore, Nekrasov, and Shatashvili [241], and others.

Hamiltonian systems related to loop groups and the description of some integrability mechanisms can be found in [4, 110, 119, 328, 330, 365]; see also [2, 308]. We will touch on certain dynamical systems related to loop groups in Sections 3 and 5.

Note that systems on loop groups help explain integrability of finite-dimensional discrete systems [77]. The study of Poisson structures of differential-geometric type on loop spaces and the corresponding integrable systems, initiated by Dubrovin and Novikov in [91, 92], is another direction of active current research; see [272]. It also prompted the theory of Frobenius manifolds [89, 251]. Finally, we mention the construction of a universal symplectic form for a large class of soliton equations in [221, 218], and the relation between integrable Hamiltonian systems and the Riemann–Hilbert problem [76].

For manifolds M of higher dimensions the corresponding current groups, i.e., the spaces of maps from M to a finite-dimensional Lie group G, along with their central extensions, are studied in more detail, for example, in [264, 322, 248]. We discuss some of these groups in more detail in Section 5, where we deal with the case of a two-dimensional manifold M. The case of current groups on noncompact manifolds is treated in [288].

2 Diffeomorphisms of the Circle and the Virasoro–Bott Group

This section deals with the Lie group $\mathrm{Diff}(S^1)$ of orientation-preserving diffeomorphisms of the circle and its Lie algebra $\mathrm{Vect}(S^1)$ of smooth vector fields on the circle, as well as with their central extensions. We start by showing that the Lie algebra of vector fields on the circle admits a unique nontrivial central extension, the so-called Virasoro algebra. This central extension gives rise to a central extension of the Lie group of circle diffeomorphisms, which is called the Virasoro–Bott group. Similar to the case of loop groups, the Virasoro–Bott group has a "nicer" coadjoint representation than the nonextended group of circle diffeomorphisms. Its coadjoint orbits can be classified in terms of conjugacy classes of the finite-dimensional group $\mathrm{SL}(2,\mathbb{R})$. Finally, we describe the Euler equations corresponding to right-invariant metrics on the Virasoro–Bott group and encounter the KdV and related partial differential equations among them.

2.1 Central Extensions

Let us consider the Lie algebra $\mathrm{Vect}(S^1)$ of smooth vector fields on the circle. After fixing a coordinate θ on the circle, any vector field can be written as $f(\theta)\partial_\theta$, where f is a smooth function on S^1 and ∂_θ stands for $\frac{\partial}{\partial\theta}$. Under this identification, the commutator of two elements in $\mathrm{Vect}(S^1)$ is given by

$$[f(\theta)\partial_\theta, g(\theta)\partial_\theta] = (f'(\theta)g(\theta) - g'(\theta)f(\theta))\partial_\theta,$$

where f' denotes the derivative in θ of the function f.[11]

Definition / Proposition 2.1 *The map* $\omega : \mathrm{Vect}(S^1) \times \mathrm{Vect}(S^1) \to \mathbb{R}$ *given by*

$$\omega(f(\theta)\partial_\theta, g(\theta)\partial_\theta) = \int_{S^1} f'(\theta)g''(\theta)d\theta \qquad (2.5)$$

is a nontrivial 2-cocycle on $\mathrm{Vect}(S^1)$*, called the* Gelfand–Fuchs cocycle*. The corresponding central extension of* $\mathrm{Vect}(S^1)$ *is called the* Virasoro algebra *and is denoted by* $\mathfrak{vir}$*.*

Exercise 2.2 Prove the cocycle identity for ω.

The following proposition shows that the Virasoro algebra is the unique (up to isomorphism) nontrivial central extension of the Lie algebra $\mathrm{Vect}(S^1)$.

Proposition 2.3 *The second continuous cohomology group* $H^2(\mathrm{Vect}(S^1), \mathbb{R})$ *is one-dimensional and is generated by the Gelfand–Fuchs cocycle* ω*.*

---

[11] Note that this Lie bracket is the negative of the commonly assumed commutator of vector fields, as the calculations in Exercise 2.3 of Chapter I shows; see [24].

PROOF.   The proof of this proposition (see [322]) is similar to that of Proposition 1.6. Our goal is to show that, up to a coboundary, any continuous 2-cocycle ω on the Lie algebra $\text{Vect}(S^1)$ is a multiple of the Gelfand–Fuchs cocycle. First, let us extend the cocycle ω from $\text{Vect}(S^1)$ to a complex bilinear form on the complexification $\text{Vect}(S^1)_{\mathbb{C}} = \text{Vect}(S^1) \otimes \mathbb{C}$ of the Lie algebra $\text{Vect}(S^1)$. An element $f(\theta)\partial_\theta \in \text{Vect}(S^1)_{\mathbb{C}}$ can be expanded into a Fourier series

$$f(\theta) = \sum f_n e^{in\theta} \,.$$

By continuity, the cocycle ω is completely determined by its values on the basis fields $L_n = ie^{in\theta}\partial_\theta$. Note that the commutator of the fields L_n and L_m is given by

$$[L_n, L_m] = (m - n)L_{n+m} \,.$$

The cocycle identity for ω and the triple L_0, L_m, L_n gives

$$\omega([L_0, L_m], L_n) + \omega(L_m, [L_0, L_n]) = \omega(L_0, [L_m, L_n]) \,,$$

which implies that the cohomology class of the cocycle ω is unchanged under rotations of S^1 that are generated by the vector field L_0. Indeed, the right-hand side of the equation above is an *exact cocycle* (i.e., coboundary) $d\alpha$, where α is the linear functional on $\text{Vect}(S^1)$ defined by $\alpha(L_m) := \omega(L_0, L_m)$. (Here by definition $d\alpha(L_n, L_m) := \alpha([L_n, L_m])$.) In particular, the cocycle obtained from ω by averaging over S^1 belongs to the same cohomology class as ω. Therefore, we can assume ω to be *rotation invariant*, i.e.,

$$\omega([L_0, L_m], L_n) + \omega(L_m, [L_0, L_n]) = 0 \,. \tag{2.6}$$

Set $\omega_{n,m} := \omega(L_n, L_m)$. Then the commutator relation of the fields L_n and L_m and equation (2.6) imply

$$m\omega_{m,n} + n\omega_{m,n} = 0 \,.$$

This implies that $\omega_{m,n} = 0$ for $m + n \neq 0$. Antisymmetry of the cocycle ω implies $\omega_{n,-n} = \omega_{-n,n}$, so that it is enough to determine $\omega_{n,-n}$ for $n \in \mathbb{N}$.

The cocycle identity for ω evaluated on the triple L_m, L_n, L_{-m-n} implies

$$(m - n)\omega_{m+n,-n-m} + (2m + n)\omega_{n,-n} - (2n + m)\omega_{m,-m} = 0 \,.$$

In particular, for $m = 1$ the equation above reads as follows:

$$(-n + 1)\omega_{n+1,-n-1} + (n + 2)\omega_{n,-n} - (2n + 1)\omega_{1,-1} = 0 \,.$$

Hence $\omega_{n,-n}$ is defined recursively once $\omega_{1,-1}$ and $\omega_{2,-2}$ are fixed. This shows that the space of the bilinear forms ω that satisfy the 2-cocycle condition is at most two-dimensional. Two linear independent elements of this space are given by $\omega_{n,-n} = n^3$ and $\omega_{n,-n} = n$. But the 2-cocycle defined by $\omega_{n,-n} = n$

is exact, since it coincides with $d\widetilde{\alpha}$, where $\widetilde{\alpha}$ is the linear functional defined by $\widetilde{\alpha}(L_n) = -\frac{1}{2}\delta_{n,0}$.

So up to a 2-coboundary, any 2-cocycle ω has the "cubic" form

$$\omega(L_n, L_m) = c\delta_{n,-m}n^3$$

for some $c \in \mathbb{C}$.

It remains to show that the "cubic" cocycle ω is nontrivial. Suppose that $\omega = d\beta$ for some 1-cocycle β. This means that β is a linear map and $\omega(L_n, L_m) = \beta([L_n, L_m])$. In particular, we have $\beta([L_n, L_{-n}]) = 2ni\beta(L_0)$, which shows that in this case $\omega(L_n, L_{-n})$ would have to depend linearly on n. This contradiction completes the proof. $\qquad\square$

Our next goal is to show that the central extension of the Lie algebra of vector fields $\mathrm{Vect}(S^1)$ defined by the Gelfand–Fuchs cocycle ω can be lifted to a central extension of the group of circle diffeomorphisms $\mathrm{Diff}(S^1)$. It turns out that the situation here is much simpler than that in the case of the loop groups. The central extension of the group $\mathrm{Diff}(S^1)$ corresponding to the Lie algebra $\mathfrak{vir}$ is topologically trivial and hence can be defined by a continuous group 2-cocycle.

Let $\varphi : \theta \mapsto \varphi(\theta)$ be a diffeomorphism of the circle, and φ' stands for its derivative in θ.

Definition / Proposition 2.4 *The map* $B : \mathrm{Diff}(S^1) \times \mathrm{Diff}(S^1) \to S^1$ *given by*

$$(\varphi, \psi) \mapsto \frac{1}{2}\int_{S^1} \log(\varphi \circ \psi)' d\log\psi'$$

is a continuous 2-cocycle on the group $\mathrm{Diff}(S^1)$. *The Lie algebra of the corresponding central extension* $\widehat{\mathrm{Diff}}(S^1)$ *is the Virasoro algebra* $\mathfrak{vir}$. *The 2-cocycle B is called the* Bott cocycle, *and the corresponding central extension of the group* $\mathrm{Diff}(S^1)$ *is called the* Virasoro–Bott group.

PROOF. To show that the map B defines a group 2-cocycle, we have to check the identity

$$B(\varphi \circ \psi, \eta) + B(\varphi, \psi) = B(\varphi, \psi \circ \eta) + B(\psi, \eta).$$

It is provided by the chain rule, which immediately gives

$$B(\varphi \circ \psi, \eta) = \frac{1}{2}\int_{S^1} \log(\varphi \circ \psi \circ \eta)' \, d\log\eta' = \frac{1}{2}\int_{S^1} \log(\varphi' \circ \psi \circ \eta) \, d\log\eta' + B(\psi, \eta)$$

and

$$B(\varphi, \psi \circ \eta) = \frac{1}{2} \int_{S^1} \log(\varphi \circ \psi \circ \eta)' \, d\log(\psi \circ \eta)'$$

$$= B(\varphi, \psi) + \frac{1}{2} \int_{S^1} \log(\varphi' \circ \psi \circ \eta) \, d\log \eta' \,.$$

Now we verify that the infinitesimal version of the Bott group cocycle B coincides with the Gelfand–Fuchs Lie algebra 2-cocycle ω. Let $f\partial_\theta$ and $g\partial_\theta$ be two smooth vector fields on S^1 and consider the corresponding flows φ_s and ψ_t on S^1, starting at the identity diffeomorphism: $\varphi_0 = \psi_0 = \mathrm{id}$.

We have to check that

$$\omega(f\partial_\theta, g\partial_\theta) = \frac{d^2}{dt\,ds}\Big|_{t=0,s=0} B(\varphi_t, \psi_s) - \frac{d^2}{dt\,ds}\Big|_{t=0,s=0} B(\psi_s, \varphi_t)$$

(see Proposition 3.14 of Chapter I). The latter holds, since

$$\frac{d}{dt}\Big|_{t=0} B(\varphi_t, \psi_s) = \frac{1}{2} \int_{S^1} \left(\log'(\varphi_0 \circ \psi_s)'\right) (f \circ \psi_s)' d\log \psi_s'$$

$$= \frac{1}{2} \int_{S^1} (f' \circ \psi_s) \, d\log \psi_s'$$

and

$$\frac{d}{ds}\Big|_{s=0} \frac{1}{2} \int_{S^1} (f' \circ \psi_s) \, d\log \psi_s' = \frac{1}{2} \int_{S^1} f' dg' \,.$$

Similarly, we obtain

$$\frac{d^2}{dt\,ds}_{t=0,s=0} B(\psi_s, \varphi_t) = \frac{1}{2} \int_{S^1} g' df' = -\frac{1}{2} \int_{S^1} f' dg' \,,$$

which, combined with the equation above, yields the assertion. $\square$

2.2 Coadjoint Orbits of the Group of Circle Diffeomorphisms

Before we start classifying the coadjoint orbits of the Virasoro group, let us take a look at the coadjoint representation of the nonextended group of orientation-preserving diffeomorphisms of the circle. Observe that the dual spaces to the infinite-dimensional Lie algebras considered below are always understood as smooth duals, i.e., they are identified with appropriate spaces of smooth functions.

Let $\mathrm{Diff}(S^1)$ be the group of all orientation-preserving diffeomorphisms of S^1 and let $\mathrm{Vect}(S^1)$ be its Lie algebra.

Proposition 2.5 ([202]) *The (smooth) dual space* $\mathrm{Vect}(S^1)^*$ *is naturally identified with the space of quadratic differentials* $\Omega^{\otimes 2}(S^1) = \{u(\theta)(d\theta)^2\}$ *on the circle. The pairing is given by the formula*

$$\langle u(\theta)(d\theta)^2, \; v(\theta)\partial_\theta\rangle\rangle = \int_{S^1} u(\theta)v(\theta)\, d\theta$$

for any vector field $v(\theta)\partial_\theta \in \mathrm{Vect}(S^1)$. The coadjoint action coincides with the action of a diffeomorphism on the quadratic differential: for a diffeomorphism $\varphi \in \mathrm{Diff}(S^1)$ the action is

$$\mathrm{Ad}^*_{\varphi^{-1}}: \quad u\,(d\theta)^2 \mapsto u(\varphi)\cdot(\varphi')^2\,(d\theta)^2 = u(\varphi)\cdot(d\varphi)^2.$$

It follows from this proposition that the square root $\sqrt{u(\theta)(d\theta)^2}$ (when it makes sense) transforms under a diffeomorphism as a differential 1-form. In particular, if the function $u(\theta)$ does not have any zeros on the circle (say, $u(\theta) > 0$), then $\Phi(u(\theta)(d\theta)^2) := \int_{S^1} \sqrt{u(\theta)}\, d\theta$ is a Casimir function, i.e., an invariant of the coadjoint action. One can see that there is only one Casimir function in this case, since the corresponding orbit has codimension 1 in the dual space $\mathrm{Vect}(S^1)^*$. Indeed, there exists a diffeomorphism that sends the quadratic differential $u(\theta)(d\theta)^2$ without zeros to the constant quadratic differential $u_0(d\theta)^2$, where the constant u_0 is such that $\sqrt{u_0}$ is the average value of the 1-form $\sqrt{u(\theta)}\, d\theta$ on the circle:

$$2\pi\sqrt{u_0} = \int_{S^1} \sqrt{u(\theta)}\, d\theta.$$

The value u_0 parametrizes the orbits close to $u(\theta)(d\theta)^2$, and hence all these orbits have codimension 1 in $\Omega^{\otimes 2}(S^1)$. The stabilizer of a constant quadratic differential is the group S^1 of rigid rotations, so that the orbit through $u(\theta)(d\theta)^2$ is diffeomorphic to $\mathrm{Diff}(S^1)/S^1$.

On the other hand, if a differential $u(\theta)(d\theta)^2$ changes sign on the circle, then the integrals

$$\int_a^b \sqrt{|u(\theta)|}\, d\theta,$$

evaluated between any two consecutive zeros a and b of the function $u(\theta)$, are invariant. In particular, since $u(\theta)$ has at least two zeros, the coadjoint orbit of such a differential $u(\theta)(d\theta)^2$ necessarily has codimension higher than 1, and there exist coadjoint orbits of the group $\mathrm{Diff}(S^1)$ of arbitrarily high codimension. The classification of orbits in $\mathrm{Vect}(S^1)^*$ was described in [201, 203].

Remark 2.6 One can show that if the function $u(\theta)$ has two simple zeros, changing sign exactly twice on the circle, then the corresponding coadjoint orbit of the group $\mathrm{Diff}(S^1)$ has codimension 2; see [203]. (The corresponding two Casimirs are the integrals of $\sqrt{|u(\theta)|}\, d\theta$ over two different parts of the circle between these two zeros, while there are no extra local invariants at zeros themselves: quadratic differentials with simple zeros are all locally diffeomorphic to $\pm\theta\,(d\theta)^2$.)

In other words, in a family of quadratic differentials $\bar{u}^\epsilon := u^\epsilon(\theta)\,(d\theta)^2$, where the function u^ϵ is everywhere positive for $\epsilon > 0$, has a double zero for $\epsilon = 0$, and has two simple zeros for $\epsilon < 0$ (e.g., $u^\epsilon = \cos\theta + 1 + \epsilon$) the codimension of the coadjoint orbit of $\bar{u}^\epsilon = u^\epsilon(d\theta)^2$ changes from 1 for $\epsilon > 0$ to 2 for $\epsilon \leq 0$, since the number of Casimirs jumps from 1 to 2. (Note that for $\epsilon = 0$ the orbit codimension of $\bar{u}^0$ is also 2, since the existence of a double zero imposes an extra constraint on a quadratic differential.)

This change of "codimension parity" of the (infinite-dimensional) coadjoint orbits is rather surprising, since in finite dimensions the existence of a symplectic structure on each coadjoint orbit forces all of them to be even-dimensional, and hence codimensions of coadjoint orbits for a given (finite-dimensional) group are always of the same parity: either all even or all odd. However, for $\mathrm{Vect}(S^1)^* = \Omega^{\otimes 2}(S^1)$ we observe that there exist orbits of both codimensions 1 and 2!

In particular, this shows that the Weinstein theorem [384] on the existence of the transverse Poisson structure to symplectic leaves does not hold for infinite-dimensional Poisson manifolds; cf. Remark I.4.7. Indeed, on a two-dimensional transversal to $\bar{u}^0$ in $\mathrm{Vect}(S^1)^*$, neighboring coadjoint orbits of $\bar{u}^\epsilon$ have traces of both codimensions 1 and 2. One can consider the following example, clarifying how the change of parity can occur for infinite-dimensional symplectic leaves. Define the Poisson structure in an infinite-dimensional vector space $\{(x_0, x_1, x_2, \dots)\}$ by the bivector field

$$\Pi = x_0\,\partial_{x_1} \wedge \partial_{x_2} + \partial_{x_2} \wedge \partial_{x_3} + \partial_{x_3} \wedge \partial_{x_4} + \dots .$$

Its symplectic leaves are hyperplanes $\{x_0 = \mathrm{const} \neq 0\}$, while for $x_0 = 0$ the symplectic leaves are planes $\{x_0 = 0,\, x_1 = \mathrm{const}\}$ of codimension 2.

In the next section, however, we shall see that coadjoint orbits of the Virasoro group, the central extension of the diffeomorphism group $\mathrm{Diff}(S^1)$, do respect the codimension parity and behave much more like finite-dimensional coadjoint orbits.

2.3 The Virasoro Coadjoint Action and Hill's Operators

Let $\mathfrak{vir}$ be the Virasoro algebra, whose elements are pairs $(f(\theta)\partial_\theta, c)$, where $f(\theta)\partial_\theta$ is a vector field and c is a real number. We can think of its (smooth) dual space as the space of pairs $\mathfrak{vir}^* = \{(u(\theta)(d\theta)^2, a)\}$ consisting of a quadratic differential and a real number (the cocentral term). The pairing between $\mathfrak{vir}$ and $\mathfrak{vir}^*$ is given by

$$\langle (f(\theta)\partial_\theta, c), (u(\theta)(d\theta)^2, a) \rangle = \int_{S^1} f(\theta)u(\theta)d\theta + c \cdot a .$$

Our goal in this section is to derive a classification of the coadjoint orbits of the Virasoro–Bott group $\widehat{\mathrm{Diff}}(S^1)$. This classification turns out to be similar to that of the coadjoint orbits of the centrally extended loop groups in Section 1.2.

We begin by noticing that the center of the group $\widehat{\mathrm{Diff}}(S^1)$ acts trivially on the dual space $\mathfrak{vir}^*$. This is why to describe the coadjoint representation of the Virasoro–Bott group we need the action of (nonextended) diffeomorphisms only.

Definition / Proposition 2.7 *The coadjoint action of a diffeomorphism* $\varphi \in \mathrm{Diff}(S^1)$ *on the dual* $\mathfrak{vir}^*$ *of the Virasoro algebra is given by the following formula:*

$$\mathrm{Ad}^*_{\varphi^{-1}} : \quad \left(u\,(d\theta)^2,\, a\right) \mapsto \left(u(\varphi)\cdot(\varphi')^2\,(d\theta)^2 + aS(\varphi)\,(d\theta)^2,\, a\right), \qquad (2.7)$$

where

$$S(\varphi) = \frac{\varphi'\varphi''' - \frac{3}{2}(\varphi'')^2}{(\varphi')^2}$$

is the Schwarzian derivative *of the diffeomorphism* φ.

The same formula can be used to define the Schwarzian derivative $S(\phi)$ for a smooth map $\phi : \mathbb{R} \to \mathbb{R}$ or $\mathbb{R} \to \mathbb{R}P^1 \simeq S^1$.

PROOF.   The coadjoint action of the Virasoro algebra is defined by the identity

$$\langle \mathrm{ad}^*_{(v\partial_\theta, b)}(u(d\theta)^2,\, a),\, (w\partial_\theta, c)\rangle = -\langle (u(d\theta)^2,\, a),\, [(v\partial_\theta, b),\, (w\partial_\theta, c)]\rangle.$$

Using the definition of the Virasoro commutator and integrating by parts we obtain that the right-hand side is equal to

$$\int_{S^1} -w(2uv' + u'v + av''')\, d\theta\,.$$

Thus the coadjoint operator is

$$\mathrm{ad}^*_{(v\partial_\theta, b)}(u(d\theta)^2,\, a) = -((2uv' + u'v + av''')(d\theta)^2,\, 0). \qquad (2.8)$$

It remains to check that equation (2.7) indeed defines a representation of the group $\mathrm{Diff}(S^1)$ on the space $\mathfrak{vir}^*$ and that the infinitesimal version of this action is given by equation (2.8). Both assertions can be checked by direct calculations, which we leave to the reader. $\square$

Exercise 2.8 Prove the following transformation law for the Schwarzian derivative:

$$S(\varphi \circ \psi) = (S(\varphi) \circ \psi) \cdot (\psi')^2 + S(\psi)\,. \qquad (2.9)$$

Check that the formula (2.7) defines a group representation by using this law.

It turns out to be more convenient to regard the dual Virasoro space $\mathfrak{vir}^*$ not as the space of pairs $\{(u(\theta)(d\theta)^2, a)\}$, but as the space of *Hill's operators*, i.e., differential operators $a\partial_\theta^2 + u(\theta)$, where ∂_θ^2 stands for the second derivative $d^2/d\theta^2$. Indeed, the group action on Hill's operators

$$\mathrm{Ad}^*_{\varphi^{-1}} : \quad a\partial_\theta^2 + u(\theta) \mapsto a\partial_\theta^2 + u(\varphi) \cdot (\varphi')^2 + aS(\varphi) \tag{2.10}$$

has the following nice geometric interpretation (see, e.g., [342, 202, 205, 304]).

Look at a hyperplane $a = \mathrm{const}$ corresponding to nonzero a in the dual space $\mathfrak{vir}^*$. For instance, we fix $a = 1$ and consider Hill's operators of the form $\partial_\theta^2 + u(\theta)$, where θ is a coordinate on S^1. Let f and g be two independent solutions of the corresponding Hill differential equation

$$(\partial_\theta^2 + u(\theta))y = 0 \tag{2.11}$$

for an unknown function y. Although this equation has periodic coefficients, the solutions need not necessarily be periodic, but instead are defined over $\mathbb{R}$. Consider the ratio $\eta := f/g : \mathbb{R} \to \mathbb{R}P^1$. (Below we use the same notation θ for the coordinate on the circle $S^1 = \mathbb{R}/2\pi\mathbb{Z}$ and on its cover $\mathbb{R}$.)

Proposition 2.9 *The potential u is (one-half) the Schwarzian derivative of the ratio η:*

$$u = \frac{S(\eta)}{2}.$$

PROOF.   First we note that the *Wronskian*

$$W(f, g) := \det \begin{pmatrix} f & g \\ f' & g' \end{pmatrix} = fg' - f'g$$

is constant, since it satisfies $W' = 0$. For two independent solutions the Wronskian does not vanish, and we normalize W by setting $W = -1$.

This additional condition allows one to find the potential u from the ratio r. Indeed, first one reconstructs the solutions f and g from the ratio η by differentiating:

$$\eta' = \frac{f'g - fg'}{g^2} = \frac{-W}{g^2} = \frac{1}{g^2}.$$

Therefore,

$$g = \frac{1}{\sqrt{\eta'}}, \qquad f = g \cdot \eta = \frac{\eta}{\sqrt{\eta'}}.$$

Given two solutions f and g, one immediately finds the corresponding differential equation they satisfy by writing out the following 3×3 determinant:

$$\det \begin{bmatrix} y & f & g \\ y' & f' & g' \\ y'' & f'' & g'' \end{bmatrix} = 0.$$

Since f and g satisfy the equation $y'' + u \cdot y = 0$, one obtains from the determinant above that

$$u = -\det \begin{bmatrix} f' & g' \\ f'' & g'' \end{bmatrix}.$$

The explicit formula for u expressed in terms of η turns out to be one-half the Schwarzian derivative of η.   □

Corollary 2.10 *The Schwarzian derivative $S(\eta)$ is invariant with respect to a Möbius transformation $\eta \mapsto (a\eta+b)/(c\eta+d)$, where a, b, c, d are real numbers such that $ad - bc = 1$.*

In particular, if η itself is a Möbius transformation $\eta : \theta \mapsto (a\theta+b)/(c\theta+d)$, then $S(\eta) = S(\mathrm{id}) = 0$, where $\mathrm{id} : \theta \mapsto \theta$.

PROOF.   Indeed, for a given potential u the solutions f and g of the corresponding differential equation are not defined uniquely, but up to a transformation of the pair (f, g) by a matrix from $\mathrm{SL}(2, \mathbb{R})$. Then the ratio η changes by a Möbius transformation. Thus Möbius equivalent ratios η correspond to the same potential $u = S(\eta)/2$. For the identity diffeomorphism $\mathrm{id} : \theta \mapsto \theta$, the explicit formula for the Schwarzian derivative gives $S(\mathrm{id}) = 0$.   □

Proposition 2.11 *The Virasoro coadjoint action of a diffeomorphism φ on the potential $u(\theta)$ gives rise to a diffeomorphism change of coordinate in the ratio η:*

$$\varphi : \quad \eta(\theta) \to \eta(\varphi(\theta)).$$

PROOF.   We look at the corresponding infinitesimal action on the solutions of the differential equation $(\partial_\theta^2 + u(\theta))y = 0$. For a diffeomorphism $\varphi^{-1}(\theta) = \theta + \epsilon v(\theta)$ close to the identity, consider the infinitesimal Virasoro action of φ^{-1} on the potential $u(\theta)$:

$$u \mapsto u + \epsilon \cdot \delta u, \quad \text{where} \quad \delta u = 2uv' + u'v + \frac{1}{2}v'''$$

(cf. formula (2.8) for $a = \frac{1}{2}$ and note that we are considering the action of φ^{-1}). It is consistent with the following action on a solution y of the above differential equation:

$$y \mapsto y + \epsilon \cdot \delta y, \quad \text{where} \quad \delta y = -\frac{1}{2}yv' + y'v.$$

The consistency means that $(\partial_\theta^2 + u + \epsilon \cdot \delta u)(y + \epsilon \cdot \delta y) = 0 + \mathcal{O}(\epsilon^2)$.

Note that the action $\epsilon \cdot \delta y = \epsilon \cdot (-\frac{1}{2}yv' + y'v)$ is an infinitesimal version of the following action of the diffeomorphism $\varphi^{-1}(\theta) = \theta + \epsilon v(\theta)$ on y:

$$\varphi^{-1}: \quad y(\theta) \mapsto y(\varphi(\theta))(\varphi'(\theta))^{-1/2}.$$

Thus solutions to Hill's equation transform as densities of degree $-1/2$. Therefore the ratio η of two solutions transforms as a function under a diffeomorphism action. $\qquad\square$

In short, to calculate the coadjoint action on the potential u one can first pass from this potential to the ratio of two solutions of the corresponding Hill equation, then change the variable in the ratio, and finally take the Schwarzian derivative of the new ratio to reconstruct the new potential $\mathrm{Ad}^*_{\varphi^{-1}}u$ (see Figure 2.1).

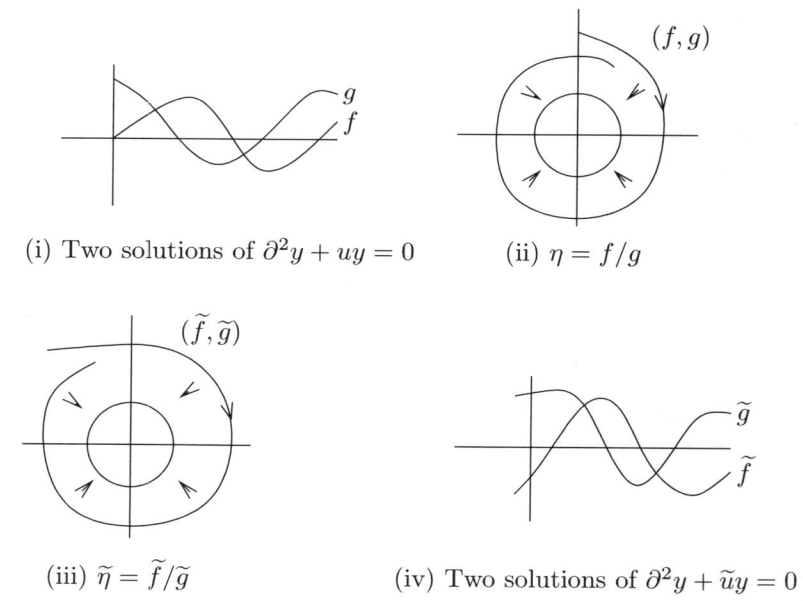

(i) Two solutions of $\partial^2 y + uy = 0$ (ii) $\eta = f/g$

(iii) $\widetilde{\eta} = \widetilde{f}/\widetilde{g}$ (iv) Two solutions of $\partial^2 y + \widetilde{u}y = 0$

Fig. 2.1. Schematic picture of the action of a diffeomorphism of S^1 on Hill's operators and their solutions: To two solutions of the equation $(\partial^2 + u)y = 0$, one associates their ratio $\eta : \mathbb{R} \to \mathbb{R}P^1$. A diffeomorphism φ acts on the ratio η by reparametrization, and one reconstructs the corresponding solutions and Hill's operator $\partial^2 + \widetilde{u}$ from the new ratio $\widetilde{\eta}$.

Now we return to our goal, the classification of the Virasoro coadjoint orbits. Any ($a = \mathrm{const}$)-hyperplane in the Virasoro dual $\mathfrak{vir}^*$ is invariant under the coadjoint action (see equation (2.10)), and identified with Hill's operators for $a \neq 0$. While all of the above considerations of Hill's operators were of local nature (local in θ), now we will make use of the fact that Hill's operators are periodic: $u(\theta)$ is defined on a circle.

Consider the universal covering $\widetilde{\mathrm{SL}}(2,\mathbb{R})$ of the group $\mathrm{SL}(2,\mathbb{R})$. The group $\widetilde{\mathrm{SL}}(2,\mathbb{R})$ admits an outer automorphism of order 2, taking the inverse of a matrix. One can see that the identity is its only fixed point on the universal covering group. Consider the set $(\widetilde{\mathrm{SL}}(2,\mathbb{R}) \setminus \{\mathrm{id}\})/\mathbb{Z}_2$, where we first dropped the identity element from $\widetilde{\mathrm{SL}}(2,\mathbb{R})$ before taking the quotient. The main result of this section is the following theorem.

Theorem 2.12 ([342, 202]) *Given $a \neq 0$, there is a one-to-one correspondence between the set of coadjoint orbits of the Virasoro–Bott group in the hyperplane $\{a\partial_\theta^2 + u(\theta)\} \subset \mathfrak{vir}^*$ and the set of conjugacy classes in the quotient $(\widetilde{\mathrm{SL}}(2,\mathbb{R}) \setminus \{\mathrm{id}\})/\mathbb{Z}_2$*

PROOF.   Consider a pair (f, g) of linearly independent solutions of the Hill equation $(a\partial_\theta^2 + u(\theta))y = 0$. For a periodic potential $u(\theta)$ these solutions are quasiperiodic, i.e., the values $(f(\theta), g(\theta))$ and $(f(\theta+2\pi), g(\theta+2\pi))$ are related by a monodromy matrix $M \in \mathrm{SL}(2,\mathbb{R})$:

$$(f(\theta + 2\pi), g(\theta + 2\pi)) = (f(\theta), g(\theta))\, M \,. \tag{2.12}$$

Recall that we use the same notation θ for the coordinate on the circle $S^1 = \mathbb{R}/2\pi\mathbb{Z}$ and on its cover $\mathbb{R}$. Note that the monodromy matrix M can be viewed as an element in the universal cover $\widetilde{\mathrm{SL}}(2,\mathbb{R})$, where the lift to the cover is provided by the fundamental solution $(f(\theta), g(\theta))$, starting at the identity: $\begin{pmatrix} f & g \\ f' & g' \end{pmatrix}|_{\theta=0} = \mathrm{id}$.

Similarly, the values of the "projective solution," the ratios $\eta(\theta) := f(\theta)/g(\theta)$ and $\eta(\theta + 2\pi) := f(\theta + 2\pi)/g(\theta + 2\pi)$, are related by a Möbius transformation $\mathcal{M} \in \mathrm{PSL}(2,\mathbb{R}) = \mathrm{SL}(2,\mathbb{R})/\{\pm\,\mathrm{id}\}$. The monodromy matrix M (respectively $\mathcal{M}$) changes to a conjugate matrix if we pick a different pair of solutions (f, g) for the same differential equation.

Now regard the ratio $\eta = f/g$ for $\theta \in [0, 2\pi]$ as a map $\eta : [0, 2\pi] \to \mathbb{R}P^1$ describing a motion ("rotation") along the circle $\mathbb{R}P^1 \simeq S^1$. One can see that the condition $W \neq 0$ on the Wronskian is equivalent to the condition $\eta' = -W/g^2 \neq 0$, i.e., that the rotation "does not stop." Choosing the positive sign of the Wronskian, $W > 0$, we can assume that the rotation always goes in the negative direction: $\eta' < 0$.

Recall that the Virasoro action on η is, in fact, a circle reparametrization for the coordinate θ. By a diffeomorphism change of the coordinate $\theta \mapsto \varphi(\theta)$, one can always turn the map $\eta : [0, 2\pi] \to \mathbb{R}P^1$ into a *uniform* rotation along $\mathbb{R}P^1$, while keeping the boundary values of $\eta(\theta)$ on the segment $[0, 2\pi]$ satisfying the monodromy relation $\eta(\theta + 2\pi) = \eta(\theta)\mathcal{M}$. Furthermore, the number of rotations (the "winding number") for the map $\eta : [0, 2\pi] \to \mathbb{R}P^1$ does not change under a reparametrization by a circle diffeomorphism φ. In other words, the orbits of the maps η (or, equivalently, of the potentials $\{u(x)\}$) are described by one continuous parameter (the conjugacy class of M) and one discrete parameter (the winding number). One can see that these two parameters together encode nothing else but the conjugacy class of the monodromy matrices M in the universal covering of $SL(2, \mathbb{R})$.

Note that the choice in the sign of the Wronskian reflects the $\mathbb{Z}_2$-action on the universal covering $\widetilde{SL}(2, \mathbb{R})$. Indeed, with this choice $(W > 0)$ the path η always goes in the negative direction $(\eta' < 0)$, so one can reach only the "negative half" of the conjugacy classes in the universal cover $\widetilde{SL}(2, \mathbb{R})$.

Finally, note that the identity matrix in the universal covering $\widetilde{SL}(2, \mathbb{R})$ (or in its projectivization $\widetilde{SL}(2, \mathbb{R})/\mathbb{Z}_2$) cannot be obtained as a monodromy matrix for the maps $\eta : [0, 2\pi] \to \mathbb{R}P^1$. Indeed, any map η starting at the identity has to move out from it, since $\eta'(0) \neq 0$. $\square$

Corollary 2.13 *The Virasoro orbits in the hyperplane* $\{a\partial_\theta^2 + u(\theta) \mid a = a_0\} \subset \mathfrak{vir}^*$ *with fixed* $a_0 \neq 0$ *are classified by the Jordan normal form of matrices in* $SL(2, \mathbb{R})$ *and by a positive integer parameter, the winding number. In this hyperplane* $\{a = a_0\}$ *of the dual* $\mathfrak{vir}^*$ *the orbit containing Hill's operator* $a\partial_\theta^2 + u(\theta)$ *has codimension equal to the codimension in* $SL(2, \mathbb{R})$ *of the conjugacy class of the monodromy matrix* M *corresponding to this Hill's operator.*

Matrices in the group $SL(2, \mathbb{R})$ split into three classes, whose normal forms are the exponentials of the following three classes in the Lie algebra $\mathfrak{sl}(2, \mathbb{C})$, the complexification of $\mathfrak{sl}(2, \mathbb{R})$:

$$(i) \begin{pmatrix} \mu & 0 \\ 0 & -\mu \end{pmatrix}, \quad (ii) \begin{pmatrix} 0 & \pm 1 \\ 0 & 0 \end{pmatrix}, \text{ and } \quad (iii) \begin{pmatrix} 0 & 0 \\ 0 & 0 \end{pmatrix}; \tag{2.13}$$

see Figure 2.2. The codimensions of the corresponding conjugacy classes in $SL(2, \mathbb{R})$ are 1 in cases (i) and (ii), and 3 in case (iii). Note that the set of real matrices that are exponentials of (i)-type matrices consists of the elliptic and hyperbolic parts: rotation matrices (for $\mu \in i\mathbb{R}$) and hyperbolic rotations (for $\mu \in \mathbb{R}$). Furthermore, hyperbolic rotations correspond to one-sheeted hyperboloids. Rotations in the clockwise and counterclockwise directions correspond to different sheets of two-sheeted hyperboloids, and they belong to different conjugacy classes in $SL(2, \mathbb{R})$. (The rotation by $180°$ has a three-dimensional stabilizer and corresponds to a one-point conjugacy class.) The group $SL(2, \mathbb{R})$ is topologically a solid torus, and the adjacency of conjugacy classes described in Figure 2.2 is observed near both id and $-$ id in this group.

The equality of the codimensions of the Virasoro coadjoint orbits in $\mathfrak{vir}^*$ and the codimensions of (the conjugacy classes of) the corresponding

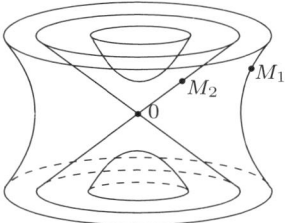

Fig. 2.2. The points M_1, M_2, and 0 in $\mathfrak{sl}(2,\mathbb{R})$ (which is a local picture of the group $\mathrm{SL}(2,\mathbb{R})$) correspond to Virasoro orbits of types (i), (ii), and (iii) respectively.

monodromy matrices in $\mathrm{SL}(2,\mathbb{R})$ follows from the smooth dependence on a parameter in the above classification. (The versal deformations of the Virasoro orbits can be defined in terms of the Jordan–Arnold normal forms of the monodromy matrices depending on a parameter; cf. [15, 233, 306].) Alternatively, one can describe the dimensions of the corresponding stabilizers; see [202, 342]. To visualize (the three-dimensional transversal to) the set of the Virasoro orbits, one can imagine the universal covering of $\mathrm{SL}(2,\mathbb{R})$ as a cylinder filled with an infinite number of copies of Figure 2.2, stacked one on top of another, while the $\mathbb{Z}_2$-quotient keeps only "half" of this infinite cylinder.

Remark 2.14 Regarded as homogeneous spaces, the orbits of type (i) are often denoted by $\mathrm{Diff}(S^1)/S^1$, the notation $\mathrm{Diff}(S^1)/\mathbb{R}^1$ stands for (ii) (and sometimes for the case $\mu \in \mathbb{R}$ in (i)), and $\mathrm{Diff}(S^1)/\mathrm{SL}(2,\mathbb{R})$ corresponds to (iii).

To see the reasoning for this, we describe the stabilizers for coadjoint orbits containing constant elements, i.e., Hill's operators $\partial_\theta^2 + u(\theta)$ with constant potentials $u(\theta) \equiv p = \mathrm{const}$. For such an operator, the corresponding monodromy matrix $M_p \in \mathrm{SL}(2,\mathbb{R})$ is given explicitly:

$$M_p = \begin{pmatrix} \cos(2\pi\sqrt{p}) & \frac{1}{\sqrt{p}}\sin(2\pi\sqrt{p}) \\ -\sqrt{p}\sin(2\pi\sqrt{p}) & \cos(2\pi\sqrt{p}), \end{pmatrix} \qquad \text{for } p > 0,$$

$$M_p = \begin{pmatrix} \cosh(2\pi\sqrt{-p}) & \frac{1}{\sqrt{-p}}\sinh(2\pi\sqrt{-p}) \\ \sqrt{-p}\sinh(2\pi\sqrt{-p}) & \cosh(2\pi\sqrt{-p}) \end{pmatrix} \qquad \text{for } p < 0,$$

and

$$M_0 = \begin{pmatrix} 1 & 2\pi \\ 0 & 1 \end{pmatrix} \qquad \text{for } p = 0.$$

One can see that for Hill's differential operators with potentials $p < 0$ or $p = 0$ the stabilizer is $\mathbb{R}$. In the case $p > 0$ it is the group S^1 of rigid rotations, provided that $M_p \neq \pm\,\mathrm{id} \in \mathrm{SL}(2,\mathbb{R})$. Finally, the stabilizer is three-dimensional, once the monodromy M_p is plus or minus the identity matrix, i.e., the exponential of type (iii) in the list (2.13). This can be the case only if $p = m^2/4$

for some $m \in \mathbb{N}$. In the latter case, the stabilizer of the corresponding Hill's equation is the m-fold covering of $\mathrm{PSL}(2, \mathbb{R})$ in $\mathrm{Diff}(S^1)$.

Finally, we note that the trace function $\mathrm{tr} : \mathrm{SL}(2, \mathbb{R}) \to \mathbb{R}$ is invariant on conjugacy classes in $\mathrm{SL}(2, \mathbb{R})$ and it gives rise to a Casimir function on the hyperplane $\{a = a_0\} \subset \mathfrak{vir}^*$ for the Lie–Poisson bracket: $\mathrm{tr}\{a_0 \partial_\theta^2 + u(\theta)\} :=$ $\mathrm{tr}(M)$. where M is a monodromy matrix of the given Hill operator. Along with the value of a, this is the only Casimir for generic Virasoro orbits, since their codimension in the hyperplane $\{a = a_0\}$ is equal to 1.

2.4 The Virasoro–Bott Group and the Korteweg–de Vries Equation

The *Korteweg–de Vries* (or KdV) equation is the nonlinear evolution equation

$$u_t = -3uu' - au''' , \tag{2.14}$$

which describes traveling waves in a shallow canal. Here, u is a function of the time variable t and one space variable θ, u_t and u' denote the corresponding partial derivatives in t and θ, and a is a nonzero constant. A brief history of this equation can be found in [294].

In this section we show how the Korteweg–de Vries equation appears as the Euler equation with respect to a certain right-invariant metric on the Virasoro–Bott group. Recall that the Euler equation with respect to a right-invariant metric on a Lie group G is a dynamical system on the corresponding Lie algebra $\mathfrak{g}$ describing the evolution of the tangent vector along a geodesic on G, where this vector is pulled back to the Lie algebra of G by right translation; see Section I.4.

Consider the L^2-inner product on the Virasoro algebra $\mathfrak{vir} = \mathrm{Vect}(S^1) \oplus \mathbb{R}$ defined by

$$\langle (v(\theta)\partial_\theta, a), (w(\theta)\partial_\theta, c) \rangle = \int_{S^1} v(\theta)w(\theta)d\theta + a \cdot c . \tag{2.15}$$

Extend this quadratic form to every tangent space on the Virasoro–Bott group by right translations to define a (weak) right-invariant L^2-metric on the group.

Theorem 2.15 ([305]) *The Euler equation for the right-invariant L^2-metric on the Virasoro group is (the family of) the KdV equation:*

$$u_t = -3uu' - au''' , \tag{2.16}$$

$$a_t = 0 . \tag{2.17}$$

PROOF.   According to Arnold's Theorem I.4.14, the Euler equation on $\mathfrak{g}^*$ for the *right-invariant* metric on the group G has the form

$$\frac{d}{dt}m(t) = \mathrm{ad}^*_{A^{-1}m(t)}\, m(t)\,, \tag{2.18}$$

where $m(t)$ is a point in $\mathfrak{g}^*$, and $A : \mathfrak{g} \to \mathfrak{g}^*$ is the inertia operator defined by the metric (using the left-invariant metric would give the minus sign in the equation; cf. Remark I.4.16).

In the Virasoro coadjoint action, an element $(v\partial_\theta, c) \in \mathfrak{vir}$ of the Virasoro algebra acts on an element $(u(d\theta)^2, a) \in \mathfrak{vir}^*$ of the dual space via

$$\mathrm{ad}^*_{(v\partial_\theta,c)}(u(d\theta)^2, a) = ((-2v'u - vu' - av''')(d\theta)^2, 0)\,;$$

see equation (2.8). Furthermore, the L^2-inner product gives rise to the "identity" inertia operator $A : \mathfrak{vir} \to \mathfrak{vir}^*$:

$$(u\partial_\theta, a) \mapsto (u(d\theta)^2, a)\,,$$

mapping a vector field $u(\theta)\partial_\theta$ to the quadratic differential $u(\theta)(d\theta)^2$ with the same function $u(\theta)$.

Then, substituting $(u(d\theta)^2, a)$ for m, we see that the Euler equation (2.18) becomes

$$\frac{d}{dt}(u(d\theta)^2, a) = ((-3uu' - au''')(d\theta)^2, 0)\,,$$

from which one immediately reads off the KdV equation (2.16), (2.17). $\qquad\square$

The component a does not change with time (see equation (2.17)) and plays the role of a constant parameter in the KdV equation. It has the physical meaning of the characteristic thickness of the shallow-water approximation (see, e.g., [228], p. 169).

Remark 2.16 One can study more general metrics on the Virasoro algebra, some of which are of particular interest in mathematical physics. Consider, for instance, the following two-parameter family of weighted $H^1_{\alpha,\beta}$-inner products on $\mathfrak{vir}$:

$$\langle (v\partial_\theta, b), (w\partial_\theta, c) \rangle_{H^1_{\alpha,\beta}} = \int_{S^1} (\alpha\, vw + \beta\, v'w')\, d\theta + bc\,.$$

The case $\alpha = 1, \quad \beta = 0$ corresponds to the L^2 inner product above, while $\alpha = \beta = 1$ corresponds to the H^1 product.

Theorem 2.17 ([192]) *The Euler equations for the right-invariant $H^1_{\alpha,\beta}$-metric (with $\alpha \neq 0$) on the Virasoro group are given by the following system:*

$$\alpha(u_t + 3uu') - \beta((u'')_t + 2u'u'' + uu''') + au''' = 0\,, \tag{2.19}$$

$$a_t = 0\,. \tag{2.20}$$

Exercise 2.18 Give a proof of the latter theorem along the lines of the proof of the L^2-case above. (Hint: The inertia operator for the weighted H^1 metric is

$$A : (v\partial_\theta, a) \mapsto ((\Lambda v)(d\theta)^2, a),$$

where $\Lambda := \alpha - \beta\partial_\theta^2$ is a second-order differential operator. Verify that in terms of $v = \Lambda^{-1}u$ the Euler equation has the form

$$\frac{d}{dt}(\Lambda v) = -2(\Lambda v)v' - (\Lambda v')v + av''',$$

which is equivalent to equation (2.19).)

Remark 2.19 For $\alpha = 1$, $\beta = 0$, equation (2.19) is the KdV equation (2.16). For $\alpha = \beta = 1$, one recovers the *Camassa–Holm equation* (see [268]). For $\alpha = 0$, $\beta = 1$, equation (2.19) becomes the *Hunter–Saxton equation*. We note that in the case of $\alpha = 0$, the $H^1_{\alpha,\beta}$-metric becomes the homogeneous $\dot{H}^1$-metric, which is degenerate. Therefore, to define the Euler equations one has to pass to the homogeneous space $\widehat{\mathrm{Diff}}(S^1)/S^1$ (or $\mathrm{Diff}(S^1)/S^1$) and define the geodesic flow on it; see details in [192]. It turns out that the space $\mathrm{Diff}(S^1)/S^1$ equipped with the $\dot{H}^1$-metric is isometric to an open subset of an L^2-sphere; see [237, 238]. This isometry, in particular, allows one to extend solutions of the Hunter–Saxton equation beyond breaking time and interpret them after wave-breaking in an appropriate weak sense.

   We also note that the case $a = 0$ corresponds to the nonextended Lie algebra $\mathrm{Vect}(S^1)$ of vector fields on the circle, rather than to the Virasoro algebra $\mathfrak{vir}$. In the nonextended case, depending on the values of α and β, one obtains the Hopf (or inviscid Burgers) equation $u_t + 3uu' = 0$ or the nonextended Camassa–Holm equation [69, 127]

$$u_t + 3uu' + 2u'u'' + uu''' + (u'')_t = 0.$$

2.5 The Bi-Hamiltonian Structure of the KdV Equation

The KdV equation is not only a Hamiltonian system; it also exhibits strong integrability properties. As we discussed before, there are various definitions of what an integrable infinite-dimensional system is: one can require from the system either the existence of action-angle coordinates, or the existence of "sufficiently many" independent integrals of motion, or some other properties, which may differ substantially in infinite dimensions. In this section we show that the KdV equation is not only Hamiltonian, but in fact bi-Hamiltonian, thus exhibiting one of the "strongest forms" of integrability. More precisely, in addition to being Hamiltonian with respect to the Lie–Poisson bracket on the dual space $\mathfrak{vir}^*$, this equation turns out to be Hamiltonian with respect to another compatible Poisson structure on the same space.

   Recall that for any Lie algebra $\mathfrak{g}$, every point $m_0 \in \mathfrak{g}^*$ gives rise to a "constant" Poisson bracket $\{\ ,\ \}_0$ on $\mathfrak{g}^*$ by "freezing" the usual Lie–Poisson bracket $\{\ ,\ \}_{LP}$ at the point m_0. The constant Poisson bracket is defined for smooth functions f, g on $\mathfrak{g}^*$ by

$$\{f, g\}_0(m) := \langle [df_m, dg_m], m_0 \rangle\,;$$

see Section I.4.4. Furthermore, the Poisson brackets $\{\ ,\ \}_{LP}$ and $\{\ ,\ \}_0$ are compatible for all choices of the point m_0 (see Lemma I.4.21). The main goal of this section is to show that for a certain choice of the "freezing point" $m_0 \in \mathfrak{vir}^*$, the KdV equation is Hamiltonian with respect to the constant Poisson structure $\{\ ,\ \}_0$. Note that other equations discussed above (Camassa–Holm, Hunter–Saxton) have a similar bi-Hamiltonian structure, but related to different choices of the "freezing point"; see [192].

Theorem 2.20 *The KdV equation (2.14) is Hamiltonian with respect to the constant Poisson bracket on $\mathfrak{vir}^*$ with the "freezing point" $m_0 = (\frac{1}{2}(d\theta)^2, 0) \in \mathfrak{vir}^*$.*

PROOF.   Let $F(u, a)$ be a function on $\mathfrak{vir}^*$ and let $(v\partial_\theta, b) :=$ $(\delta F/\delta u, \delta F/\delta a) \in \mathfrak{vir}$ be the (variational) derivative $dF_{(u(d\theta)^2, a)}$ of F at $(u(d\theta)^2, a)$. Then the Hamiltonian equation with the Hamiltonian function F, computed with respect to the constant Poisson structure "frozen" at $(u_0(dx)^2, a_0)$, has the form

$$\frac{d}{dt}(u(d\theta)^2, a) = \mathrm{ad}^*_{(v\partial_\theta, b)}(u_0(d\theta)^2, a_0) = -\big((2u_0 v' + (u_0)'v + a_0 v''')(d\theta)^2, 0\big)\,.$$

(Here we use Remark I.4.22 and the explicit form (2.8) of the coadjoint action ad^* for the Virasoro algebra.)

Now specifying the "freezing" point to $(u_0(d\theta)^2, a_0) = (\frac{1}{2}(d\theta)^2, 0) \in \mathfrak{vir}^*$, we come to

$$\frac{d}{dt}(u(d\theta)^2, a) = -(v'(d\theta)^2, 0)\,, \tag{2.21}$$

where v is defined as the "partial derivative" of the functional F, i.e., $v\partial_\theta = \delta F/\delta u$.

Next, consider the functional F of the form

$$F(u, a) = \int_{S^1} \left(\frac{1}{2}u^3 - \frac{a}{2}(u')^2 \right) d\theta\,.$$

By definition, the *variational derivative* $(\delta F/\delta u, \delta F/\delta a) \in \mathfrak{vir}$ of the functional F is determined by the following identity satisfied for any $(\xi(d\theta)^2, c) \in \mathfrak{vir}^*$:

$$\left\langle (\xi(d\theta)^2, c), \left(\frac{\delta F}{\delta u}, \frac{\delta F}{\delta a} \right) \right\rangle = \frac{d}{d\epsilon}\Big|_{\epsilon=0} F(u + \epsilon\xi, a + \epsilon c).$$

For equation (2.21) we need only the partial derivative $\delta F/\delta u$, and we find it as follows:

$$\frac{d}{d\epsilon}\bigg|_{\epsilon=0} F(u + \epsilon\xi, a) = \frac{d}{d\epsilon}\bigg|_{\epsilon=0} \int_{S^1} \left(\frac{1}{2}(u + \epsilon\xi)^3 - \frac{a}{2}(u' + \epsilon\xi')^2 \right) d\theta$$

$$= \int_{S^1} \left(\frac{3}{2}u^2\xi - au'\xi' \right) d\theta = \int_{S^1} \left(\frac{3}{2}u^2\xi + au''\xi \right) d\theta$$

$$= \left\langle \left(\left(\frac{3}{2}u^2 + au'' \right) \partial_\theta, a \right), (\xi(d\theta)^2, 0) \right\rangle.$$

Hence we obtain the derivative $\delta F/\delta u = (\frac{3}{2}u^2 + au'')\partial_\theta$. Since $v\partial_\theta = \delta F/\delta u$, now we substitute $v = \frac{3}{2}u^2 + au''$ into the equation (2.21), which gives

$$u_t = -3uu' - au''',$$

the KdV equation.   □

Corollary 2.21 *The KdV equation is bi-Hamiltonian with respect to the compatible Poisson structures* $\{\ ,\ \}_{LP}$ *and* $\{\ ,\ \}_0$ *on* $\mathfrak{vir}^*$*, where* $\{\ ,\ \}_0$ *denotes the constant Poisson structure, frozen in the point* $(\frac{1}{2}(d\theta)^2, 0) \in \mathfrak{vir}^*$.

The constant bracket for the KdV is usually called the *first KdV structure*, or the Gardner–Faddeev–Zakharov bracket [140, 392], while the linear Lie–Poisson structure is called the *second KdV*, or the Magri *bracket* [246]. The analogues of these structures for higher-order differential operators are called the first and second Adler–Gelfand–Dickey structures; see Section 4.

Remark 2.22 Similarly, one can show that both the Camassa–Holm and Hunter–Saxton equations are bi-Hamiltonian with respect to the Lie–Poisson structure and a constant structure on $\mathfrak{vir}^*$. The respective "freezing" points of the constant Poisson bracket on $\mathfrak{vir}^*$ are $m_1 = (\frac{1}{2}(d\theta)^2, 1)$ for the Camassa–Holm equation and $m_2 = (0, 1)$ for the Hunter–Saxton equation (see Figure 2.3).

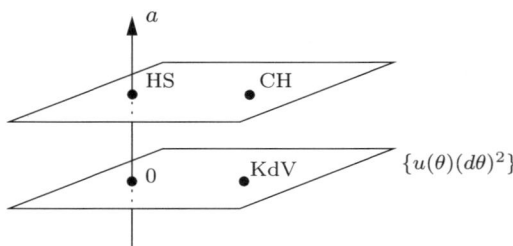

Fig. 2.3. The freezing points of the constant Poisson bracket on $\mathfrak{vir}^*$ that give rise to the bi-Hamiltonian structures for the KdV, the Camassa–Holm, and the Hunter–Saxton equations respectively.

Remark 2.23 The bi-Hamiltonian nature of the KdV equation allows one to obtain the whole hierarchy of the KdV Hamiltonians via the Lenard–Magri scheme discussed in the introduction.

Namely, consider the linear combination of the Poisson brackets for the KdV: the Virasoro Lie–Poisson and the constant ones:

$$\{ \, , \, \}_\lambda := \{ \, , \, \}_{LP} + \lambda^2 \{ \, , \, \}_0 \, .$$

Since the corresponding brackets are compatible, $\{ \, , \, \}_\lambda$ defines a Poisson bracket for all $\lambda \in \mathbb{R}$, which is the usual Lie–Poisson bracket on $\mathfrak{vir}^*$ shifted in the direction of $m_0 = (\frac{1}{2}(d\theta)^2, 0) \in \mathfrak{vir}^*$. (Note that one uses here the parameter λ^2 rather than λ for the expansion below to have a simpler form.)

Recall that we think of the dual space $\mathfrak{vir}^* = \{a\partial_\theta^2 + u(\theta)\}$ as the space of Hill's operators. The monodromy of a differential operator $\partial_\theta^2 + u(\theta)$, which is a matrix in $SL(2,\mathbb{R})$, changes to a conjugate matrix under the Virasoro coadjoint action (see Theorem 2.12). Therefore, the trace of the monodromy is a Casimir function for the Poisson bracket $\{ \, , \, \}_{LP}$ on (the hyperplane $a = 1$ in) the dual space $\mathfrak{vir}^*$. The same reasoning allows one to obtain the following result.

Lemma 2.24 *Let M_λ denote the monodromy of the differential operator $\frac{d^2}{d\theta^2} + u(\theta) - \lambda^2$. Then the function*

$$h_\lambda := \log(\mathrm{tr}(M_\lambda))$$

is a Casimir function for the Poisson bracket $\{ \, , \, \}_\lambda$ on the space $\mathfrak{vir}^*$.

Indeed, $\{ \, , \, \}_\lambda$ is the usual Lie–Poisson bracket on $\mathfrak{vir}^*$ shifted in the direction of $m_0 = (0 \cdot \partial_\theta^2 + 1/2) \in \mathfrak{vir}^*$, and so, instead of the Lie–Poisson Casimir $\mathrm{tr}(M)$, we can use the shifted Casimir $\mathrm{tr}(M_\lambda)$ or any function of it, in particular, $\log(\mathrm{tr}(M_\lambda))$.

Finally, by expanding h_λ into a power series in λ^{-1} one produces first integrals of the KdV equation:

$$h_\lambda = 2\pi\lambda - \sum_{n=1}^{\infty} h_{2n-1} \lambda^{1-2n} \, ,$$

where

$$h_1 = \frac{1}{2} \int_{S^1} u \, d\theta, \quad h_3 = \frac{1}{8} \int_{S^1} u^2 \, d\theta, \quad h_5 = \frac{1}{16} \int_{S^1} \left(u^3 - \frac{1}{2}(u')^2\right) d\theta, \quad \dots;$$

see, e.g., [31, 37].

2.6 Bibliographical Notes

The cohomology of the Lie algebra of vector fields on the circle was computed by Gelfand and Fuchs in [143], where the term Virasoro algebra was coined after the paper [376] (see also [118, 138]). The Bott cocycle on the group $\mathrm{Diff}(S^1)$ first appeared in [53].

The group $\mathrm{Diff}(S^1)$ (and hence the Virasoro group) does not admit a natural complexification, i.e., a group corresponding to the complexified Lie algebra $\mathrm{Vect}(S^1)_{\mathbb{C}}$; see [205, 322] and Example I.1.25. However, for a cone in this complex Lie algebra consisting of those vector fields that "point outside of the circle," there exists a semigroup. This *annulus semigroup* appeared in the papers of Neretin [291] and Segal [343]. For more details on representations and applications of the Virasoro group and algebra see [292, 322, 153].

The classification of the coadjoint orbits of the Virasoro group can be found in the literature under different guises: as a classification of projective structures on the circle by Kuiper [223], as a classification of Hill's operators by Lazutkin and Pankratova [233], and in the present form, as Virasoro orbits, in the papers by Kirillov [202, 205] and Segal [342]; see also [164, 304, 388, 32]. The adjoint orbits of the diffeomorphism group of the circle were described in [164].

The Virasoro coadjoint orbit $\mathrm{Diff}(S^1)/S^1$ can also be understood as the universal Teichmüller space [326]; cf. [284, 285]. Curvatures of a Kähler metric on this orbit were described in [208], while its complex Hilbert manifold structure is discussed in [338, 363]. The Virasoro group itself admits a complex structure and can be viewed as a holomorphic $\mathbb{C}^*$-bundle over its orbit $\mathrm{Diff}(S^1)/S^1$ [236].

There is a vast literature related to the geometry and Hamiltonian properties of the KdV equation, which is one of the key examples in any book on soliton theory. The description of the KdV equation, as well as its super-analogue, as an Euler–Arnold equation on the Virasoro group with respect to the L^2-metric can be found in [305, 344, 346]. More general, H^1-type, metrics on this group were considered in [268, 189, 192], and we followed the latter paper in our exposition. Regularity properties of the Riemannian exponential maps for these and other Sobolev metrics on the Virasoro and the diffeomorphism groups are described in [73, 74].

The Adler–Gelfand–Dickey structures [3, 141, 142] are the generalizations of the KdV Poisson structures from Hill's operators to linear differential operators of higher order, and we discuss them in Section 4.

For the algebro-geometric approach to constructing solutions of the KdV equation we address the reader to [216]; the description of the corresponding infinite-dimensional Grassmann manifolds can be found in [347] and the references therein. Various analytical aspects of the KdV theory, its spectral theory, and the angle-action variables are discussed in the book [182]. The geometry related to the KAM theory for near-integrable Hamiltonian systems, applied in the infinite-dimensional context, e.g., to the KdV-type equations, is discussed in [224, 182].

3 Groups of Diffeomorphisms

Diffeomorphism groups constitute one of the most intriguing classes of infinite-dimensional Lie groups. They often serve as configuration spaces for various equations of fluid and gas dynamics. In this section, our main objects of study are the groups of volume-preserving diffeomorphisms, in particular their geometry and coadjoint orbits. The Euler geodesic equations for such groups and for their generalizations deliver the Euler equations for ideal incompressible fluids on manifolds, as well as the equations of compressible fluids and magnetohydrodynamics. We also discuss how invariants of the coadjoint action of certain diffeomorphism groups are related to knot theory and to the symplectic structure on the space of immersed curves in $\mathbb{R}^3$.

3.1 The Group of Volume-Preserving Diffeomorphisms and Its Coadjoint Representation

Let M be an n-dimensional compact manifold (possibly with boundary) and let μ be a volume form on M.

Definition 3.1 The *group of volume-preserving diffeomorphisms* of the manifold M consists of all diffeomorphisms of M preserving the volume form μ:

$$S\mathrm{Diff}(M) = \{\varphi \in \mathrm{Diff}(M) \mid \varphi^*\mu = \mu\},$$

where the group multiplication is the composition of diffeomorphisms. (Whenever this group of diffeomorphisms is not connected, our notation $S\mathrm{Diff}$ stands for the connected component of the identity of this group.)

Exercise 3.2 The corresponding Lie algebra $S\mathrm{Vect}(M)$ consists of divergence-free vector fields on M:

$$S\mathrm{Vect}(M) = \{\xi \in \mathrm{Vect}(M) \mid L_\xi\mu = 0 \text{ and } \xi \text{ is tangent to } \partial M\},$$

where the algebra commutator is the negative of the Lie bracket of two vector fields: $[\xi, \eta] = -L_\xi\eta$; cf. Example I.2.2.

It turns out that the (smooth part of the) dual of the Lie algebra $S\mathrm{Vect}(M)$ has a natural geometric description.

Proposition 3.3 *The smooth part of the dual of the Lie algebra $S\mathrm{Vect}(M)$ is naturally identified with the space $\Omega^1(M)/d\Omega^0(M)$ of smooth 1-forms on M modulo exact 1-forms on M. The pairing is as follows:*

$$\langle \xi, [u] \rangle = \int_M \iota_\xi u \wedge \mu,$$

where $u \in \Omega^1(M)$ is any 1-form from the coset $[u]$.

In the coadjoint representation, a diffeomorphism $\varphi \in \mathrm{SDiff}(M)$ acts by "change of coordinates": it sends a coset $[u] \in \Omega^1(M)/d\Omega^0(M)$ to the coset $[(\varphi^{-1})^*u]$. Accordingly, the coadjoint action of $S\mathrm{Vect}(M)$ on $\Omega^1(M)/d\Omega^0(M)$ is given by the negative of the Lie derivative:

$$\mathrm{ad}^*_\xi \ : \ [u] \mapsto -[L_\xi u]$$

for any divergence-free vector field ξ on M.

PROOF.   Assume first that the manifold M has no boundary. Note that divergence-free vector fields on M are in one-to-one correspondence with the space of all closed $(n-1)$-forms on M. Indeed, any vector field ξ on M defines an $(n-1)$-form ν_ξ via interior product with the volume form on M: $\nu_\xi = \iota_\xi\mu$. Moreover, this form ν_ξ is closed for a divergence-free field ξ, since $0 = L_\xi\mu = d\iota_\xi\mu + \iota_\xi d\mu = d\nu_\xi$.

The pairing between $\Omega^1(M)/d\Omega^0(M)$ and $S\mathrm{Vect}(M)$ can now be rewritten as

$$\langle \xi, [u] \rangle = \int_M u \wedge \nu_\xi$$

for any representative 1-form u. Two different representatives u and $\tilde{u}$ differ by a differential df and give the same pairing. Indeed, since the form ν_ξ is closed, so is $df \wedge \nu_\xi$. Then the Stokes formula gives

$$\langle \xi, df \rangle = \int_M df \wedge \nu_\xi = \int_M d(f \wedge \nu_\xi) = 0$$

for M without boundary.

In the case of a manifold M with boundary ∂M, the restriction of the closed $(n-1)$-form ν_ξ vanishes on ∂M. Then the application of the Stokes formula gives the same result as above. Thus, the pairing between $S\mathrm{Vect}(M)$ and $\Omega^1(M)/d\Omega^0(M)$ is indeed well defined.

Finally, recall that in the adjoint representation, the group $\mathrm{Diff}(M)$ acts on its Lie algebra $\mathrm{Vect}(M)$ by coordinate changes: $\mathrm{Ad}_\varphi(\xi) = \varphi_*\xi \circ \varphi^{-1}$. Hence we obtain

$$\langle \xi, \mathrm{Ad}^*_\varphi[u] \rangle = \langle \mathrm{Ad}_{\varphi^{-1}}\xi, [u] \rangle = \int_M u \wedge \iota_{\varphi_*^{-1}\xi}\mu = \int_M (\varphi^{-1})^*u \wedge \iota_\xi \left((\varphi^{-1})^*\mu \right) .$$

Since φ preserves the volume form μ, we have $\mathrm{Ad}^*_\varphi[u] = [(\varphi^{-1})^*u]$. The statement on infinitesimal coadjoint action of the Lie algebra $S\mathrm{Vect}(M)$ readily follows.

Note also that the actions Ad^* and ad^* are well defined on the cosets, since changes of coordinates commute with taking d.   □

Exercise 3.4 Complete the proof of Proposition 3.3 in the case of a manifold M with boundary. (Hint: in order to show that the restriction of the closed $(n-1)$-form ν_ξ vanishes on ∂M, use the fact that the field ξ is tangent to ∂M.)

3.2 The Euler Equation of an Ideal Incompressible Fluid

Imagine an ideal incompressible fluid occupying a domain M in $\mathbb{R}^n$.

Definition 3.5 The fluid motion is described by a velocity field $v(t, x)$ and a pressure function $p(t, x)$ that satisfy the classical *Euler equation of an ideal incompressible fluid*:

$$\partial_t v + (v, \nabla)v = -\nabla p, \tag{3.22}$$

where div $v = 0$ and the field v is tangent to the boundary of M. The function p is defined uniquely modulo an additive constant by the conditions that v have zero divergence and be tangent to the boundary.

In Euclidean coordinates, equation (3.22) reads

$$\partial_t v_i + \sum_{j=1}^n v_j \frac{\partial v_i}{\partial x_j} = -\frac{\partial p}{\partial x_i}.$$

The same equation describes the motion of an ideal incompressible fluid filling an arbitrary Riemannian manifold M equipped with a volume form μ. In the latter case, v is a divergence-free vector field on M, the notation $(v, \nabla)v$ stands for the Riemannian covariant derivative $\nabla_v v$ of the field v in the direction of itself, and div v is taken with respect to the volume form μ.

Remark 3.6 Equation (3.22) has a natural interpretation as a geodesic equation [12, 18]. Indeed, consider the flow $(t, x) \mapsto \phi(t, x)$ defined by the velocity field $v(t, x)$, which describes the motion of fluid particles:

$$\partial_t \phi(t, x) = v(t, \phi(t, x)), \quad \phi(0, x) = x$$

for all $x \in M$. The chain rule immediately gives

$$\partial_t^2 \phi(t, x) = (\partial_t v + (\partial_x v)(\partial_t \phi))(t, \phi(t, x)) = (\partial_t v + (v, \nabla)v)(t, \phi(t, x)),$$

and hence the Euler equation (3.22) is equivalent to

$$\partial_t^2 \phi(t, x) = -(\nabla p)(t, \phi(t, x)),$$

while the incompressibility condition div $v = 0$ becomes $\det(\partial_x \phi(t, x)) = 1$. The latter form of the Euler equation (for a smooth flow $\phi(t, x)$) says that the acceleration of the flow is given by the gradient of a function. Therefore this acceleration is L^2-orthogonal to the set of volume-preserving diffeomorphisms (or, rather, to its tangent space, the space of divergence-free fields):

$$\int_M (\nabla p, \xi)\mu = 0$$

for all divergence-free vector fields ξ and all smooth functions p on M. In other words, the fluid motion $\phi(t, .)$ is a geodesic line on the set of volume-preserving diffeomorphisms of the domain M with respect to the induced L^2-metric on the group $SDiff(M)$. This L^2-metric is the kinetic energy of the fluid flow: for a velocity field v its energy is $E(v) = \frac{1}{2}\int_M (v, v)\mu$. Fluid flows describe geodesics on the group $SDiff(M)$, since they extremize this energy according to the least action principle.

In the case of an arbitrary Riemannian manifold M, the Euler equation defines the geodesic flow on the group of volume-preserving diffeomorphisms of M with respect to the right-invariant L^2-metric on $SDiff(M)$; see [12, 96, 24].

3.3 The Hamiltonian Structure and First Integrals of the Euler Equations for an Incompressible Fluid

The geodesic properties of the Euler equation can also be described within the Hamiltonian framework. The latter turns out to be useful in establishing numerous conservation laws for flows of an ideal fluid.

Consider the Lie algebra $SVect(M)$ of divergence-free vector fields on a compact manifold M and its dual space $SVect(M)^* = \Omega^1(M)/d\Omega^0(M)$ of 1-forms modulo exact 1-forms. This algebra is equipped with the L^2 quadratic form

$$(\xi_1, \xi_2)_{SVect} = \int_M (\xi_1, \xi_2)\mu, \tag{3.23}$$

which is induced by fixing a Riemannian metric (,) on the manifold M. The corresponding inertia operator

$$A : SVect(M) \to \Omega^1(M)/d\Omega^0(M)$$

sends a vector field ξ to the coset $[u]$ of the 1-form u that is obtained from the field ξ by "raising indices" with the help of the metric: $u(.) = (\xi, .)$. (In $\mathbb{R}^n$ with the standard Euclidean metric, the inertia operator A maps a vector field $\xi = \sum \xi_i \partial/\partial x_i$ to the 1-form $u_\xi = \sum \xi_i dx_i$.) Since the Riemannian metric is nondegenerate, so is the map $A : \xi \mapsto [u]$. Note that in each coset $[u]$ there is a unique representative 1-form, which is obtained by raising indices from a *divergence-free* field.

Lemma 3.7 *Let H be the quadratic Hamiltonian on $\Omega^1(M)/d\Omega^0(M)$ defined by the inertia operator A:*

$$H([u]) = \frac{1}{2}\langle A^{-1}[u], [u]\rangle .$$

Then the Hamiltonian equation on $\Omega^1(M)/d\Omega^0(M)$ for the Hamiltonian function $-H$ with respect to the Lie–Poisson bracket is the following:

$$\frac{d}{dt}[u] = -L_\xi[u] , \tag{3.24}$$

where $[u]$ is the coset of the 1-form u related to the divergence-free field ξ with the help of the metric, $u(\,.\,) = (\xi, \,.\,)$.

PROOF.   Recall from Corollary I.4.11 that the Hamiltonian equation with respect to the function $-H$ is given by

$$\frac{d}{dt}[u] = -\operatorname{ad}^*_{(-A^{-1}[u])}[u] = \operatorname{ad}^*_{A^{-1}[u]}[u] .$$

The coadjoint action of the Lie algebra $S\mathrm{Vect}(M)$ on its dual is given by the negative of the Lie derivative, so that we get equation (3.24). $\square$

Corollary 3.8 *The Hamiltonian equation (3.24) is equivalent to the Euler equation for an ideal incompressible fluid on the manifold M.*

PROOF.   Rewriting equation (3.24) in terms of the coset representative, the 1-form $u \in [u]$, gives
$$\partial_t u = -L_\xi u - d\tilde{p} \tag{3.25}$$
for a time-dependent function $\tilde{p}$. Passing back to vector fields, i.e., applying A^{-1} to both sides of equation (3.25), we get our final equation

$$\partial_t \xi = -(\xi, \nabla)\xi - \nabla p .$$

(Here we use the fact that "raising indices" sends the covariant derivative $(\xi, \nabla)\xi$ to the Lie derivative $L_\xi u$ up to a full differential; see, e.g., [24].) The latter is the Euler equation for an ideal incompressible fluid on M, where p is interpreted as the pressure function. $\square$

   This description of the dual space and the Euler hydrodynamics equation allows one to describe certain first integrals of the fluid motion practically without calculations. Indeed, start with the following proposition, which provides certain Casimir functions for the coadjoint action of the group $S\mathrm{Diff}(M)$. Let M be a compact manifold equipped with a volume form μ.

Proposition 3.9 ([307]) *If the manifold M is of odd dimension $2m+1$, then the function*

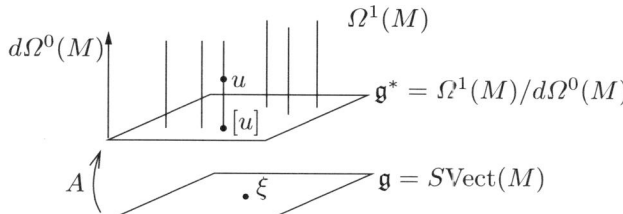

Fig. 3.1. Projection of $u \in \Omega^1(M)$ to the coset $[u] \in \Omega^1(M)/d\Omega^0(M)$. The inertia operator A sends a vector field ξ to the coset $[u]$ of 1-forms.

$$I([u]) = \int_M u \wedge (du)^m$$

on the space of cosets, $\Omega^1(M)/d\Omega^0(M) = S\mathrm{Vect}(M)^*$, is invariant under the coadjoint action of the group $S\mathrm{Diff}(M)$.

If $\dim(M) = 2m$, then for each $k \in \mathbb{N}$, the function

$$I_k([u]) = \int_M \left(\frac{(du)^m}{\mu} \right)^k \mu$$

on the space of cosets is invariant under the action of the group $S\mathrm{Diff}(M)$.

Here, $(du)^m/\mu$ denotes the ratio of a $2m$-form over the volume form on M. This ratio is a function, which can be integrated against the volume form μ on M.

PROOF. First we have to check that I and I_k are well-defined functionals on $\Omega^1(M)/d\Omega^0(M)$. Note that for any exact 1-form df we have $I(df) = 0$ and $I_k(df) = 0$. Similarly, one can see that each of the functionals I and I_k depends not on a representative, but just on the corresponding coset, e.g., $I(u) = I(u + df) = I([u])$. Furthermore, the group $S\mathrm{Diff}(M)$ acts on $\Omega^1(M)/d\Omega^0(M)$ by change of coordinates. Since the integrals I and I_k are defined in a coordinate-free way, they are indeed invariant under this action. □

Corollary 3.10 *The functionals I and, respectively, I_k on $\Omega^1(M)/d\Omega^0(M)$ are first integrals of the Euler equation for an ideal incompressible fluid filling M of odd and, respectively, even dimension.*

PROOF. As we have seen in Corollary 3.8, the Euler equation is Hamiltonian with respect to the usual Lie–Poisson bracket on the dual space of the Lie algebra $S\mathrm{Vect}(M)$. Therefore its flow lines are always tangent to coadjoint orbits of the group $S\mathrm{Diff}(M)$. But by Proposition 3.9, the functions I and I_k are constant on the coadjoint orbits and hence constant along the flow lines. □

Remark 3.11 Being Casimirs, the functionals I and I_k give conservation laws for any Hamiltonian equation on $S\mathrm{Vect}(M)^* = \Omega^1(M)/d\Omega^0(M)$ with respect to the Lie–Poisson bracket. In particular, the functionals I and I_k are first integrals of the Euler equation for an *arbitrary metric* on M. These integrals manifest the "kinematic symmetries" of the hydrodynamical system, while the energy is an invariant related to the system's "dynamics."

Example 3.12 Specifying to the case of a three-dimensional domain $M \subset \mathbb{R}^3$, we see that the function

$$I(\xi) = \int_M u \wedge du\,,$$

where u and ξ are related by means of the Euclidean metric, is a first integral of the Euler equation. A short direct calculation allows one to rewrite $I(\xi)$ in the form

$$I(\xi) = \int_M (\xi, \mathrm{curl}\ \xi)\, d^3x\,.$$

The latter integral has a natural geometric meaning of *helicity* of the vector field $\mathrm{curl}\,\xi$:

Definition 3.13 Let M be a three-dimensional compact manifold with a volume form μ and suppose that the second de Rham cohomology group of M vanishes: $H^2(M, \mathbb{R}) = 0$. Then, if η is a divergence-free vector field on M tangent to the boundary of M, the corresponding 2-form $\omega_\eta = \iota_\eta \mu$ is closed, and hence exact. So we can find a 1-form α on M such that $\omega_\eta = d\alpha$. The *helicity* of the vector field η is defined by

$$\mathrm{Hel}(\eta) = \int_M \alpha \wedge d\alpha = \int_M d^{-1}(\omega_\eta) \wedge \omega_\eta\,.$$

Exercise 3.14 Show that the helicity does not depend on the ambiguity in the definition of the 1-form $\alpha = d^{-1}\omega_\eta$.

The helicity of a vector field η has a topological interpretation as the asymptotic Hopf invariant, or "average linking number" of the trajectories of the vector field η (see [16, 24, 269] for details).

Example 3.15 Similarly, specifying to the case of a two-dimensional domain $M \subset \mathbb{R}^2$, we find infinitely many first integrals of the Euler equation, namely

$$I_k(\xi) = \int_M (\mathrm{curl}\ \xi)^k d^2x\,,$$

where $\mathrm{curl}\ \xi := \frac{\partial \xi_1}{\partial x_2} - \frac{\partial \xi_2}{\partial x_1}$ is the *vorticity function* on $M \subset \mathbb{R}^2$.

Remark 3.16 The functions I and I_k on $\Omega^1(M)/d\Omega^0(M)$ are invariant under the coadjoint action of the group $S\mathrm{Diff}(M)$. However, it should be noted that they do *not* form a *complete* set of invariants of the coadjoint representation. One can construct parametrized families of orbits on which these functions take the same values.

Exercise 3.17 Prove that for an odd-dimensional manifold M not only is the integral $I([u])$ over the entire manifold M invariant under the coadjoint action of $S\mathrm{Diff}(M)$, but so are the integrals

$$I_C([u]) = \int_C u \wedge (du)^m$$

over every set C invariant with respect to the *vorticity vector field* $\eta = \mathrm{curl}\,\xi$. This field is defined for any odd $n = 2m+1$ as the kernel field of the $2m$-form $(du)^m$ on the manifold M: $\iota_\eta \mu = (du)^m$, where the 1-form u is related to the vector field ξ with the help of the Riemannian metric.

Certain nonanalytic invariants of the coadjoint action on $S\mathrm{Vect}(M)^*$ can be extracted from the ergodic description of the helicity functional as the average of pairwise linkings of the trajectories of the field n; see [16]. In the even-dimensional case, slightly more general invariants than I_k are obtained by replacing $((du)^m/\mu)^k$ with $f((du)^m/\mu)$ for any function $f : \mathbb{R} \to \mathbb{R}$.

The full classification of the coadjoint orbits of the diffeomorphism groups is still an open problem [24, 205]. Below we show that a complete set of such invariant functionals defined on singular coadjoint orbits of this group would include all knot and link invariants!

3.4 Semidirect Products: The Group Setting for an Ideal Magnetohydrodynamics and Compressible Fluids

It is curious to note that the similarity pointed out by V. Arnold between the Euler top on the group $SO(3)$ and the Euler ideal fluid equations on $S\mathrm{Diff}(M)$ has a "magnetic analogue": a parallelism between the Kirchhoff and magnetohydrodynamics (MHD) equations; see the table in Example 4.18 of Chapter I.

The latter equations are both related to the groups that are *semidirect products*. The Kirchhoff equations for the motion of a rigid body in a fluid are associated with the group $E(3) = SO(3) \ltimes \mathbb{R}^3$ of Euclidean motions of the three-dimensional space. In this group, a Euclidean motion is described by a pair (a, b), a rotation $a \in SO(3)$ and a translation by a vector $b \in \mathbb{R}^3$, so that the group multiplication law between two such pairs,

$$(a_2,\, b_2) \circ (a_1,\, b_1) = (a_2 a_1,\, b_2 + a_2 b_1)\,,$$

is determined by the consecutive application of motions.

Similarly, magnetohydrodynamics is governed by the group $S\mathrm{Diff}(M) \ltimes S\mathrm{Vect}(M)$. Elements of the corresponding Lie algebra are pairs of divergence-free vector fields (v, B), which can be interpreted as the velocity field v and the magnetic field B.

In the idealized setting, an inviscid incompressible fluid obeying hydro-dynamical principles transports a magnetic field. In turn, the medium itself experiences a reciprocal influence of the magnetic field. The evolution is de-scribed by the following system of Maxwell's equations.

Definition 3.18 Consider an electrically conducting incompressible fluid that fills some domain $M \subset \mathbb{R}^3$ and transports a divergence-free magnetic field B. Then the evolution of the field B and of the fluid velocity field v is described by the system of ideal *magnetohydrodynamics (MHD) equations*

$$\partial_t v = -(v, \nabla)v + (\mathrm{curl}\, B) \times B - \nabla p \,,$$
$$\partial_t B = -L_v B \,,$$

where $\mathrm{div}\, B = \mathrm{div}\, v = 0$ with respect to the standard volume form $\mu = d^3 x$ in M and the coefficients are normalized by a suitable choice of units.

Here, the second equation is the definition of the "frozenness" of the mag-netic field B into the medium, and $L_v B$ denotes the Lie bracket of two vector fields. (Note that a frozenness-type condition usually indicates the structure of a semidirect product in the corresponding symmetry group; cf. [253].) The term $(\mathrm{curl}\, B) \times B$ represents the Lorentz force $\mathbf{j} \times B$, which acts on a unit charge moving with velocity $\mathbf{j}$ in the magnetic field B. The motion of electric charges produces the electrical current field $\mathbf{j}$ proportional to $\mathrm{curl}\, B$.

Theorem 3.19 ([377, 253]) *The magnetohydrodynamics equations are the Euler equations corresponding to the right-invariant metric on the group $S\mathrm{Diff}(M) \ltimes S\mathrm{Vect}(M)$ generated by the sum of the kinetic and magnetic energies*

$$E(v, B) := \frac{1}{2} \int_M [(v, v) + (B, B)] \, \mu$$

on its Lie algebra of pairs (v, B).

A somewhat similar description is available for compressible fluids as well. Consider, for instance, *barotropic gases* or *fluids*, whose pressure is a fixed function of density.

Definition 3.20 The equations of motion of a barotropic gas (or fluid) on a Riemannian manifold M are given by

$$\begin{cases} \rho\, \partial_t v = -\rho\, (v, \nabla)v - \nabla h(\rho) \,, \\ \partial_t \rho + \mathrm{div}(\rho \cdot v) = 0 \,, \end{cases} \tag{3.26}$$

where $\rho \in C^\infty(M)$ is the density of the gas, the time-dependent vector field v is the velocity vector field, and $h : C^\infty(M) \to C^\infty(M)$ is the fixed correspondence between the density function ρ and the pressure function $p = h(\rho)$. The equations of *barotropic gas dynamics* usually correspond to a specific choice of $h(\rho) = \mathrm{const} \cdot \rho^a$. For an ideal gas, $a = 1 + \frac{2}{D}$, where D is the number of degrees of freedom of a molecule. For instance, for monatomic gases (argon, krypton) $D = 3$ and $a = 5/3$, while for diatomic gases (such as nitrogen, oxygen, and hence approximately for air) $D = 5$ and $a = 7/5$.

To describe these equations as Euler equations we note first that the main player here will be the group $\mathrm{Diff}(M)$ of all (not only volume-preserving) diffeomorphisms of a manifold M. Its Lie algebra $\mathrm{Vect}(M)$ consists of all smooth vector fields on M.

For the barotropic gas dynamics we need a certain extension of this group. Namely, consider the semidirect product group $G = \mathrm{Diff}(M) \ltimes C^\infty(M)$. This is the group consisting of pairs (φ, f), where φ is a diffeomorphism of the manifold M, and f is a smooth function. The product on the group $\mathrm{Diff}(M) \ltimes C^\infty(M)$ is given by

$$(\varphi, f) \cdot (\psi, g) = (\varphi \circ \psi, \varphi_* g + f) \,,$$

where $\varphi_* g = g \circ \varphi^{-1}$ denotes the *pushforward* of the function g by the diffeomorphism φ (or, equivalently, the pullback by φ^{-1}). The smoothness of the group product on $\mathrm{Diff}(M) \ltimes C^\infty(M)$ follows from that of the product on $\mathrm{Diff}(M)$ and that of the action of $\mathrm{Diff}(M)$ on $C^\infty(M)$.

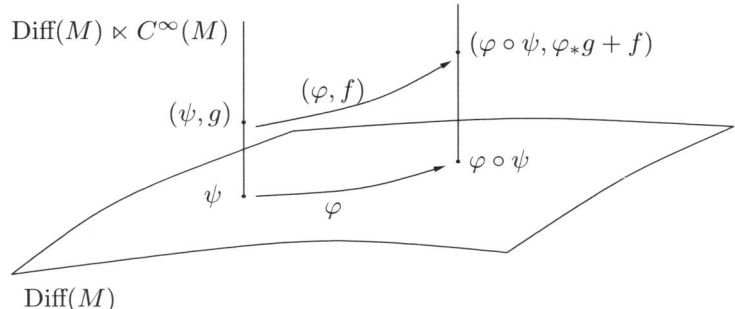

Fig. 3.2. As a topological space, the semidirect product $\mathrm{Diff}(M) \ltimes C^\infty(M)$ is the product of the spaces $\mathrm{Diff}(M)$ and $C^\infty(M)$. The group structure is not a direct product, but is twisted by the action of $\mathrm{Diff}(M)$ on $C^\infty(M)$ by pushforwards.

The Lie algebra $\mathfrak{g}$ of the group $\mathrm{Diff}(M) \ltimes C^\infty(M)$ is also a semidirect product, the Lie algebra $\mathrm{Vect}(M) \ltimes C^\infty(M)$. As a vector space, it is the direct sum $\mathrm{Vect}(M) \oplus C^\infty(M)$, while the Lie algebra structure on this space encompasses the natural action of vector fields on functions

Exercise 3.21 Find explicit formulas for the adjoint representation of the group $\text{Diff}(M) \ltimes C^\infty(M)$ and the Lie bracket of its algebra. (Hint: for more details on this and the following exercises, see Appendix A.7.)

To describe the corresponding dual space, we return first to the group $\text{Diff}(M)$ of all diffeomorphisms, whose Lie algebra $\text{Vect}(M)$ consists of all smooth vector fields.

Exercise 3.22 Show that the (smooth) dual of the Lie algebra $\text{Vect}(M)$ can be naturally identified with the space $\Omega^1(M) \otimes_{C^\infty(M)} \Omega^n(M)$. Elements of this space are linear combinations of tensor products of 1-forms with n-forms on M, where given a 1-form α and an n-form μ we identify the elements $f \cdot \alpha \otimes \mu$ with $\alpha \otimes f \cdot \mu$ for any smooth function f. The pairing between vector fields on M and elements of this space $\Omega^1(M) \otimes_{C^\infty(M)} \Omega^n(M) =: \text{Vect}(M)^*$ is given by

$$\langle \xi, \alpha \otimes \mu \rangle = \int_M (\iota_\xi \alpha)\mu \,.$$

The naturality of the pairing means that the coadjoint action of $\text{Diff}(M)$ is the change of coordinates by a diffeomorphism.

Now the smooth part of the dual of the semidirect product Lie algebra $\text{Vect}(M) \ltimes C^\infty(M)$ can be identified with the space $(\Omega^1(M) \otimes_{C^\infty(M)} \Omega^n(M)) \oplus \Omega^n(M)$ via the pairing

$$\langle (\xi, f), (\alpha \otimes \mu, \nu) \rangle = \int_M (\iota_\xi \alpha)\mu + \int_M f\nu \,.$$

Exercise 3.23 Show that the coadjoint action of the group $\text{Diff}(M) \ltimes C^\infty(M)$ on the space $(\Omega^1(M) \otimes_{C^\infty(M)} \Omega^n(M)) \oplus \Omega^n(M)$ is given by

$$\text{Ad}^*_{(\varphi, f)^{-1}}(\alpha \otimes \eta, \nu) = (\varphi^*\alpha \otimes \varphi^*\eta + \varphi^*df \otimes \varphi^*\nu, \varphi^*\nu) \,. \tag{3.27}$$

In order to define a dynamical system related to this group, we need some additional data. Fix a Riemannian metric $(\ , \ )$ and a volume form μ on the manifold M. Furthermore, fix a function $h : C^\infty(M) \to C^\infty(M)$ that assigns to a density ρ of the fluid its pressure $h(\rho)$.

Define the Hamiltonian functional $H : \text{Vect}(M) \oplus C^\infty(M) \to \mathbb{R}$ by

$$H(\xi, \rho) = \int_M \left(\frac{1}{2}(\xi, \xi)\,\rho + \rho\,\Phi(\rho) \right) \mu \,,$$

where $\Phi(\rho)$ is chosen such that $\rho^2 \Phi'(\rho) = h(\rho)$.

Exercise 3.24 Show that the Euler equation corresponding to the lift of the Hamiltonian $-H$ to the dual of the Lie algebra is exactly the Euler equation of a barotropic fluid. (The lift of H to the dual space is given by the inertia operator defined by the Riemannian metric, see details in Appendix A.7.)

Exercise 3.25 Find Casimir functions similar to the ones we discussed in the case of the Euler equation of an ideal incompressible fluid. (Hint: in the coordinate system "moving with the fluid" the equations of motion can be thought of as those for an incompressible fluid.)

Remark 3.26 Thus it turns out that the barotropic fluid equations also have an infinite number of conservation laws in the even-dimensional case, and at least one first integral in the odd-dimensional case; see [24, 307].

   The Hamiltonian approach allows one to apply the technique of Casimir functions to the stability study of barotropic fluids and ideal MHD systems: their dynamics take place on coadjoint orbits of the corresponding groups, and the Casimir functions help in describing the corresponding conditional extrema of the Hamiltonians.

   More general equations of compressible fluids include also the function of entropy. The corresponding Lie group can be identified with the semidirect product $G = \mathrm{Diff}(M) \ltimes (C^\infty(M) \oplus C^\infty(M))$: the first summand of the extension stands for the density, while the second one represents the entropy function; see [90].

3.5 Symplectic Structure on the Space of Knots and the Landau–Lifschitz Equation

Consider an embedding of the (oriented) circle S^1 into Euclidean space $\mathbb{R}^3$, i.e., a knot in $\mathbb{R}^3$. Associate to this curve a linear functional on the space $S\mathrm{Vect}(\mathbb{R}^3)$ of divergence-free fields in $\mathbb{R}^3$, that is, an element of the dual space $S\mathrm{Vect}(\mathbb{R}^3)^*$, as follows. Take a compact oriented surface $S \subset \mathbb{R}^3$ such that the oriented boundary of S coincides with the knot $\gamma \colon \partial S = \gamma$. Such a surface is called a *Seifert surface* for γ. Then to any vector field $\xi \in \mathrm{Vect}(\mathbb{R}^3)$ we can associate its flux through the surface S. If the vector field ξ is divergence-free, the flux does not depend on the choice of the surface S, so that the curve γ defines a functional on the Lie algebra $S\mathrm{Vect}(\mathbb{R}^3)$:

$$\langle \gamma, \xi \rangle = \mathrm{Flux}(\xi)|_S = \int_S \iota_\xi \mu \,, \qquad (3.28)$$

where $\mu = d^3x$ is the standard volume form in $\mathbb{R}^3$; see Figure 3.3.

Remark 3.27 Obviously, the functional on the Lie algebra $S\mathrm{Vect}(\mathbb{R}^3)$ defined by the curve γ does not lie in the smooth part of the dual of $S\mathrm{Vect}(\mathbb{R}^3)$. But we can still associate to γ a coset of *singular*, rather than smooth, 1-forms on $\mathbb{R}^3$. Namely, given a Seifert surface S we consider the "δ-type" 1-form u_S supported on S, whose integral over any closed curve σ in $\mathbb{R}^3$ intersecting the surface S transversally counts (with signs) the intersections of the curve σ with this surface S.

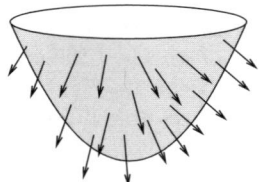

Fig. 3.3. A Seifert surface for a knot defines the flux functional on vector fields.

Exercise 3.28 Verify that although the 1-form u_S depends on the choice of the surface S, its coset $[u_S]$ does not: the choice of another Seifert surface $\widetilde{S}$ changes the 1-form u_S by a complete differential.

This coset $[u_S]$ belongs to a closure $\overline{\Omega^1(\mathbb{R}^3)/d\Omega^0(\mathbb{R}^3)}$ of the smooth dual space $S\mathrm{Vect}(\mathbb{R}^3)^* = \Omega^1(\mathbb{R}^3)/d\Omega^0(\mathbb{R}^3)$. Note that the latter space of all smooth 1-forms modulo exact 1-forms is isomorphic to the space $Z^2(\mathbb{R}^3)$ of all smooth closed 2-forms in $\mathbb{R}^3$: The exterior derivative d takes any such coset of 1-forms to a closed 2-form without any loss of information, since $H^1(\mathbb{R}^3) = 0$.

Alternatively, one can directly associate to the curve γ a singular closed 2-form ω_γ in $\mathbb{R}^3$. This is the δ-type 2-form (called the de Rham current) supported on γ. (More precisely, *de Rham currents* are, by definition, continuous linear functionals on differential forms. The curve γ defines the current ω_γ, whose value on any smooth 1-form α on $\mathbb{R}^3$ is $\langle \omega_\gamma, \alpha \rangle = \int_\gamma \alpha$. Similarly, the current u_S is defined by prescribing its value $\langle u_S, \beta \rangle = \int_S \beta$ for any smooth 2-form β on $\mathbb{R}^3$.) The relation $du_S = \omega_\gamma$ between these currents exactly manifests the relation $\partial S = \gamma$ between a knot γ and its Seifert surface S.

In this way, a knot γ in $\mathbb{R}^3$ can be seen as an element of the (full) dual space to the Lie algebra of divergence-free vector fields on $\mathbb{R}^3$. From this viewpoint, the coadjoint orbit through γ is the equivalence class of the knot γ under various isotopies. (Evidently, any isotopy of a knot $\gamma \subset \mathbb{R}^3$ can be extended to a volume-preserving diffeomorphism of the ambient space $\mathbb{R}^3$.) This leads to the curious observation that knot invariants constitute a part of the coadjoint invariants of the whole Lie group $S\mathrm{Diff}(\mathbb{R}^3)$; see, e.g., [16, 24].

Let us extend our consideration from embedded to *immersed* closed curves in $\mathbb{R}^3$. The space $\mathcal{C}$ of such curves can be regarded as an infinite-dimensional symplectic manifold. Let $\gamma = \gamma(S^1) \subset \mathbb{R}^3$ be a closed immersed curve in $\mathbb{R}^3$, and for now fix a parametrization θ of the circle S^1. A tangent vector α to $\mathcal{C}$ at γ is an infinitesimal variation of the curve γ, that is, a normal vector field attached to $\gamma(S^1)$.

Definition 3.29 The *Marsden–Weinstein symplectic structure* is the following 2-form ω_{MW} on the space $\mathcal{C}$ of immersed closed curves in $\mathbb{R}^3$ with a fixed volume form μ. Its value on the two tangent vectors a and b to $\mathcal{C}$ at γ is given by the oriented volume of the collar spanned along γ by the vectors $a(\theta)$ and $b(\theta)$; see Figure 3.4. Explicitly, for the chosen parametrization θ,

$$\omega_{MW}(a,b) := \int_{S^1} \mu(a(\theta), b(\theta), \gamma'(\theta))d\theta = \int_\gamma \iota_b \iota_a \mu \,.$$

The latter integral shows that the value of ω does not depend on the parametrization, i.e., ω_{MW} is indeed a symplectic structure on the space of nonparametrized curves.

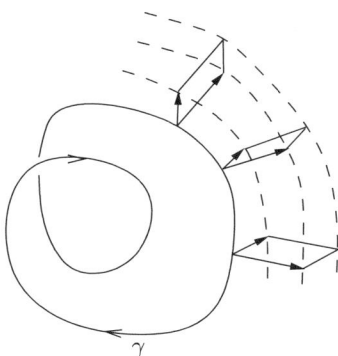

Fig. 3.4. The value of the symplectic structure ω_{MW} evaluated on two variations of the curve γ is the volume of the collar spanned by the variations.

Proposition 3.30 *The Marsden–Weinstein symplectic structure on the space of embedded curves containing γ coincides with the natural Kirillov–Kostant symplectic structure on the coadjoint orbit of the group $S\mathrm{Diff}(\mathbb{R}^3)$ through the curve γ, regarded as an element of the dual space to $S\mathrm{Vect}(\mathbb{R}^3)$.*

Exercise 3.31 Prove this proposition. (Hint: see [255, 24].)

Remark 3.32 The same construction works for any three-dimensional manifold equipped with a volume form. In higher dimensions, a similar construction allows one to define a symplectic structure on the space of closed immersed submanifolds of codimension 2. One can also prove its equivalence with the Kirillov–Kostant symplectic structure on ("singular") coadjoint orbits for $S\mathrm{Diff}(M^n)$, which are linear functionals on divergence-free vector fields in M, represented by "fluxes through" hypersurfaces in M bounded by these submanifolds; cf. [156].

Returning to the 3D case, we are going to make use of not only the volume form, but also the Euclidean structure in $\mathbb{R}^3$. Define the following "energy" function H on the space of immersed closed curves that assigns to a loop $\gamma \in \mathcal{C}$ its length:

$$H(\gamma) = \int_{S^1} \|\gamma'(\theta)\| \, d\theta = \int_{S^1} \sqrt{(\gamma'(\theta), \gamma'(\theta))} \, d\theta \, .$$

Proposition 3.33 *The Hamiltonian equation corresponding to this energy function H with respect to the Marsden–Weinstein symplectic structure is given by the* binormal *(or* filament*) equation*

$$\partial_t \gamma = \gamma' \times \gamma'' \tag{3.29}$$

in the arc-length parameter θ, see Figure 3.5. (Here, γ' denotes the derivative $\frac{\partial}{\partial \theta} \gamma$, and in the arc-length parametrization $\|\gamma'(\theta)\| = 1$ for all $\theta \in S^1$.)

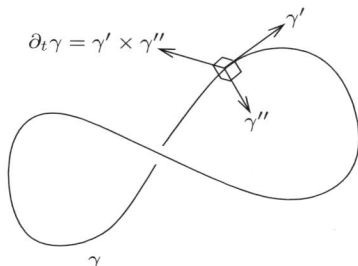

Fig. 3.5. The binormal equation describes the movement of a curve in $\mathbb{R}^3$ in the direction $\partial_t \gamma = \gamma' \times \gamma''$.

Exercise 3.34 Prove this proposition. Show also that if θ is not arc-length, then the corresponding Hamiltonian equation becomes

$$\partial_t \gamma = k(\theta, t) \gamma' \times \gamma'' \, , \tag{3.30}$$

where $k(\theta, t)$ is the curvature of the curve at the point θ at time t.

(Hint: For the functional $H(\gamma)$ in the arc-parametrization, the variational derivative (i.e., the "gradient" of H in the space of curves $\mathcal{C}$) is $\delta H/\delta\gamma = -(\text{length of } \gamma)^{-1} \gamma''$. This can be thought of as a vector field normal to γ. In turn, the Marsden–Weinstein symplectic structure at $\gamma \in \mathcal{C}$ can be regarded as the symplectic structure averaged from all two-dimensional planes normal to the loop γ. Then to obtain the corresponding Hamiltonian field v_H, which is the skew-gradient of H, we have to rotate the above gradient field $\delta H/\delta\gamma$ by $\pi/2$ around each tangent vector $\gamma'(\theta)$: if $J_\gamma(*) = \gamma' \times *$ is the corresponding rotation operator, we get

$$v_H = J_\gamma(\delta H/\delta\gamma) = -(\text{length of } \gamma)^{-1} \gamma' \times \gamma'' \, . \,)$$

Notice that the binormal vector $\gamma' \times \gamma''$ is, in particular, orthogonal to γ', and hence the evolution of the curve γ does not change the length of γ, i.e., the Hamiltonian is indeed preserved.

This equation is also called the *da Rios* or *localized induction approximation (LIA) equation*; see the nice survey [331]. It can also be obtained from the three-dimensional Euler equation (3.22) by keeping only "local terms" in the corresponding evolution law for a vortex filament; see, e.g., [139].

Remark 3.35 Equation (3.29) is equivalent to the *Heisenberg magnetic chain* equation. Namely, set $L = \gamma'$. Then equation (3.29) becomes

$$\partial_t L = \partial_t \gamma' = \gamma'' \times \gamma'' + \gamma' \times \gamma''' = \gamma' \times \gamma''',$$

so that we get

$$\partial_t L = L \times L'', \tag{3.31}$$

which is also the simplest form of the *Landau–Lifschitz* equation.

Remark 3.36 The Heisenberg magnetic chain (or Landau–Lifschitz) equation has another interpretation as the Euler equation for a certain *left-invariant sub-Riemannian* metric on the loop group $LSO(3) = C^\infty(S^1, SO(3))$. The Hamiltonian formulation is as follows. Let us identify the Lie algebra $L\mathfrak{so}(3) = C^\infty(S^1, \mathfrak{so}(3))$ with the smooth part of its dual $L\mathfrak{so}(3)^*$ via the pairing

$$\langle X, Y \rangle = -\int_{S^1} \operatorname{tr}(X(\theta)Y(\theta))\, d\theta.$$

Now instead of defining an inertia operator $A : L\mathfrak{so}(3) \to L\mathfrak{so}(3)^*$ we define the following noninvertible self-adjoint operator $B : L\mathfrak{so}(3)^* \to L\mathfrak{so}(3)$ acting in the opposite direction: $B(Y) = -Y''$. (If B were invertible, it would have the meaning of the inverse of the corresponding inertia operator: $B = A^{-1}$.) The corresponding Hamiltonian function on the dual space $L\mathfrak{so}(3)^*$ requires only the operator B for its definition and it is given by

$$H(Y) := \frac{1}{2}\langle Y, B(Y) \rangle = -\frac{1}{2}\langle Y, Y'' \rangle = \frac{1}{2}\langle Y', Y' \rangle$$

for $Y \in L\mathfrak{so}(3)^*$.

The image of B in $L\mathfrak{so}(3)$ is the subspace $\mathfrak{n}_0$ of $\mathfrak{so}(3)$-valued functions on the circle with zero mean. On this hyperplane $\mathfrak{n}_0 \subset L\mathfrak{so}(3)$ the operator B can be inverted, and this gives rise to the so called H^{-1}-metric

$$E(X) = \frac{1}{2}\langle \partial_\theta^{-1} X, \partial_\theta^{-1} X \rangle,$$

since it is given by the squared L^2-norm of the antiderivative $\partial_\theta^{-1} X$ for functions X with zero mean, $X \in \mathfrak{n}_0 \subset L\mathfrak{so}(3)$.

Note that the quadratic form on the subspace $\mathfrak{n}_0$ does not extend to a left-invariant Riemannian metric on a subgroup of $LSO(3)$. Indeed, this subspace $\mathfrak{n}_0 \subset L\mathfrak{so}(3)$ does not form a Lie subalgebra: the bracket of two loops with zero mean does not necessarily have zero mean. The subspace $\mathfrak{n}_0$ of the tangent space at the identity id $\in LSO(3)$ generates a left-invariant distribution on the group $LSO(3)$, and we can extend the quadratic form $E(X)$ from $\mathfrak{n}_0$ to a metric on this distribution. This is an example of an *infinite-dimensional non-integrable* distribution on a group with a left-invariant sub-Riemannian metric. Normal geodesics for this metric are described by the same Hamiltonian picture as for a left-invariant Riemannian metric on the group, i.e., by the Heisenberg magnetic chain (or Landau–Lifschitz) equation (3.31).

Note also that the same Hamiltonian equation on $L\mathfrak{so}(3)^*$ can be obtained from an invertible operator $\bar{B} := \mathrm{id} + B$, i.e., for $\bar{B}(Y) := Y - Y''$, which defines a left-invariant Riemannian metric on the group $LSO(3)$; see [7]. Indeed, the addition of the identity inertia operator does not change the Hamiltonian dynamics on the orbits, since the latter operator corresponds to the Killing form, and hence on each coadjoint orbit the new Hamiltonian differs from the old one by a constant.

Exercise 3.37 Show that the Landau–Lifschitz equation (3.31) is the Euler equation for the energy E (or, equivalently, for the Hamiltonian function H).

Derive the Landau–Lifschitz equation with the same Hamiltonian H for the loops in any semisimple Lie algebra $\mathfrak{g}$, where $-\mathrm{tr}(XY)$ is replaced by the Killing form on $\mathfrak{g}$.

Remark 3.38 One can rewrite the binormal equations in the Frenet frame for γ as evolution equations for the curvature $k(\theta, t)$ and torsion $\tau(\theta, t)$. Curiously, by passing to the new functions "velocity" $v := \tau$ and "energy density" $\rho := k^2$ one obtains exactly the equations of a 1-dimensional barotropic fluid for a specific function h relating density and pressure; see [370].

Exercise 3.39 Find this function $h(\rho)$ corresponding to the binormal equation.

Another important feature of the binormal equation was observed by Hasimoto [158]. He found the transformation

$$\psi(\theta, t) = k(\theta, t) \exp\left(i \cdot \int_0^\theta \tau(\eta, t) d\eta \right),$$

which sends the binormal equation to the nonlinear Schrödinger equation

$$-i\, \partial_t \psi = \psi'' + \frac{1}{2}|\psi|^2 \psi$$

for a complex-valued wave function $\psi : S^1 \to \mathbb{C}$. The latter equation is known to be a completely integrable (bi-Hamiltonian) infinite-dimensional system (see, e.g., [90]). And so is the binormal equation, since the Hasimoto transformation respects the corresponding Poisson and symplectic structures; see [232, 65]. A direct relation between the Schrödinger equation and the barotropic fluid equation is described in [360].

3.6 Diffeomorphism Groups as Metric Spaces

In this section we sidestep slightly from our main theme, the coadjoint orbits of the group of diffeomorphisms, and take a look at the geometry of the diffeomorphism groups themselves. We discuss without proofs various peculiar properties of the groups of volume-preserving and Hamiltonian diffeomorphisms.

Consider a volume-preserving diffeomorphism of a bounded domain and regard it as a final fluid configuration for a fluid flow starting at the identity diffeomorphism. In order to reach the position prescribed by this diffeomorphism, every fluid particle has to move along some path in the domain. The distance of this diffeomorphism from the identity in the diffeomorphism group is the averaged characteristic of the path lengths of the particles.

It turns out that the geometry of the groups of volume-preserving diffeomorphisms of two-dimensional manifolds differs drastically from that of higher-dimensional ones. This difference is due to the fact that in three (and more) dimensions there is enough space for particles to move to their final positions without hitting each other. On the other hand, the motion of the particles in the plane might necessitate their rotations about one another. The latter phenomenon of "braiding" makes the system of paths of particles in 2D necessarily long, in spite of the boundedness of the domain. The distinction between different dimensions can be formulated in terms of properties of $S\mathrm{Diff}(M)$ as a *metric space*.

Recall that on a Riemannian manifold M the group $S\mathrm{Diff}(M)$ of volume-preserving diffeomorphisms is equipped with the right-invariant L^2-metric, which is defined at the identity by the L^2-energy of vector fields.

Definition 3.40 To any smooth path $\{\phi(t,.) \mid 0 \le t \le 1\}$ on the group $S\mathrm{Diff}(M)$, i.e., to a family of volume preserving diffeomorphisms, we associate its (L^2-) *length:*

$$\ell\{\phi(t,.)\} := \int_0^1 \left(\int_{M^n} |\partial_t \phi(t,x)|^2 \mu \right)^{1/2} dt.$$

Then the *distance* between two fluid configurations $\varphi, \psi \in S\mathrm{Diff}(M)$ is the infimum of the lengths of all paths in $S\mathrm{Diff}(M)$ connecting them:

$$\mathrm{dist}_{S\mathrm{Diff}}(\varphi, \psi) = \inf \ell\{\phi(t,.)\},$$

where the $\phi(0,.) = \varphi$ and $\phi(1,.) = \psi$. It is natural to define the (L^2-) *diameter* of the group $S\mathrm{Diff}(M)$ as the supremum of distances between any two of its elements:

$$\mathrm{diam}\ (\ S\mathrm{Diff}(M)\ ) = \sup_{\varphi,\psi\in S\mathrm{Diff}(M)} \mathrm{dist}_{S\mathrm{Diff}}(\varphi,\psi).$$

Theorem 3.41 ([355, 357]) *For a unit n-dimensional cube M^n where $n \geq 3$, the diameter of the group of smooth volume-preserving diffeomorphisms $S\mathrm{Diff}(M^n)$ is finite in the right-invariant metric $\mathrm{dist}_{S\mathrm{Diff}}$ and it is bounded above as follows:*

$$\mathrm{diam}\ (\ S\mathrm{Diff}(M^n)\ ) \leq 2\sqrt{\frac{n}{3}}.$$

Finiteness of the diameter holds for an arbitrary simply connected manifold M of dimension three or higher. The two-dimensional case is completely different:

Theorem 3.42 ([100]) *For an arbitrary manifold M of dimension $n = 2$, the diameter of the group $S\mathrm{Diff}(M)$ is infinite.*

Note that the diameter in the case $n \geq 3$ can become infinite if the fundamental group of M is nontrivial [100]. On the other hand, in the two-dimensional case the infiniteness of the diameter is of "local" nature. The main difference between the geometries of the groups of diffeomorphisms in two and three dimensions is based on the observation that for a long path on $S\mathrm{Diff}(M^3)$ that twists the particles in space, there always exists a "shortcut" untwisting them by making use of the third coordinate. One can compare this with the corresponding linear problems: $\pi_1(\mathrm{SL}(2,\mathbb{R})) = \mathbb{Z}$, while $\pi_1(\mathrm{SL}(n,\mathbb{R})) = \mathbb{Z}/2\mathbb{Z}$ for $n \geq 3$. (Such a linear problem arises if we associate to a diffeomorphism in $S\mathrm{Diff}(M)$ and some fixed point in M the Jacobi matrix of the diffeomorphism at that point. Then for a path in $S\mathrm{Diff}(M)$ this gives a path in $\mathrm{SL}(n,\mathbb{R})$.)

Remark 3.43 More precisely, for an n-dimensional cube ($n \geq 3$) the distance between two volume-preserving diffeomorphisms $\varphi, \psi \in S\mathrm{Diff}(M)$ is bounded above by some power of the L^2-norm of the "difference" between them:

$$\mathrm{dist}_{S\mathrm{Diff}}(\varphi,\psi) \leq C \cdot ||\varphi - \psi||^{\alpha}_{L^2(M)},$$

where the exponent α in this inequality is at least $2/(n+4)$, and, presumably, this estimate is sharp [357]. This property means that the embedding of the group $S\mathrm{Diff}(M^n)$ into the vector space $L^2(M,\mathbb{R}^n)$ for $n \geq 3$ is "Hölder-regular" (the greater α, the more regular is the embedding, although apparently, it is far from being smooth). Certainly, this Hölder property implies the finiteness of the diameter of the diffeomorphism group. A similar estimate exists for a simply connected higher-dimensional M.

Fig. 3.6. Profile of the Hamiltonian function (left) whose flow (right) for sufficiently long time provides a "long path" on the group $SDiff(B^2)$ of area-preserving diffeomorphisms.

However, no such estimate can hold for $n = 2$: one can find a pair of volume-preserving diffeomorphisms arbitrarily far from each other on the group $SDiff(M^2)$, but close in the L^2-metric on the square or a disk. For instance, let M be the unit disk $B^2 \subset \mathbb{R}^2$ with the standard volume form. An explicit example of a long path on the group $SDiff(B^2)$ is given by the following flow for sufficiently long time t: in polar coordinates it is defined by

$$(r, \phi) \mapsto (r, \phi + t \cdot v(r)),$$

where the angular velocity $v(r)$ is *nonconstant*; see Figure 3.6. One can show that the distance of this diffeomorphism from the identity in the group grows linearly in time.

Remark 3.44 This also allows one to give the following example of an *unattainable diffeomorphism* of the square [356], i.e., a diffeomorphism that cannot be connected to the identity in $SDiff(M)$ by a piecewise-smooth path of finite length. The corresponding Hamiltonian has "hills" of infinitely increasing heights with supports on a sequence of disks converging to the boundary of the square (see Figure 3.7). In contrast, if M is an n-dimensional cube and $n \geq 3$, then all volume-preserving diffeomorphisms of M are attainable [356].

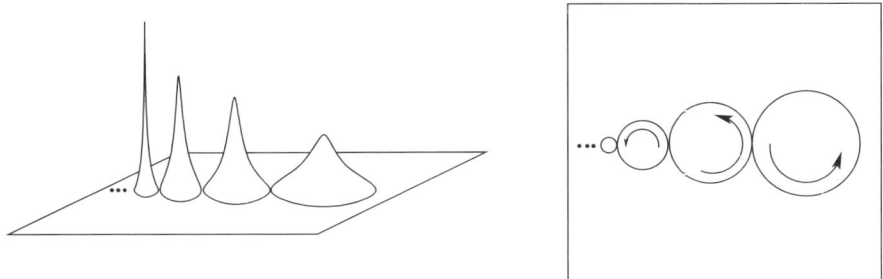

Fig. 3.7. An unattainable diffeomorphism of the square.

Furthermore, the above 2D results hold in the more general context of Hamiltonian diffeomorphisms of symplectic manifolds. Let M^{2n} be a compact manifold with a symplectic form ω that is exact: $\omega = d\alpha$ for some 1-form α on M. Note that such a manifold M has to have a nonempty boundary, since otherwise we would have

$$\int_M \omega^n = \int_M d(\alpha \wedge \omega^{n-1}) = 0\,,$$

which contradicts the nondegeneracy of the symplectic form.

Definition 3.45 Let $\mathrm{Ham}(M)$ be the group of Hamiltonian diffeomorphisms of M, i.e., the group of time-one flows of Hamiltonian vector fields on M. The Lie algebra of the group $\mathrm{Ham}(M)$ is the Lie algebra of Hamiltonian vector fields on M tangent to ∂M.

Theorem 3.42 on the infinite diameter in 2D admits the following generalization to the case of the group $\mathrm{Ham}(M)$ for an exact symplectic $2n$-dimensional manifold M.

Theorem 3.46 ([100]) *The L^p-diameter $\mathrm{diam}_p(\mathrm{Ham}(M))$ of the group of Hamiltonian diffeomorphisms of M is infinite in any right-invariant L^p-metric*

Here we define the right-invariant L^p-metric on the group $\mathrm{Ham}(M)$ or $S\mathrm{Diff}(M)$ by the same formula as above, but using the L^p-norm instead of L^2: the L^p-length of a path is

$$\ell_p\{\phi(t,.)\} := \int_0^1 \|\partial_t\phi(t,x)\|_{L^p(M)}\, dt = \int_0^1 \left(\int_M |\partial_t\phi(t,x)|^p \mu\right)^{1/p} dt\,,$$

where $\mu = \omega^n$ is the volume form. Note that the strongest result is in the L^1-metric, since

$$\ell_p\{\,.\,\} \geq C_{M,p} \cdot \ell_1\{\,.\,\}$$

for some constant $C_{M,p} > 0$.

In Appendix A.6.2 we give a proof of a simplified version of this theorem, following [99]. It turns out that the L^1-lengths of paths on the group of Hamiltonian diffeomorphisms stationary on the boundary of M are bounded below by the Calabi invariant.

Remark 3.47 Yet another manifestation of the difference between the geometries of the groups of volume-preserving diffeomorphisms in the two- and three-dimensional cases is the geometry of their Riemannian exponential maps. The corresponding problem of description of conjugate points along geodesics in $S\mathrm{Diff}(M)$ was posed by Arnold back in the 1960s in the paper [12] on the geometry of the Euler equation. It is shown in [97] that for a compact two-dimensional surface M without boundary the exponential map of the L^2-metric on $SDiff(M)$ is a nonlinear Fredholm map of index zero. In particular, this implies that conjugate points are isolated and of finite multiplicity along finite geodesic segments. In other words, in this case the Riemannian exponential map on $SDiff(M)$ has the same structure of singularities as that on a finite-dimensional manifold. For a three-dimensional M the situation changes drastically: the set of conjugate points is not discrete. In particular, conjugate points cluster to the first one, while the exponential operator is not Fredholm, as its range is not closed [97, 323].

Remark 3.48 Finally, it is interesting to relate the geometry of the group of volume-preserving diffeomorphisms to that of all diffeomorphisms. In a sense, the dynamics of an ideal fluid are dual to the Monge–Kantorovich mass transport problem, which asks for the most economical way to move, say, a pile of sand to a prescribed location. Mass (or density) is transported most effectively by gradient vector fields. The latter are L^2-orthogonal to divergence-free ones, which, in contrast, preserve the volume (or mass). The corresponding transportation (also called Kantorovich or Wasserstein) metric on the space of densities and the L^2-metric on volume-preserving diffeomorphisms can be viewed as natural extensions of each other within the framework of the group of all diffeomorphisms, see [303, 374, 193] and Appendix A 5.

3.7 Bibliographical Notes

V. Arnold's seminal paper [12], which described the Euler hydrodynamics equation for an ideal fluid [108] as the equation of geodesics on the group of volume-preserving diffeomorphisms, generated a great deal of interest in the group-theoretical and Hamiltonian aspects of hydrodynamics. The analytical questions of this approach were treated in [96]; see also more on analysis on diffeomorphism groups in [95, 157]. The classical treatise on fluid dynamics is the Landau–Lifschitz book [229].

The Hamiltonian nature of the Euler equation was described in [13, 18]. The geometry of coadjoint orbits of the Lie algebra of divergence-free vector fields was studied in [255, 205, 24]. Finite-dimensional mechanical prototypes of ideal fluid dynamics were discussed in [13, 18, 24, 82, 214].

The frozenness of the vorticity in the two- and three-dimensional cases was known to Helmholtz and Kelvin. The invariance of the helicity for the Euler equation was observed in [269, 278]. The fact that the helicity and energy are the only invariants of the first order (i.e., whose local density depends only on the field and on its first partial derivatives) for the three-dimensional Euler equation was established in [351]. In higher dimensions the above-mentioned first integrals of the Euler hydrodynamics equation were found in [352] for the case of $\mathbb{R}^n$ and in [307, 187] for the general case.

The general two-point variational problem for finding a shortest geodesic between two diffeomorphisms in $S\mathrm{Diff}(M)$ does not always have a solution, as was shown by Shnirelman in [355, 357]. Brenier proved that the solution, however, always exists in a wider class of generalized flows [59]. Peculiar weak solutions to the Euler equation are discussed in [358]. The group-theoretic meaning of various equations of mathematical physics and, in particular, those related to semidirect products and extensions, is discussed in [253]; see also [24, 42, 90, 295, 167, 187, 245, 377, 378].

The symplectic structure on curves in $\mathbb{R}^3$ appeared in [255]. The book [62] is a good source of references for various geometric structures on such curves and for the corresponding evolution equations. The interpretation of the Landau–Lifschitz equation as the Euler equation on a loop group was obtained in [7]; see [244]. For finite-dimensional examples of invariant sub-Riemannian metrics on Lie groups, similar to the infinite-dimensional Example 3.36, see [44, 113, 114, 273].

The results on the diameter of the groups of volume-preserving diffeomorphisms in right-invariant metrics can be found in [355, 357] and [99, 100]. We also discuss the bi-invariant Hofer metric on the group of Hamiltonian diffeomorphisms in Appendix A.6.1 and address the reader to the book [317] for more detail. Curvature calculations for diffeomorphism groups can be found in [12, 243]. The structure of the exponential map on the diffeomorphism group was treated in [97, 323]. Appendix A.5 discusses the Riemannian geometry of the full diffeomorphism group and its relation to problems of optimal mass transport; see the references therein.

The structure theory of various diffeomorphism groups is discussed in [368, 33, 172]. Semigroups of polymorphisms (or stochastic kernels) can the thought of as natural completions of the groups of diffeomorphisms; see [293] on their geometry and representation theory.

4 The Group of Pseudodifferential Symbols

The Lie algebra of vector fields on the circle embeds naturally into the Lie algebra of differential operators on the circle, where the Lie bracket of two differential operators of arbitrary degrees is given by their commutator: $[A, B] = A \circ B - B \circ A$. In this section, we study yet a bigger Lie algebra of *pseudodifferential symbols* on the circle, which contains the algebra of differential operators as a subalgebra.

We have seen in Section 2 that the Lie algebra of vector fields on the circle admits a unique nontrivial central extension, the Virasoro algebra $\mathfrak{vir}$. It turns out that both the Lie algebras of differential operators DO and pseudodifferential symbols ψ DS on the circle also admit central extensions, $\widehat{\mathrm{DO}}$ and $\widehat{\psi \mathrm{DS}}$ respectively, which, upon restriction to the subalgebra of vector fields, give the Virasoro algebra; see the following diagram:

$$
\begin{array}{ccccc}
\mathrm{Vect}(S^1) & \hookrightarrow & \mathrm{DO} & \hookrightarrow & \psi \mathrm{DS} \\
\uparrow & & \uparrow & & \uparrow \\
\mathfrak{vir} & \hookrightarrow & \widehat{\mathrm{DO}} & \hookrightarrow & \widehat{\psi \mathrm{DS}}.
\end{array}
$$

From this point of view, the algebra of pseudodifferential symbols is the most general of the three Lie algebras mentioned above. Somewhat surprisingly, when passing to this big Lie algebra ψ DS, the construction of its central extension drastically simplifies and resembles the construction of the affine Lie algebras in Section 1.

Furthermore, roughly "half of the algebra $\widehat{\psi \mathrm{DS}}$" can be exponentiated to a Lie group of pseudodifferential symbols of arbitrary real (or complex) degrees. This group turns out to be a *Poisson Lie group*. Its Poisson structure is closely related to the theory of the KdV- and KP-type equations, and we describe a universal hierarchy of Hamiltonians on this group encompassing many integrable systems.

4.1 The Lie Algebra of Pseudodifferential Symbols

Definition 4.1 The associative algebra ψ DS of *pseudodifferential symbols* on the circle consists of formal semi-infinite series

$$
A(\theta, \partial) = \sum_{k=-\infty}^{n} a_k(\theta) \partial^k
$$

with coefficients $a_k \in C^\infty(S^1)$. The product of two such symbols is defined according to the Leibniz rule

$$
\partial \circ f = f \partial + f',
$$

where $f' = df/d\theta$. (This relation explains the meaning of the symbol $\partial := d/d\theta$: applying both sides to any "test" function g gives $\partial(fg) = f \partial g + f' g$, where f stands for the operator of multiplication by a function f.) The natural extension of this rule to arbitrary integers k and n gives

$$\partial^k \circ f \partial^n = f \partial^{n+k} + \sum_{i=1}^{\infty} \binom{k}{i} f^{(i)} \partial^{n+k-i}, \qquad (4.32)$$

where

$$\binom{k}{i} := \frac{k(k-1)\cdots(k-i+1)}{i!} \qquad (4.33)$$

denotes the binomial coefficient. The Lie algebra structure on ψDS is given by the usual commutator $[A, B] = A \circ B - B \circ A$.

Exercise 4.2 Check the associativity of the product $\circ$ on ψDS.

Remark 4.3 For integral $k > 0$ the sum in equation (4.32) is finite and manifests the composition of differential operators.

For $k < 0$ the binomial coefficients (4.33) are never zero, so that the sum in the right-hand side of equation (4.32) is infinite. For example, for $k = -1$ and $n = 0$, we obtain

$$\partial^{-1} \circ f = f \partial^{-1} - f' \partial^{-2} + f'' \partial^{-3} - \cdots.$$

To make sense of this equality, one can apply ∂ to both sides and get the identity

$$\partial \circ (\partial^{-1} \circ f) = \partial \circ (f \partial^{-1} - f' \partial^{-2} + f'' \partial^{-3} - \cdots)$$
$$= f + f' \partial^{-1} - f' \partial^{-1} - f'' \partial^{-2} + f'' \partial^{-2} + \cdots = f.$$

Thus ∂ and ∂^{-1} are indeed inverses of each other.

Remark 4.4 The Lie algebra ψDS has two natural subalgebras: the Lie algebra DO of differential operators on the circle

$$\text{DO} := \left\{ \sum_{k=0}^{n} a_k(\theta) \partial^k \mid a_k \in C^{\infty}(S^1) \right\}$$

and the Lie algebra INT of integral symbols, which is given by

$$\text{INT} := \left\{ \sum_{k=-\infty}^{-1} a_k(\theta) \partial^k \mid a_k \in C^{\infty}(S^1) \right\}.$$

As a vector space, the Lie algebra ψDS is a direct sum of these subalgebras: $\psi DS = DO \oplus INT$.

Furthermore, the Lie algebra of pseudodifferential symbols admits an algebraic trace, an analogue of the trace of usual matrices:

Definition 4.5 The *trace* of a pseudodifferential symbol $A = \sum a_i(\theta)\partial^i$ is defined by

$$\mathrm{tr}\left(\sum_{i=-\infty}^{n} a_i(\theta)\partial^i\right) := \int_{S^1} a_{-1}(\theta)d\theta\,.$$

Exercise 4.6 (i) Show that $\mathrm{tr}[A, B] = 0$ for all A, $B \in \psi\,\mathrm{DS}$. This implies that the bilinear form $\langle A, B\rangle := \mathrm{tr}(A \circ B)$ is invariant, i.e., it satisfies $\langle[A, B], C\rangle = \langle A, [B, C]\rangle$.

(ii) Show that the bilinear form $\langle A, B\rangle = \mathrm{tr}(A \circ B)$ is nondegenerate on $\psi\,\mathrm{DS}$ and that the subalgebras DO and INT are isotropic subspaces of $\psi\,\mathrm{DS}$, i.e., that the restrictions of this form to both DO and INT vanish.

Remark 4.7 One can define the pseudodifferential symbols on $\mathbb{C}^*$, rather than on the circle S^1, using the variable $z = e^{i\theta}$, similarly to Remark 1.5. Here one considers the series $\sum_{k=-\infty}^{n} a_k(z)\partial_z^k$, where a_k is a Laurent polynomial in z. The relation of the corresponding Lie algebras is provided by the chain rule: $\partial := \partial_\theta = (\partial z/\partial\theta)\partial_z = iz \cdot \partial_z$. Then the trace for symbols on $\mathbb{C}^*$ can be defined by $\mathrm{tr}\left(\sum_{k=-\infty}^{n} a_k(z)\partial_z^k\right) := \mathrm{res}|_{z=0}(a_{-1}(z)/z)\,.$

4.2 Outer Derivations and Central Extensions of $\psi\,\mathrm{DS}$

It is well known that the exponent of the derivative is the shift operator on functions, as given by the Taylor formula:

$$\exp(t\partial) f(\theta) = \left(\sum_{k=0}^{\infty} \frac{t^k}{k!}\partial^k\right) f(\theta) = f(\theta) + tf'(\theta) + \frac{t^2}{2}f''(\theta) + \cdots = f(\theta + t)\,.$$

Below we are going to define the *logarithm of the derivative operator* ∂. This notion turns out to have a completely different flavor: this will be an operator not on functions, but on pseudodifferential symbols, and it appears rather useful in describing their central extensions

To introduce this notion we first remark that in the definition of the product of two pseudodifferential symbols in formula (4.32) one can replace the integer k by an arbitrary *real* (or *complex*) number α to obtain

$$[\partial^\alpha, f\partial^n] = \sum_{i=1}^{\infty} \binom{\alpha}{i} f^{(i)}\partial^{n+\alpha-i}\,, \tag{4.34}$$

where the binomial coefficients on the right-hand side are defined just as in the integer case:

$$\binom{\alpha}{i} := \frac{\alpha(\alpha - 1)\cdots(\alpha - i + 1)}{i!}\,. \tag{4.35}$$

Let us formally write the identity $\partial^\alpha = e^{\alpha\log\partial}$, which implies

$$\frac{d}{d\alpha}\Big|_{\alpha=0}\partial^\alpha = (\log \partial)\, \partial^\alpha|_{\alpha=0} = \log \partial\,.$$

Hence, differentiating both sides of equation (4.34) at $\alpha = 0$ gives

$$[\log \partial, f\partial^n] = \sum_{i=1}^{\infty} \frac{(-1)^{i+1}}{i} f^{(i)} \partial^{n-i}. \qquad (4.36)$$

We use this formula to define the map $[\log \partial, \,.\,] : \psi\,\mathrm{DS} \to \psi\,\mathrm{DS}$, where $\log \partial$ can be thought of as the "velocity vector" to the family of differentiations ∂^α.

Definition / Proposition 4.8 *The linear map* $[\log \partial, \,.\,] : \psi\,\mathrm{DS} \to \psi\,\mathrm{DS}$ *given by formula (4.36) is an outer derivation of the Lie algebra* $\psi\,\mathrm{DS}$ *of pseudodifferential symbols.*

Exercise 4.9 Verify that the map $[\log \partial, \,.\,]$ is indeed a derivation of the associative algebra $\psi\,\mathrm{DS}$:

$$[\log \partial, A \circ B] = [\log \partial, A] \circ B + A \circ [\log \partial, B]\,,$$

which implies the statement about the Lie algebra.

We postpone the explanation why this derivation is outer until Proposition 4.12 and Remark 4.13. Now we employ the derivation $[\log \partial, \,.\,]$ to define a central extension of the Lie algebra $\psi\,\mathrm{DS}$.

Theorem 4.10 ([215]) *The map* $\varpi : \psi\,\mathrm{DS} \times \psi\,\mathrm{DS} \to \mathbb{R}$ *given by*

$$\varpi(A, B) := \mathrm{tr}([\log \partial, A] \circ B)$$

defines a Lie algebra 2-cocycle on $\psi\,\mathrm{DS}$.

Exercise 4.11 Prove the theorem.

It turns out that the cocycle ϖ on the Lie algebra ψDS generalizes some of the Lie algebra cocycles we have come across before:

Proposition 4.12 *The restriction of the 2-cocycle* ϖ *to the subalgebra* $\mathrm{Vect}(S^1) \subset \psi\,\mathrm{DS}$ *gives the Gelfand–Fuchs 2-cocycle*

$$\varpi(f\partial, g\partial) = \frac{1}{6} \int_{S^1} f'' g'\, d\theta$$

for any vector fields $f(\theta)\partial$ *and* $g(\theta)\partial$ *on* S^1.

PROOF.   This is a direct calculation. We have

$$\varpi(f\partial, g\partial) = \mathrm{tr}([\log\partial, f\partial]\circ g\partial) = \mathrm{tr}(g\partial \circ [\log\partial, f\partial])$$

$$= \mathrm{tr}\left(g\partial \circ \left(f'\partial^0 - \frac{f''}{2}\partial^{-1} + \frac{f'''}{3}\partial^{-2} - \cdots\right)\right)$$

$$= \mathrm{tr}\left(\cdots + \left(-g\frac{f''}{2} + g\frac{f'''}{3}\right)\partial^{-1} + \cdots\right)$$

$$= \mathrm{tr}\left(\cdots - \frac{1}{6}gf'''\partial^{-1} + \cdots\right),$$

where the dots denote terms other than ∂^{-1}. Hence by the trace definition, we have

$$\varpi(f\partial, g\partial) = -\frac{1}{6}\int_{S^1} f'''g\, d\theta = \frac{1}{6}\int_{S^1} f''g'\, d\theta,$$

which is a multiple of the Gelfand–Fuchs cocycle. $\square$

Remark 4.13 The logarithmic cocycle ϖ on the Lie algebra $\psi\,\mathrm{DS}$ is nontrivial, since even its restriction to the subalgebra $\mathrm{Vect}(S^1) \subset \psi\,\mathrm{DS}$ is nontrivial. Furthermore, the nontriviality of the cocycle ϖ also implies that the derivation $[\log\partial, .]$ is outer, since otherwise the 2-cocycle ϖ would be cohomologous to a 2-coboundary on $\psi\,\mathrm{DS}$. Indeed, if $\log\partial$ were a pseudodifferential symbol, we could rewrite the cocycle $\varpi(A, B) = \mathrm{tr}([\log\partial, A]\circ B) = \mathrm{tr}(\log\partial\circ[A, B]) = \langle\log\partial, [A, B]\rangle$ as a linear functional on the commutator, i.e., as a 2-coboundary on $\psi\,\mathrm{DS}$. Of course, $\log\partial \notin \psi\,\mathrm{DS}$, and the above line is not justified. (The above "transformations" are useful to keep in mind, however, when checking the cocycle identity for ϖ; cf. Exercise 4.11.)

Exercise 4.14 Show that the restriction of the 2-cocycle ϖ to the subalgebra of differential operators $\mathrm{DO} \subset \psi\,\mathrm{DS}$ is a multiple of the Kac–Peterson cocycle:

$$\varpi(f\partial^n, g\partial^m) = \frac{n!\,m!}{(n + m + 1)!}\int_{S^1} f^{(m+1)}g^{(n)}\, d\theta.$$

(The corresponding central extension $\widehat{\mathrm{DO}}$ of differential operators is often called the $W_{1+\infty}$-algebra in the physics literature. The reason for this funny notation is that the algebra W_∞, which appeared first, was the subalgebra of differential operators generated by the Virasoro generators, i.e., by vector fields, and did not include functions. The above notation $W_{1+\infty}$ emphasizes that this larger algebra of all differential operators includes functions as well.)

In particular, the restriction of the cocycle ϖ to the abelian subalgebra $C^\infty(S^1)$ of smooth functions on the circle gives the 2-cocycle that defines the infinite-dimensional Heisenberg algebra:

$$\varpi(f, g) = \int_{S^1} f' g \, d\theta$$

for any smooth functions $f = f(\theta)$ and $g = g(\theta)$ on S^1.

Remark 4.15 One can easily generalize the cocycle ϖ to the Lie algebra of $\mathfrak{gl}(n)$-valued pseudodifferential symbols. Then the restriction to the subalgebra of Lie-algebra-valued functions gives back the cocycle defining the affine Lie algebra $\widehat{L\mathfrak{gl}(n)}$.

But rather than viewing the cocycle of the affine algebra as a particular case of the logarithmic cocycle, one can see the following parallelism in the two constructions. The 2-cocycle on the loop algebra $L\mathfrak{g}$ was defined by

$$\omega(X, Y) = \int_{S^1} \mathrm{tr}(XY') \, d\theta = \langle X, [\partial, Y] \rangle,$$

where $\partial = \frac{d}{d\theta}$; see Section 1. So the outer derivation $[\partial, \,.\,]$ of the loop algebra $L\mathfrak{g}$ is "replaced" by the outer derivation $[\log \partial, \,.\,]$ of the Lie algebra $\psi \mathrm{DS}$.

Remark 4.16 Consider the linear function $f : \mathbb{R} \to \mathbb{R}$, $f(\theta) = \theta$. Even though the function θ is not periodic and hence not defined on the circle S^1, its commutator $[\theta, A]$ with any pseudodifferential symbol $A \in \psi \mathrm{DS}$ is again a pseudodifferential symbol on S^1. This allows one to define another 2-cocycle ϖ' on the Lie algebra $\psi \mathrm{DS}$ by

$$\varpi'(A, B) := \mathrm{tr}([\theta, A] \circ B).$$

The similarity between the cocycles ϖ and ϖ' becomes more apparent if we consider the Lie algebra of pseudodifferential symbols on $\mathbb{C}^*$. By identifying $e^{i\theta}$ with z, one rewrites θ as $-i \log z$, and hence the cocycle ϖ' assumes the form

$$\varpi'(A, B) = -i \cdot \mathrm{tr}([\log z, A] \circ B).$$

On the other hand, an analogue of the above cocycle ϖ in the coordinate z has a similar form,

$$\varpi(A, B) = \mathrm{tr}([\log \partial_z, A] \circ B), .$$

cf. Remark 4.7. The cocycles ϖ and ϖ' span the second cohomology of the Lie algebra $\psi \mathrm{DS}$, since this cohomology group is known to be two-dimensional [116, §4]. For us, however, the cocycle ϖ will play a more important role, since its restriction to vector fields gives the Virasoro algebra, while the other cocycle ϖ' vanishes there.

Exercise 4.17 Show that the restriction of the cocycle ϖ' to the subalgebra $\mathrm{DO} \subset \psi \mathrm{DS}$ vanishes: $\varpi'(A, B) = 0$ for any two differential operators A and B on the circle.

4.3 The Manin Triple of Pseudodifferential Symbols

The 2-cocycle ϖ discussed in the preceding section defines a central extension $\widehat{\psi\,\mathrm{DO}}$ of the Lie algebra $\psi\,\mathrm{DS}$. In this section we are concerned with yet a bigger Lie algebra, obtained by extending $\psi\,\mathrm{DO}$ by adding both the central extension and the derivation $[\log\partial,\,.\,]$.

Definition 4.18 The *doubly extended* Lie algebra of $\psi\,\mathrm{DS}$ is the semidirect product of the (centrally extended) Lie algebra $\widehat{\psi\,\mathrm{DS}}$ of pseudodifferential symbols with the one-dimensional space of derivations $\{\lambda\log\partial \mid \lambda\in\mathbb{R}\}$. As a vector space, this Lie algebra can be written as

$$\widehat{\psi\,\mathrm{DS}} = \widehat{\psi\,\mathrm{DO}} \oplus \mathbb{R}\log\partial = \psi\,\mathrm{DO} \oplus \mathbb{R}\cdot\varpi \oplus \mathbb{R}\cdot\mathrm{lcg}\,\partial\,,$$

while the Lie bracket between $\log\partial$ and $\psi\,\mathrm{DO}$ is given by the derivation $[\log\partial,\,.\,]$, and the center $\mathbb{R}\cdot\varpi$ commutes with everything: $[\varpi,\,.\,]=0$. (Here, slightly abusing notation, we denote by $\mathbb{R}\cdot\varpi$ the central direction in the algebra $\widehat{\psi\,\mathrm{DS}}$.)

Remark 4.19 Furthermore, as a vector space, the Lie algebra $\widehat{\psi\,\mathrm{DS}}$ has a direct sum decomposition $\widehat{\psi\,\mathrm{DS}} = \widehat{\mathrm{DO}} \oplus \widetilde{\mathrm{INT}}$, where $\widehat{\mathrm{DO}}$ denotes the *centrally* extended Lie algebra of differential operators on S^1:

$$\widehat{\mathrm{DO}} = \Big\{ b\,\varpi + \sum_{i=0}^{n} a_i(\theta)\partial^i \mid a_i\in C^\infty(S^1),\ b\in\mathbb{R} \Big\}\,,$$

while $\widetilde{\mathrm{INT}}$ denotes the Lie algebra of integral symbols INT extended by the *"cocentral"* direction $\log\partial$:

$$\widetilde{\mathrm{INT}} = \Big\{ \alpha\log\partial + \sum_{i=-\infty}^{-1} a_i(\theta)\partial^i \mid a_i\in C^\infty(S^1),\ \alpha\in\mathbb{R} \Big\}$$

(see Figure 4.1).

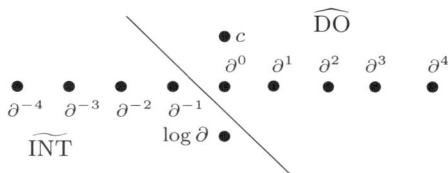

Fig. 4.1. $\widehat{\mathrm{DO}}$ and $\widetilde{\mathrm{INT}}$ as subalgebras of the Lie algebra $\widehat{\psi\,\mathrm{DS}}$.

The advantage of the double extension $\widetilde{\psi\,\mathrm{DS}}$ is that this Lie algebra admits a nondegenerate invariant bilinear form (the "Killing form"):

Exercise 4.20 Check that the bilinear form $\langle\ ,\ \rangle$ on the Lie algebra $\widetilde{\psi\,\mathrm{DS}}$ given by

$$\langle A_1 + b_1\cdot\varpi + \alpha_1\cdot\log\partial\,,\ A_2 + b_2\cdot\varpi + \alpha_2\cdot\log\partial\rangle := \mathrm{tr}(A_1\circ A_2) + b_1\alpha_2 + b_2\alpha_1 \quad (4.37)$$

is nondegenerate and invariant. Show that the subalgebras $\widehat{\mathrm{DO}}$ and $\widehat{\mathrm{INT}}$ are isotropic subspaces of $\widetilde{\psi\,\mathrm{DS}}$ with respect to this form.

Corollary 4.21 *The bilinear form* $\langle\ ,\ \rangle$ *on* $\widetilde{\psi\,\mathrm{DS}}$ *identifies the subalgebra* $\widehat{\mathrm{INT}}$ *with the (smooth part of) the dual to the subalgebra* $\widehat{\mathrm{DO}}$ *and vice versa.*

These observations can be summarized in the language of Manin triples.

Definition 4.22 Three Lie algebras $\bar{\mathfrak{g}}$, $\mathfrak{g}_+$, and $\mathfrak{g}_-$ form a *Manin triple* if the following two conditions are satisfied:

1. The Lie algebras $\mathfrak{g}_\pm$ are Lie subalgebras of $\bar{\mathfrak{g}}$ such that $\bar{\mathfrak{g}} = \mathfrak{g}_+ \oplus \mathfrak{g}_-$ as vector spaces.
2. There exists a nondegenerate invariant bilinear form on $\bar{\mathfrak{g}}$ such that $\mathfrak{g}_+$ and $\mathfrak{g}_-$ are isotropic subspaces.

Example 4.23 (*i*) The algebras $(\psi\,\mathrm{DS}, \mathrm{DO}, \mathrm{INT})$ form a Manin triple with respect to the bilinear form $\langle A\,,\ B\rangle = \mathrm{tr}(A\circ B)$; cf. Exercise 4.6.

(*ii*) The algebras $(\widetilde{\psi\,\mathrm{DS}}, \widehat{\mathrm{DO}}, \widehat{\mathrm{INT}})$ form a Manin triple with respect to the bilinear form (4.37), as Exercise 4.20 shows.

The importance of Manin triples comes from the fact that there is a one-to-one correspondence between Manin triples and Lie bialgebras:

Definition 4.24 A Lie algebra $\mathfrak{g}$ is called a *Lie bialgebra* if its dual space $\mathfrak{g}^*$ comes equipped with a Lie algebra structure $[\ ,\ ]^*$ such that the map $\alpha : \mathfrak{g} \to \mathfrak{g} \wedge \mathfrak{g}$ dual to the commutator map $[\ ,\ ]^* : \mathfrak{g}^* \wedge \mathfrak{g}^* \to \mathfrak{g}^*$ satisfies

$$\mathrm{ad}_X\,\alpha(Y) = \alpha([X, Y])\,. \quad (4.38)$$

Here ad denotes the adjoint action $\mathrm{ad}_X(Z \wedge W) = \mathrm{ad}_X(Z) \wedge W + Z \wedge \mathrm{ad}_X(W)$ of $\mathfrak{g}$ on $\mathfrak{g} \wedge \mathfrak{g}$.

The condition on the map $\alpha : \mathfrak{g} \to \mathfrak{g} \wedge \mathfrak{g}$ can be stated in terms of Lie algebra cohomology: the map α has to be a Lie algebra 1-cocycle on $\mathfrak{g}$ relative to the adjoint representation of $\mathfrak{g}$ on $\mathfrak{g} \wedge \mathfrak{g}$.

Theorem 4.25 *Let* $(\bar{\mathfrak{g}}, \mathfrak{g}_+, \mathfrak{g}_-)$ *be a Manin triple. Then the Lie algebra* $\mathfrak{g}_-$ *is naturally identified with* $\mathfrak{g}_+^*$ *and it endows* $\mathfrak{g}_+$ *with the structure of a Lie bialgebra. On the other hand, for any Lie bialgebra* $\mathfrak{g}$ *there is a natural Lie algebra structure on* $\bar{\mathfrak{g}} = \mathfrak{g} \oplus \mathfrak{g}^*$ *such that* $(\bar{\mathfrak{g}}, \mathfrak{g}, \mathfrak{g}^*)$ *is a Manin triple.*

Namely, given a Lie bialgebra $\mathfrak{g}$, one can define a commutator on $\bar{\mathfrak{g}} = \mathfrak{g} \oplus \mathfrak{g}^*$ by

$$[X + A, Y + B]_{\bar{\mathfrak{g}}} = [X, Y]_{\mathfrak{g}} - \mathrm{ad}_B^*(X) + \mathrm{ad}_A^*(Y) + [A, B]_{\mathfrak{g}^*} + \mathrm{ad}_X^*(B) - \mathrm{ad}_Y^*(A)$$

for all $X, Y \in \mathfrak{g}$ and $A, B \in \mathfrak{g}^*$. The condition that this commutator satisfy the Jacobi identity is equivalent to the condition (4.38) on the map α. For a full proof of Theorem 4.25, see, e.g., [349, 242].

Summarizing the discussion above we come to the following result.

Corollary 4.26 ($= 4.21'$) *The Lie algebras* $\widehat{\psi \mathrm{DS}}$, $\widehat{\mathrm{DO}}$, *and* $\widetilde{\mathrm{INT}}$ *form a Manin triple. In particular,* $\widetilde{\mathrm{INT}}$ *is a Lie bialgebra.*

Although the above correspondence of Manin triples and Lie bialgebras, as well as the relation to Poisson Lie groups discussed below, is proved in a finite-dimensional context only, the related explicit formulas for the commutators, products, pairings, etc. work in many specific infinite-dimensional situations. In particular, they can be checked directly for the case of pseudodifferential symbols.

4.4 The Lie Group of α-Pseudodifferential Symbols

The Lie algebra of vector fields is the Lie algebra of a Lie group, the group of diffeomorphisms. However, when we pass from vector fields to differential operators of all degrees, there is no natural Lie group attached to the corresponding Lie algebra DO or to its central extension $\widehat{\mathrm{DO}}$.[12]

Interestingly, the Lie algebra $\widetilde{\mathrm{INT}}$ of integral symbols, which is dual to $\widehat{\mathrm{DO}}$ in the the Manin triple $(\widehat{\psi \mathrm{DS}}, \widehat{\mathrm{DO}}, \widetilde{\mathrm{INT}})$, comes with a group attached to it. Below we describe this group of α-pseudodifferential symbols $\widetilde{G}_{\mathrm{INT}}$.

Definition 4.27 Define the group of α-*pseudodifferential symbols* to be the set of monic (i.e., with the highest coefficient equal to 1) pseudodifferential symbols of degree $\alpha \in \mathbb{R}$:

$$\widetilde{G}_{\mathrm{INT}} = \left\{ \partial^\alpha \left(1 + \sum_{i=-\infty}^{-1} a_i(\theta) \partial^i \right) \mid \alpha \in \mathbb{R}, \ a_i \in C^\infty(S^1) \right\}.$$

---

[12] One can show, for instance, that the would-be group adjoint orbits in DO are dense, which is impossible for any reasonable definition of a (smooth) group action.

The group multiplication in $\widetilde{G}_{\text{INT}}$ is defined by equation (4.34):

$$\partial^\alpha \circ f = f\partial^\alpha + \sum_{i=1}^\infty \binom{\alpha}{i} f^{(i)}\partial^{\alpha-i}\,.$$

Exercise 4.28 Check that $\widetilde{G}_{\text{INT}}$ is indeed a group, i.e., that it is closed under taking products and inverses.

Remark 4.29 The group $\widetilde{G}_{\text{INT}}$ is a semidirect product

$$\widetilde{G}_{\text{INT}} = G_{\text{INT}} \rtimes \{\partial^\alpha \mid \alpha \in \mathbb{R}\}\,,$$

where

$$G_{\text{INT}} = \left\{ 1 + \sum_{i=-\infty}^{-1} a_i(\theta)\partial^i \right\} \subset \widetilde{G}_{\text{INT}}$$

denotes the group of monic pseudodifferential symbols of degree 0, while $\{\partial^\alpha \mid \alpha \in \mathbb{R}\}$ is the abelian one-parameter group of "fractional differentiations": $\partial^\alpha \circ \partial^\beta = \partial^{\alpha+\beta}$, which acts on G_{INT}.

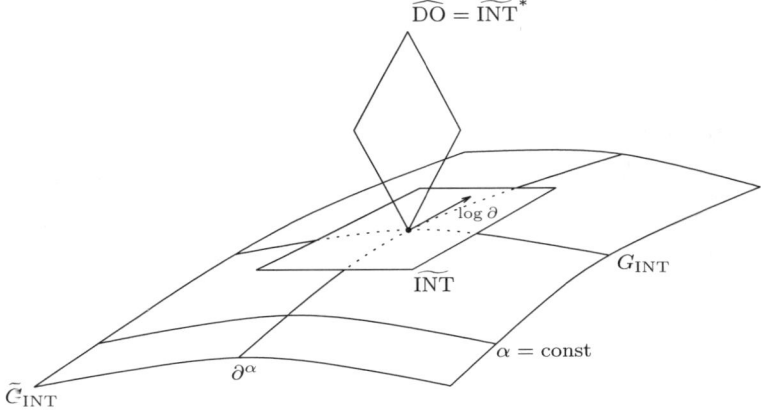

Fig. 4.2. The group $\widetilde{G}_{\text{INT}}$ of α-pseudodifferential symbols and its Lie algebra $\widetilde{\text{INT}}$.

To identify the Lie algebra for the group $\widetilde{G}_{\text{INT}}$, first we consider the finite-dimensional example of the unipotent Lie group of upper triangular matrices with 1's on the diagonal. Its Lie algebra is nilpotent and represented by strictly upper triangular matrices. Similarly, we expect the Lie algebra of the group G_{INT} to be given by the Lie algebra INT of integral symbols. On the other hand, we already know the Lie algebra of the one-parameter subgroup $\{\partial^\alpha \mid \alpha \in \mathbb{R}\} \subset \widetilde{G}_{\text{INT}}$, since it is given by $\mathbb{R} \log \partial$ by our construction. Thus we expect the group $\widetilde{G}_{\text{INT}}$ to have the Lie algebra $\widetilde{\text{INT}} = \text{INT} \rtimes \{\alpha \log \partial \mid \alpha \in \mathbb{R}\}$ (see Figure 4.2). The following theorem formalizes this argument.

Theorem 4.30 ([198]) *The exponential map* $\exp : \widetilde{\text{INT}} \to \widetilde{G}_{\text{INT}}$ *is well defined and bijective.*

The bijective property of the exponential map $\exp : \widetilde{\text{INT}} \to \widetilde{G}_{\text{INT}}$ means that the group $\widetilde{G}_{\text{INT}}$ is *quasi-unipotent*, i.e. it is an infinite-dimensional analogue of the finite-dimensional unipotent group.

Remark 4.31 The idea of the proof of Theorem 4.30 is as follows. Assume the existence of the exponential map on this group. Then for an integral symbol $A \in \text{INT}$, the one-parameter family $L_s = \exp(s(a \log \partial + A))$ should (by definition) satisfy the equation

$$\left(\frac{d}{ds} L_s \right) \circ L_s^{-1} = a \log \partial + A \tag{4.39}$$

with the initial condition $L_0 = 1$. Equation (4.39) gives a system of differential equations on the (s-dependent) coefficients of L_s, which turns out to be uniquely solvable. Hence equation (4.39) can be used to define the one-parameter subgroup L_s and therefore the exponential map. Now the proof of Theorem 4.30 follows from the following two exercises and the Campbell–Hausdorff formula; see details in the next section.

Exercise 4.32 Show that the formula

$$\exp(P) = \sum_{k \geq 0} \frac{1}{k!} P^k$$

for an integral symbol $P \in \text{INT}$ gives a well-defined map $\text{INT} \to G_{\text{INT}}$: the coefficient at each ∂^j in the right-hand side is a finite sum. Similarly, show that the inverse map $G_{\text{INT}} \to \text{INT}$ given by

$$\log(1 + P) = \sum_{k \geq 1} \frac{(-1)^k}{k} P^{k+1}$$

is well defined. (Hint: use the fact that $\deg P^k \leq -k$.)

The above exercise proves the following proposition, which is a "restricted" version of Theorem 4.30, by giving the explicit formula for the exponential map and its inverse.

Proposition 4.33 *The map* $\exp : \text{INT} \to G_{\text{INT}}$ *is well defined and bijective.*

The next exercise makes sense of the formal calculation $\frac{d}{d\alpha}|_{\alpha=0} \partial^\alpha = \log \partial$ we made before.

Exercise 4.34 Verify that the element $\log \partial$ in the Lie algebra $\widetilde{\mathrm{INT}}$ generates the one-parameter subgroup $\{\partial^\alpha \mid \alpha \in \mathbb{R}\} \subset \widetilde{G}_{\mathrm{INT}}$ by computing the limit of the commutation relations of the tangent elements to the subgroups $\{\partial^\alpha\}$ and $\bar{G}_{\mathrm{INT}}$:

$$\partial^\epsilon \circ \exp(\delta P) \circ \partial^{-\epsilon} \circ \exp(-\delta P) = 1 + \delta\epsilon[\log \partial, P] + \mathcal{O}(\epsilon^2, \delta^2)$$

for any $P \in \mathrm{INT}$ and as $\epsilon, \delta \to 0$.

(Hint: By using the approximation $\exp(\delta P) = 1 + \delta P + \mathcal{O}(\delta^2)$ show that the left-hand side, modulo the terms $\mathcal{O}(\delta^2)$, is equal to

$$1 + \delta\left[\partial^\epsilon, P\right] \circ \partial^{-\epsilon} = 1 + \delta \sum_{k=1}^\infty \binom{\epsilon}{k} a_j^{(k)} \partial^{j-k}$$

for a symbol $P = a_j(\theta)\partial^j$. Then employ the identity $\frac{d}{d\epsilon}\big|_{\epsilon=0} \binom{\epsilon}{k} = \frac{(-1)^{k+1}}{k}$ and the definition of $\log \partial$.)

Remark 4.35 Note that $\widetilde{G}_{\mathrm{INT}}$ is a Fréchet Lie group. Indeed, the decomposition $\widetilde{G}_{\mathrm{INT}} = G_{\mathrm{INT}} \rtimes \{\partial^\alpha \mid \alpha \in \mathbb{R}\}$ can be used to define a topology on this group as follows. The group G_{INT} is the inverse limit of Fréchet vector spaces (each one being the direct sum of finitely many $C^\infty(S^1)$'s, the spaces of the first n coefficients of the integral symbols, with the usual C^∞ topology). Hence it can be endowed with the inverse limit topology. Then we equip $\widetilde{G}_{\mathrm{INT}}$ with the direct product topology of G_{INT} and the usual topology on $\mathbb{R}$ in the α-direction.

4.5 The Exponential Map for Pseudodifferential Symbols

In this section we prove Theorem 4.30 on the surjectivity of the exponential map $\exp : \widetilde{\mathrm{INT}} \to \widetilde{G}_{\mathrm{INT}}$ on pseudodifferential symbols.

If P is an element in a Lie algebra (think of a matrix Lie algebra, for instance), the would-be exponential $L_s := \exp(sP)$ should satisfy the differential equation

$$\frac{dL_s}{ds} \circ (L_s)^{-1} = P. \tag{4.40}$$

We are going to use this equation to *define* the element $\exp(P) \in \widetilde{G}_{\mathrm{INT}}$.

Lemma 4.36 *For an integral symbol P, the map $P \mapsto (L_s)_{s=1}$ gives a well-defined map $\exp : \mathrm{INT} \to G_{\mathrm{INT}}$.*

Although this lemma is covered by Proposition 4.33, we prove it in a way that can be easily adapted to the extended group $\widetilde{G}_{\mathrm{INT}}$ below.

PROOF.   Fix an integral symbol $P = \sum_{i=-\infty}^{-1} a_i(\theta)\partial^i \in \mathrm{INT}$. Its would-be exponent

$$L_s = 1 + \sum_{k=-\infty}^{-1} u_k(\theta; s)\partial^k$$

has to satisfy the the differential equation (4.40) or, equivalently, $\frac{d}{ds}L_s = P \circ L_s$. Rewriting the latter equation in terms of the coefficients at ∂^k for $k \leq -1$, we obtain an infinite system of differential equations in the "triangular form"

$$\frac{d}{ds}u_k(\theta; s) = \Phi_k\left(u_1(\theta; s), \ldots, u_{k-1}(\theta; s)\right). \tag{4.41}$$

Here Φ_k is a polynomial in the $u_1, \ldots, u_{k-1}$ and their derivatives with coefficients in $C^\infty(S^1)$. (For example, we have $\frac{d}{ds}u_1(\theta; s) = a_1(\theta)$ and $\frac{d}{ds}u_2(\theta; s) = a_2(\theta) + a_1(\theta)u_1(\theta; s) + a_1(\theta)u_1'(\theta; s)$, where $u_1' := \partial u_1/\partial\theta$, etc.) This system can be solved uniquely after fixing the initial condition $u(\theta; 0) = 0$, i.e., $L_0 = 1$. Furthermore, since the functions a_i are periodic in θ, so will be the functions u_k. Hence the exponential map is well defined for all $P \in \mathrm{INT}$. $\square$

Now we are ready to prove

Theorem 4.30. *The map* $\exp : \widetilde{\mathrm{INT}} \to \widetilde{G}_{\mathrm{INT}}$ *is well defined and bijective.*

PROOF.   Take an element $\tilde{P} = \lambda \log \partial + P \in \widetilde{\mathrm{INT}}$, where $P = \sum_{i=-\infty}^{-1} a_i(\theta)\partial^i$ is an integral symbol from INT. Without loss of generality we set $\lambda = 1$. (The case $\lambda = 0$ is covered by Proposition 4.33.) Equation (4.40) now becomes

$$\frac{d\tilde{L}_s}{ds} = (\log \partial + P)\,\tilde{L}_s. \tag{4.42}$$

Let us set $\tilde{L}_s = (1 + Q_s)\partial^s \in \widetilde{G}_{\mathrm{INT}}$. Then the equation above assumes the form

$$\frac{dQ_s}{ds} = [\log \partial, Q_s] + P \circ (1 + Q_s). \tag{4.43}$$

Indeed, for the product $\tilde{L}_{s+\epsilon}\tilde{L}_s^{-1}$ we have, after some transformations,

$$\tilde{L}_{s+\epsilon}\tilde{L}_s^{-1} = (\partial^\epsilon(1 + Q_{s+\epsilon}) - [\partial^\epsilon, Q_{s+\epsilon}])(1 + Q_s)^{-1}$$

$$= 1 + \epsilon \log \partial + \epsilon\frac{dQ_s}{ds}(1 + Q_s)^{-1} - \epsilon[\log \partial, Q_s](1 + Q_s)^{-1} + \mathcal{O}(\epsilon^2),$$

as $\epsilon \to 0$. On the other hand, this product is equal to $1 + \epsilon\left(\frac{d}{ds}\tilde{L}_s\right)\tilde{L}_s^{-1} + \mathcal{O}(\epsilon^2)$, which leads to the equivalence of equations (4.42) and (4.43).

Equation (4.43) gives rise to a triangular system of ordinary differential equations similar to (4.41). Such a system can be solved uniquely, so that the exponential map is well defined on the whole Lie algebra $\widetilde{\mathrm{INT}}$.

Finally, we have to check that the exponential map is bijective. To do this we construct an inverse map to exp. Fix some $\tilde{L} \in \widetilde{G}_{\mathrm{INT}}$, where $\tilde{L} = (1+Q)\partial^s$ for an integral symbol $Q = \sum_{i=-\infty}^{-1} a_i(\theta)\partial^i$. Surjectivity of the exponential map $\mathrm{INT} \to G_{\mathrm{INT}}$ for the nonextended group (Proposition 4.33) allows us to find some $P \in \mathrm{INT}$ such that $\exp(P) = 1 + Q$. Hence we get

$$\tilde{L} = (1 + Q)\partial^s = \exp(P) \circ \exp(s \log \partial) \,.$$

The Campbell–Hausdorff formula implies that

$$\tilde{L} = \exp(P + s \log \partial + s[P, \log \partial]/2 + \cdots) \,.$$

So we can define the inverse of the exponential map by

$$\log(\tilde{L}) := P + s \log \partial + s[P, \log \partial]/2 + \cdots \,.$$

Note that each coefficient at every ∂^k in the expression for $\log(\tilde{L})$ is a finite sum (since the terms in the sum above have decreasing degree due to the increasing number of commutators). Therefore, the map $\log : \widetilde{G}_{\mathrm{INT}} \to \widetilde{\mathrm{INT}}$ is well defined, which implies that the exponential map $\exp : \widetilde{\mathrm{INT}} \to \widetilde{G}_{\mathrm{INT}}$ is indeed bijective. $\square$

4.6 Poisson Structures on the Group of α-Pseudodifferential Symbols

The fact that the Lie algebra $\widetilde{\mathrm{INT}}$ of integral symbols is actually a Lie bialgebra (or a part of a Manin triple) implies that the corresponding Lie group $\widetilde{G}_{\mathrm{INT}}$ is a Poisson Lie group:

Definition 4.37 A *Poisson Lie group* $(G, \{\ ,\ \})$ is a Lie group G equipped with a Poisson structure $\{\ ,\ \}$ such that the multiplication $G \times G \to G$ and the inverse mapping $G \to G^-$ (sending $g \mapsto g^{-1}$) are Poisson maps, where $G \times G$ carries the product Poisson structure and G^- denotes the Lie group G equipped with the opposite Poisson structure $-\{\ ,\ \}$.

Theorem 4.38 *For any connected and simply connected Lie group G, there is a one-to-one correspondence between Lie bialgebra structures on its Lie algebra $\mathfrak{g}$ and Poisson Lie group structures on G.*

For a proof of this theorem, see, e.g., [349, 242]. The explicit construction of the Poisson bracket on G from the Lie bialgebra structure on $\mathfrak{g}$ is as follows. Regard the Lie bialgebra $\mathfrak{g}$ as a part $\mathfrak{g} = \mathfrak{g}_-$ of the Manin triple $(\bar{\mathfrak{g}}, \mathfrak{g}_+, \mathfrak{g}_-)$ with corresponding Lie groups $\bar{G}$, G_+, and G_-. (We write $\mathfrak{g}_+ = \mathfrak{g}^*$ and $\mathfrak{g}_- = \mathfrak{g}$ in order to distinguish between the different duals involved.) Define the following Poisson structure on the group $G = G_-$. Let $\xi, \eta \in T_g^* G_-$ be cotangent vectors to G_- at a point $g \in G_-$ and let $\bar{\xi}$, $\bar{\eta}$ be arbitrary extensions of ξ and η to cotangent vectors of the group $\bar{G} \supset G_-$ at g. Furthermore, denote by $r_g^* \bar{\xi}$ and $l_g^* \bar{\xi}$ the pullbacks of $\bar{\xi} \in T_g^* \bar{G}$ to $T_e^* \bar{G} = \bar{\mathfrak{g}}^*$ via, respectively, the right and left translations. Finally, let $(\ )_+$ be the projection of $\bar{\mathfrak{g}}^*$ to $\mathfrak{g}_+^* = \mathfrak{g}_-$ along $\mathfrak{g}_-^*$ and define $(\ )_-$ analogously. Then the Poisson structure on the group $G = G_-$ is defined by the bivector field Π whose value on the covectors ξ and η at the point g is given by the following formula:

$$\Pi_g(\xi, \eta) = \langle (r_g^* \bar{\xi})_+, r_g^* \bar{\eta} \rangle - \langle (l_g^* \bar{\xi})_+, l_g^* \bar{\eta} \rangle . \tag{4.44}$$

Remark 4.39 Note that we do not actually need the existence of Lie groups $\bar{G}$ and G_+ corresponding to the Lie algebras $\bar{\mathfrak{g}}$ and $\mathfrak{g}_+$. For the above construction to work, it suffices to know how the Lie group G_- acts on the big Lie algebra $\bar{\mathfrak{g}} = \mathfrak{g}_+ \oplus \mathfrak{g}_-$. Indeed, for cotangent vectors ξ, $\eta \in T_g^* G$ at a point $g \in G$, let $A = r_g^* \xi$ and $B = r_g^* \eta$ denote their pullbacks to $T_e^* G_- = \mathfrak{g}_-^*$. Now, if $\bar{A}$ and $\bar{B}$ are any liftings of $A, B \in \mathfrak{g}_-^* = \bar{\mathfrak{g}} / \mathfrak{g}_+^*$ to $\bar{\mathfrak{g}}$, then formula (4.44) for the Poisson structure on the group G_- can be rewritten in the form

$$\Pi_g(\xi, \eta) = \langle (\bar{A})_+, \bar{B} \rangle - \langle (\mathrm{Ad}_g^* \bar{A})_+, \mathrm{Ad}_g^* \bar{B} \rangle . \tag{4.45}$$

This observation allows one to extend this construction to the infinite-dimensional case at hand. Indeed, set $\mathfrak{g}_- = \widetilde{\mathrm{INT}}$ and $\mathfrak{g}_+ = \widetilde{\mathrm{DO}}$, while the group $\widetilde{G}_{\mathrm{INT}}$ acts on the Lie algebra $\widetilde{\psi \mathrm{DS}}$ by conjugation. Then formula (4.45) defines a bivector field on the group $\widetilde{G}_{\mathrm{INT}}$, which endows it with the structure of a Poisson Lie group.

Corollary 4.40 *The group $\widetilde{G}_{\mathrm{INT}}$ carries a natural Poisson structure that gives it the structure of a Poisson Lie group.*

Our next goal will be to identify this Poisson structure on the group $\widetilde{G}_{\mathrm{INT}}$ explicitly. Let us express $\widetilde{G}_{\mathrm{INT}} = \bigcup_{\alpha \in \mathbb{R}} \widetilde{G}_\alpha$ as a union of the "hyperplanes"

$$\widetilde{G}_\alpha = \left\{ L \mid L = \partial^\alpha \circ \left(1 + \sum_{k=-\infty}^{-1} u_k(z) \partial^k \right) \right\}$$

of symbols of fixed degree $\alpha = \mathrm{const}$.

Definition 4.41 The *quadratic* (or *second*) *generalized Gelfand–Dickey Poisson structure* $\{\ ,\ \}_{GD}$ on the group $\widetilde{G}_{\mathrm{INT}}$ is defined as follows:

1. The degree function α is its Casimir function, i.e., the hyperplanes $\widetilde{G}_\alpha$ are Poisson submanifolds for this Poisson structure. In other words, for two functions $f, g : \widetilde{G}_{\mathrm{INT}} \to \mathbb{R}$, the value of their Poisson bracket $\{f, g\}_{GD}$ at any point $L = \partial^{\alpha_0}(1 + \cdots) \in \widetilde{G}_{\mathrm{INT}}$ depends only on the restriction of f and g to the hyperplane $\widetilde{G}_{\alpha_0}$.

2. The subsets $\widetilde{G}_\alpha$ are affine spaces, so one can identify the tangent space to the hyperplane $\widetilde{G}_{\alpha_0}$ of symbols of fixed degree α_0 with the set $\partial^{\alpha_0} \circ \mathrm{INT}$: a tangent vector at the symbol L of degree α_0 has the form $\delta L = \partial^{\alpha_0} \circ \sum_{k=-\infty}^{-1} u_k(z)\partial^k \in \partial^{\alpha_0} \circ \mathrm{INT}$.
   This allows one to identify the corresponding cotangent space with $\mathrm{DO} \circ \partial^{-\alpha_0}$. Any cotangent vector $A \in \mathrm{DO} \circ \partial^{-\alpha_0}$ defines a linear functional F_A on the tangent space $\partial^{\alpha_0} \circ \mathrm{INT}$ via the following pairing:

$$F_A(\delta L) := \langle A, \delta L \rangle = \mathrm{tr}(A \circ \delta L).$$

   Here the product $A \circ \delta L$ is a symbol of integral degree and its trace tr is well defined.

3. Finally, it is sufficient to define the Poisson bracket on linear functionals, and

$$\{F_A, F_B\}_{GD}(L) := F_B\left(V_A(L)\right),$$

   where V_A is the following Hamiltonian mapping $F_A \mapsto V_A(L)$ (from the cotangent space $\{A\}$ to the tangent space $\{\delta L\}$):

$$V_A(L) = (L \circ A)_+ \circ L - L \circ (A \circ L)_+ . \tag{4.46}$$

Here X_+ denotes the purely differential part of a pseudodifferential symbol $X \in \psi\,\mathrm{DS}$.

Note that in the above formula for the Poisson bracket we regard the functional F_B as a linear functional on the space $\widetilde{G}_{\alpha_0}$ in the left-hand side, and as a covector (on the tangent space) at L in the right-hand side. The bracket of two functionals linear in L may already be quadratic in L.

Exercise 4.42 Show that $V_A(L)$ is well defined as a vector field on $\widetilde{G}_{\alpha_0}$, i.e., show that $\deg V_A(L) \le \alpha_0 - 1$ and its degree differs from α_0 by an integer. (Hint: one can rewrite $V_A(L)$ in the form

$$V_A(L) = (L \circ A - (L \circ A)_-) \circ L - L \circ (A \circ L - (A \circ L)_-)$$
$$= -(L \circ A)_- \circ L + L \circ (A \circ L)_- ,$$

where $X_- = X - X_+$ stands for the purely integral part of a symbol X.)

Theorem 4.43 ([198, 101]) *The Poisson structure of the Poisson Lie group* $\widetilde{G}_{\mathrm{INT}}$ *coincides with the quadratic generalized Gelfand–Dickey structure.*

We start with the following lemma:

Lemma 4.44 *The function* $\deg : \widetilde{G}_{\mathrm{INT}} \to \mathbb{R}$, *which assigns to a pseudodifferential symbol its degree, is a Casimir function on* $\widetilde{G}_{\mathrm{INT}}$, *i.e., it Poisson-commutes with all differentiable functions on* $\widetilde{G}_{\mathrm{INT}}$.

PROOF. First observe that the pullback of the differential $d\deg$ from any point L to the identity in $\widetilde{G}_{\mathrm{INT}}$ corresponds to the covector $A = (0, 1) \in \widehat{\mathrm{DO}} = \mathrm{DO} \oplus \mathbb{R}\cdot\varpi$, which is an element of the center of the Lie algebra $\widehat{\mathrm{DO}}$. Extend A to an element $\bar{A} = (\bar{0}, 1, 0)$ in the center of $\widetilde{\psi\,\mathrm{DS}} = \psi\,\mathrm{DS}\oplus\mathbb{R}\cdot\varpi\oplus\mathbb{R}\cdot\log\partial$. (Here, $\bar{0}$ denotes the extension of $0 \in \mathrm{DO}$ to the zero element in $\psi\,\mathrm{DS} = \mathrm{DO} \oplus \mathrm{INT}$.) Now one can see that the expression

$$\langle (\bar{0}, 1, 0)_+, (\bar{B}, b, \beta) \rangle - \langle (L^{-1} \circ (\bar{0}, 1, 0) \circ L)_+, L^{-1} \circ (\bar{B}, b, \beta) \circ L \rangle$$

from equation (4.45) vanishes for all $(\bar{B}, b, \beta) \in \widetilde{\psi\,\mathrm{DS}}$. Hence, the function $\deg$ is indeed a Casimir function on $\widetilde{G}_{\mathrm{INT}}$. $\qquad\square$

PROOF OF THEOREM 4.43. Lemma 4.44 implies that the hyperplanes $\deg(L) = \mathrm{const}$ in $\widetilde{G}_{\mathrm{INT}}$ are Poisson submanifolds, and we can restrict our attention to their tangent spaces. Let X and Y be cotangent vectors at L to the hyperplane $\alpha = \alpha_0$. Regard X and Y as elements of $\mathrm{DO} \circ \partial^{-\alpha_0}$. Let $\bar{X}$ and $\bar{Y}$ denote arbitrary extensions of X and Y to $\psi\,\mathrm{DS} \circ \partial^{-\alpha_0}$. By the definition of the Poisson bracket on $\widetilde{G}_{\mathrm{INT}}$, one can write

$$\begin{aligned}
\Pi_L(X, Y) &= \langle (r_L^* \bar{X})_+, r_L^* \bar{Y} \rangle - \langle (\mathrm{Ad}_L^*(r_L^* \bar{X}))_+, \mathrm{Ad}_L^* r_L^* \bar{Y} \rangle \\
&= \langle (L \circ \bar{X})_+, L \circ \bar{Y} \rangle - \langle (\bar{X} \circ L)_+, \bar{Y} \circ L \rangle \\
&= \langle (L \circ \bar{X})_+ \circ L - L \circ (\bar{X} \circ L)_+, \bar{Y} \rangle ,
\end{aligned}$$

which is the quadratic generalized Gelfand–Dickey bracket. (In the calculation above, we used that right multiplication on vectors becomes left multiplication on covectors.) $\qquad\square$

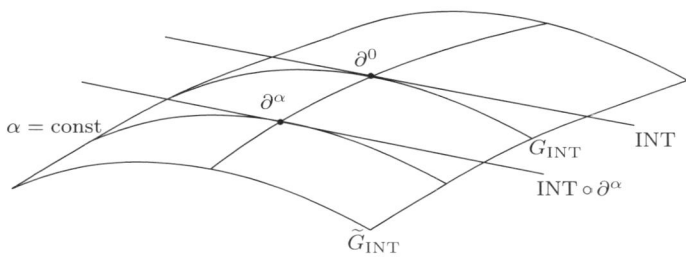

Fig. 4.3. The tangent space to the hyperplane $\alpha = \mathrm{const}$ in the group $\widetilde{G}_{\mathrm{INT}}$.

Definition 4.45 The *linear* (or *first*) *generalized Gelfand–Dickey structure* can be defined on the "integral" hyperplanes $\widetilde{G}_\alpha$ with $\alpha \in \mathbb{Z}$ in the group $\widetilde{G}_{\mathrm{INT}}$ by setting

$$V_A^1(L) := [L, A]_+ - [L, A_+].\tag{4.47}$$

Proposition 4.46 *The first and the second generalized Gelfand–Dickey structures are compatible.*

PROOF.   Indeed, from formula (4.46) for $V_A(L)$ we obtain

$$V_A(L + \lambda \partial^0) = V_A(L) + \lambda V_A^1(L),$$

so that $V_A(L) + \lambda V_A^1(L)$ is the second Gelfand–Dickey structure shifted by $\lambda \partial^0$. Note that the shift of L by $\lambda \partial^0$ makes sense only if L is of integral degree. $\qquad\square$

Remark 4.47 The classical definitions of both the first and the second Gelfand–Dickey (also called *Adler–Gelfand–Dickey*) structures are usually given in the case that α is a fixed positive integer n and L is a differential operator; cf. [3, 80].

One can see that for the second (quadratic) bracket, the set $\mathcal{L}_n := \{L \in \widetilde{G}_n \mid L_+ = L\}$ of purely differential operators is a *Poisson submanifold* in the Poisson "hyperplane" $\widetilde{G}_n$ of all monic pseudodifferential symbols of the same degree n. Indeed, for any differential operator $L = \partial^n + u_{n-1}\partial^{n-1} + \cdots + u_0$ and an *arbitrary* symbol A from the space $\mathrm{DO} \circ \partial^{-n}$, the corresponding Hamiltonian vector $V_A(L) = (LA)_+ L - L(AL)_+$ is always a differential operator of order $n-1$, and hence all Hamiltonian fields keep the submanifold $\mathcal{L}_n$ of such differential operators invariant.

The corresponding quadratic Poisson algebras of functions on the sets $\mathcal{L}_n$ (or, more generally, on the hyperplanes $\deg(L) = n$) are also called *classical W_n-algebras*.

Exercise 4.48 Show that in the Poisson Lie group $\widetilde{G}_{\mathrm{INT}}$ not only are the submanifolds $\mathcal{L}_n$ of purely differential operators Poisson, but so are the submanifolds $\mathcal{L}_n^{-1}$ of the inverses of differential operators, as well as the products $\mathcal{L}_n^{-1}\mathcal{L}_m$ for any n and m. (Hint: use the properties of a Poisson Lie group that taking products and inverses are Poisson maps.)

Remark 4.49 Another point to mention is that the linear Gelfand–Dickey bracket simplifies significantly when it is restricted to purely differential operators. Namely, linear functionals on the spaces $\mathcal{L}_n$ are given by pseudodifferential symbols A of degree $\deg(A) \leq -1$ via the pairing $\langle A, L \rangle = \mathrm{tr}(A \circ L)$. Since the degree of A is negative, we have $A_+ = 0$, whence the term $[L, A_+]$ does not appear in the expression for $V_A^1(L)$ above. Thus in this case, the first generalized Gelfand–Dickey structure reduces to the usual *first Gelfand–Dickey structure* defined as $\mathcal{V}_A^1(L) := [L, A]_+$. Finally, note that the latter Poisson structure is nothing else but the linear Lie–Poisson structure on differential operators, regarded as the dual space $\mathrm{DO} = \mathrm{INT}^*$ to the Lie algebra INT of integral symbols [234].

Remark 4.50 Both the first and the second Gelfand–Dickey Poisson structures on the spaces $\mathcal{L}_n$ can also be obtained by a Hamiltonian reduction from the dual of the affine $\widehat{L\mathfrak{gl}(n)}$-algebras in the classical Drinfeld–Sokolov reduction. The linear Lie–Poisson structure on $\widehat{L\mathfrak{gl}(n)}^*$ becomes the quadratic Gelfand–Dickey bracket on $\mathcal{L}_n$ after the reduction, while a constant Poisson structure on $\widehat{L\mathfrak{gl}(n)}^*$ becomes the linear Gelfand–Dickey bracket. We describe this construction in Appendix A.8.

It turns out that this reduction procedure can be extended to the whole hyperplanes $\deg(L) = \alpha$ and can also be defined for an arbitrary complex degree α; see [191]. The corresponding affine algebra before the reduction is constructed with the help of the algebra $\mathfrak{gl}(\alpha)$ for any complex α ("the algebra of matrices of complex size") introduced in [116].

4.7 Integrable Hierarchies on the Poisson Lie Group $\widetilde{G}_{\mathrm{INT}}$

Consider the following family of Hamiltonian functions $\{H_m\} : \widetilde{G}_{\mathrm{INT}} \to \mathbb{R}$ parametrized by an integer parameter $m \in \mathbb{N}$: the value of H_m at the pseudodifferential symbol $L \in \widetilde{G}_{\mathrm{INT}}$ of degree $\alpha \neq 0$ is

$$H_m(L) := \frac{\alpha}{m} \operatorname{tr}(L^{m/\alpha}).$$

Here any real power of L is a uniquely defined element in $\widetilde{G}_{\mathrm{INT}}$, since the group $\widetilde{G}_{\mathrm{INT}}$ is quasi-unipotent and the exponential map is one-to-one; see Theorem 4.30 and Section 4.5. One can think of the powers $L^{m/\alpha}$ as intersections of the one-parameter subgroup L^t passing through L with the "hyperplanes" of symbols of integral degree m. Since $\deg(L) = \alpha$, the degree $\deg(L^{m/\alpha}) = m$ is an integer, and hence both the corresponding trace tr and the function H_m are well defined.

Theorem 4.51 ([198, 101])   *1. The Hamiltonian equations corresponding to the functions H_m are well defined on any hyperplane $\deg(L) = \mathrm{const} \neq 0$ in the Poisson Lie group $\widetilde{G}_{\mathrm{INT}}$.*

2. Each function H_m defines the following Hamiltonian equation with respect to the (quadratic) Poisson structure on $\widetilde{G}_{\mathrm{INT}}$:

$$\frac{\partial L}{\partial t_m} = [(L^{m/\alpha})_+, L], \tag{4.48}$$

where $(L^{m/\alpha})_+$ denotes the purely differential part of the symbol $L^{m/\alpha}$.

3. All the functions H_m, $m = 1, 2, \ldots$, are first integrals for each of these Hamiltonian equations. Equivalently, these functions H_m are pairwise in involution with respect to the Poisson structure on the group $\widetilde{G}_{\mathrm{INT}}$.

PROOF.   (1) The flows of the Hamiltonians H_m do not change the degree of symbols, since $\deg(L)$ is a Casimir function on $\widetilde{G}_{\text{INT}}$.

(2) We prove part 2 for $\alpha = 1$ (and refer to [198, 101] and the exercise below for general α). In this case, $\deg(L) = 1$ and $H_m(L) = \frac{1}{m}\operatorname{tr}(L^m)$. To write out the Hamiltonian equations we need to find the *variational derivative* $\delta H_m / \delta L$ of the Hamiltonian function first. By definition,

$$\left\langle \frac{\delta H_m}{\delta L}(L), \delta L \right\rangle := \frac{d}{d\epsilon}\Big|_{\epsilon=0} H_m(L + \epsilon \cdot \delta L)$$

$$= \lim_{\epsilon \to 0} \frac{1}{\epsilon}\left(\frac{1}{m}\operatorname{tr}((L + \epsilon \cdot \delta L)^m) - \frac{1}{m}\operatorname{tr}(L^m) \right) = \operatorname{tr}(L^{m-1}\delta L).$$

Recall that tangent vectors δL to the set of symbols $\{L \mid \deg(L) = 1\}$ are symbols of degree 0. Then the cotangent vectors to the same set have the form $\text{DO} \circ \partial^{-1}$. In particular, we have $\delta H_m / \delta L = (L^{m-1})_{\geq -1}$, i.e., the variational derivative $\delta H_m / \delta L$ is the part of the symbol L^{m-1} in which we keep only the terms of degree ≥ -1. We can, however, use the whole symbol L^{m-1} as the variational derivative $\delta H_m / \delta L$, since other terms do not contribute to the pairing $\langle A, \delta L \rangle = \operatorname{tr}(A \circ \delta L)$ between the tangent and cotangent spaces.

Now substitute the value $\delta H_m / \delta L = L^{m-1}$ to the definition of the Gelfand–Dickey bracket to find the corresponding Hamiltonian field:

$$V_{\delta H_m / \delta L}(L) = (L \circ L^{m-1})_+ \circ L - L \circ (L^{m-1} \circ L)_+$$

$$= (L^m)_+ \circ L - L \circ (L^m)_+ = [(L^m)_+, L].$$

This implies the form of the Hamiltonian equations given in the theorem. The case of general α is handled similarly by showing that $\delta H_m / \delta L = L^{(m/\alpha)-1}$, but requires a bit more work (see below).

(3) The last part now follows directly, since the flow for H_m changes a symbol into a conjugate one. Thus, the ad-invariant function H_n stays invariant on the flow lines of H_m, so that the commutator $\{H_m, H_n\}$ vanishes for all m, n.   □

Exercise 4.52 Verify the last statement of the theorem above directly by obtaining zero as a result of the differentiation:

$$\{H_m, H_n\} = \frac{\partial H_n}{\partial t_m} = \frac{\alpha}{n} \frac{\partial \operatorname{tr}(L^{n/\alpha})}{\partial t_m}.$$

Exercise 4.53 Prove that for general $\alpha \neq 0$ one has $\delta H_m / \delta L = L^{(m/\alpha)-1}$. (Hint: first prove this for integral values of α; then use that the coefficients of both sides depend rationally on α; see details in [198, 101].)

Exercise 4.54 Prove that the first flow at any hierarchy has the form

$$\frac{\partial L}{\partial t_1} = \frac{\partial L}{\partial \theta} \,,$$

which expresses the fact that the space variable θ can be taken as the first time variable.

Remark 4.55 Equations (4.48) form an infinite sequence of commuting flows on the coefficients of the symbols L. These flows are defined on the hyperplanes of symbols of fixed degree α and are universal in the sense that they interpolate between several well-known hierarchies of integrable Hamiltonian systems. Here we list a few interesting cases; cf. [372]:

1. On the hyperplane $\deg(L) = 1$ one obtains the so called *Kadomtsev–Petviashvili (KP) hierarchy*, which has two equivalent and commonly used written forms:

$$\frac{\partial L}{\partial t_m} = [(L^m)_+, L] = -[(L^m)_-, L] \,.$$

The compatibility (or zero curvature) equation between the second and the third flows of this hierarchy,

$$\frac{\partial (L^2)_+}{\partial t_3} - \frac{\partial (L^3)_+}{\partial t_2} = [(L^3)_+, (L^2)_+] \,, \qquad (4.49)$$

leads to the *Kadomtsev–Petviashvili equation* in the form

$$(4u_{t_3} - 12uu_\theta - u_{\theta\theta\theta})_\theta - 3u_{t_2t_2} = 0 \qquad (4.50)$$

on the function $u(\theta, t_2, t_3)$ of two space variables $\theta = t_1$ and t_2 and one time variable t_3 (see also Exercise 4.56 below). The latter is often regarded as a two-dimensional version of the KdV equation: the KdV solutions give rise to the KP solutions independent of the space variable t_2.

2. The restriction of the universal hierarchy to the Poisson submanifolds $\mathcal{L}_n := \{L \mid L_+ = L, \deg(L) = n\}$ of purely differential operators of degree n gives the *n-KdV hierarchy*

$$\frac{\partial L}{\partial t_m} = [(L^{m/n})_+, L] \,, \qquad m = 1, 2, \ldots, n-1, n+1, n+2, \ldots \,.$$

The classical KdV hierarchy corresponds to the case $n = 2$ and $L = \partial^2 + u(\theta)$. (Notice the familiar form of a Hill's operator.) The KdV equation appears for $m = 3$ as an evolution equation on the potential $u(\theta)$.
Another classical case is the *Boussinesq equation*, which corresponds to $n = 3$, $m = 2$ and arises as a pair of equations on two functions u, v in the variables θ, t:

$$u_{\theta\theta} - 2v_\theta + u_t = 0 \,,$$
$$2u_{\theta\theta\theta} - 3v_{\theta\theta} + 3v_t + 2uu_\theta = 0 \,.$$

Using cross differentiation, it is possible to eliminate v and obtain the equation

$$3u_{tt} + u_{\theta\theta\theta\theta} + 2(u^2)_{\theta\theta} = 0 . \qquad (4.51)$$

3. Consider the space $\mathcal{K}_{n,m} := \mathcal{L}_n^{-1}\mathcal{L}_m$ of rational pseudodifferential symbols on the circle, i.e., the space of symbols of the form $L_1^{-1}L_2$, where L_1 and L_2 are (mutually prime) monic differential operators of degree n and m respectively. As we discussed in the preceding section, such spaces are Poisson submanifolds, and one can restrict to them the Hamiltonian flows. The restriction of the universal hierarchy to these spaces generates the *rational reductions of the KP hierarchy* [217].

4. Consider also pseudodifferential symbols with complex coefficients and take the Hamiltonians $H_m(L) = \frac{i}{m}\operatorname{tr}(L^m)$. Then for $m = 2$ and $L = \partial + \psi^*\partial^{-1}\psi$, the corresponding Hamiltonian flow generates the *nonlinear Schrödinger equation (NLS)* in the form

$$i\frac{\partial\psi}{\partial t} = \psi_{\theta\theta} + 2|\psi|^2\psi .$$

This equation can also be viewed as a rational KP reduction to the submanifold $\mathcal{K}_{1,2}$; see details in [217].

Exercise 4.56 Show that the compatibility equation (4.49) for a symbol $L = \partial + u\partial^{-1} + v\partial^{-2} + \cdots$ yields the following system of equations for u and v:

$$u_{t_2} = u_{\theta\theta} + 2v_\theta ,$$
$$3v_{t_2} + 3u_{\theta t_2} + 6uu_\theta = 3v_{\theta\theta} + 2u_{t_3} + u_{\theta\theta\theta} .$$

Use cross differentiation to eliminate v from the equations above and obtain the Kadomtsev–Petviashvili equation (4.50) for u.

4.8 Bibliographical Notes

Various structures related to pseudodifferential symbols are described in detail in the book [80]. The Adler–Gelfand–Dickey brackets were defined in [3, 142]. The Poisson Lie group approach to these structures was presented in [101, 198], where we refer the interested reader for more details on the questions discussed in this section. The reader can find a brief introduction to Poisson Lie groups in [87, 242, 349], for example.

The Adler–Gelfand–Dickey structures on scalar differential operators can be obtained via the Drinfeld–Sokolov reduction from matrix operators; see [88, 115] and Appendix A.8. This reduction extends to the symbols of all complex degrees [192]. In the mathematical physics literature, the Gelfand–Dickey Poisson algebras are called classical W-algebras; see [112, 41, 81]. The Lie algebra and Poisson Lie group of pseudodifferential symbols have natural q-analogues, which were described in [313, 314]; see also [131].

The 2-cocycle on differential operators on the circle appeared in [179], while the logarithmic 2-cocycle on the Lie algebra of pseudodifferential symbols on the circle was introduced in [215]. For generalizations of the logarithmic cocycle to symbols on higher-dimensional manifolds see [324, 94]. For the description of cohomology of Lie algebras of differential operators and pseudodifferential symbols see [116, 118, 94]. Representations of the centrally extended Lie algebra of differential operators were studied in [180, 130]. The logarithmic cocycle also appears in the study of the multiplicative anomaly for determinants of elliptic pseudodifferential operators [211].

One can find in [372, 250, 129] more details on the corresponding equations of mathematical physics. For rational reductions of the KP hierarchy we refer to [217, 102, 186], while higher-dimensional extensions of this hierarchy are described in [310]. Quantum versions of the corresponding equations are discussed, for example, in [117].

5 Double Loop and Elliptic Lie Groups

While loop algebras have the one-dimensional central extension, current algebras on compact manifolds admit infinite-dimensional central extensions once the manifold dimension is bigger than 1. However, if the manifold M is a two-dimensional (Riemann) surface regarded as a *complex curve*, there is a way to choose a special and, in many respects, natural *finite-dimensional* central extension for the current algebra on M. While the nonextended algebra itself is the same for all choices of the complex structure, it is the 2-cocycle on this current algebra that relies on this choice.

In this section we consider the case of an elliptic curve $M = \Sigma$, where the theory is most complete. The corresponding Lie algebra is often called the double loop or elliptic Lie algebra. It turns out that many constructions for the affine algebras have their analogues for the elliptic case. In particular, the coadjoint orbits of the corresponding elliptic Lie group are classified by equivalence classes of holomorphic G-bundles over the elliptic curve.

It turns out that the Calogero–Moser dynamical systems provide a bridge, or, rather, a ladder that unites the three classes of the Lie algebras: there is a universal construction of a Hamiltonian reduction on the dual of a Lie algebra, which for the finite-dimensional simple Lie algebras, affine Lie algebras, and elliptic Lie algebras leads to the Calogero–Moser integrable systems with, respectively, rational, trigonometric, and elliptic potentials.

5.1 Central Extensions of Double Loop Groups and Their Lie Algebras

Let $\mathfrak{g}$ be a complex semisimple Lie algebra and G the corresponding simply connected Lie group. Throughout this section we fix an elliptic curve Σ, i.e., a 2-dimensional torus endowed with a complex structure. One can always represent such a curve as a quotient $\Sigma = \mathbb{C}/(\mathbb{Z} + \tau Z)$, where τ is a complex number with $\operatorname{Im} \tau > 0$. Fix a holomorphic 1-form dz on Σ, which is canonical up to a complex factor.

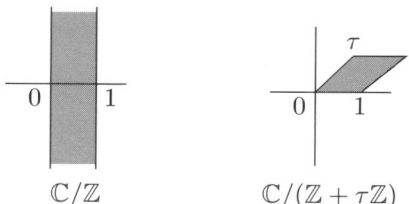

Fig. 5.1. A cylinder and an elliptic curve as quotients of $\mathbb{C}$.

The *double loop algebra* $\mathfrak{g}^{\Sigma}$ is the Lie algebra of smooth maps from the torus Σ to the Lie algebra $\mathfrak{g}$ with the pointwise bracket, $\mathfrak{g}^{\Sigma} = C^{\infty}(\Sigma, \mathfrak{g})$. Note that neither the elements nor the commutator of the double loop algebra depends on the complex structure on Σ. However, the complex structure enters the definition of the central extension of $\mathfrak{g}^\Sigma$.

Definition 5.1 The *elliptic Lie algebra* corresponding to the Lie algebra $\mathfrak{g}$ and the elliptic curve Σ is the complex one-dimensional central extension $\widehat{\mathfrak{g}}^\Sigma$ of the double loop algebra $\mathfrak{g}^\Sigma$ defined as the vector space $\widehat{\mathfrak{g}}^\Sigma = \mathfrak{g}^\Sigma \oplus \mathbb{C}$ endowed with the following commutator:

$$[(X(z,\bar{z}),a),(Y(z,\bar{z}),b)] = \left([X(z,\bar{z}),Y(z,\bar{z})],\int_\Sigma dz \wedge \langle X, dY \rangle\right),$$

where $\langle\,,\,\rangle$ denotes the Killing form on the Lie algebra $\mathfrak{g}$.

By decomposing the operator $d = \partial + \bar{\partial}$ into its Dolbault components ∂ and $\bar{\partial}$, one can rewrite the cocycle ω^Σ that defines the central extension as

$$\omega^\Sigma(X,Y) := \int_\Sigma dz \wedge \langle X, \bar{\partial} Y \rangle. \tag{5.52}$$

Note that the complex structure of the elliptic curve Σ appears only in the cocycle ω^Σ, and conformal equivalence of curves induces the isomorphism of the elliptic algebras.

The Lie group corresponding to the double loop algebra $\mathfrak{g}^\Sigma$ is given by the double loop group G^Σ, where G is the simply connected complex Lie group corresponding to the Lie algebra $\mathfrak{g}$. Similarly to the case of the loop group, the central extension of the double loop algebra gives rise to an extension of the corresponding group. However, the group extension is now given by means of a "complex analogue of the circle." Namely, the elliptic curve Σ itself can be regarded as such an analogue, while the identification $\Sigma = \mathbb{C}/(\mathbb{Z} + \tau\mathbb{Z})$ gives it the structure of an abelian group.

Theorem 5.2 *The central extension $\widehat{\mathfrak{g}}^\Sigma$ of the Lie algebra $\mathfrak{g}^\Sigma$ lifts to a central extension $\widehat{G}^\Sigma$ of the current group G^Σ, where the extension is by means of the elliptic curve Σ itself:*

$$e \to \Sigma \to \widehat{G}^\Sigma \to G^\Sigma \to e.$$

The central extension $\widehat{G}^\Sigma$, called the *elliptic Lie group*, can be constructed similarly to the central extension of the loop group LG (see [322, 106]). More explicit (quotient) constructions either use ingenious generalizations [241] of the construction of the affine group (see [263]; cf. Section 1.3) or are purely complex and involve the higher-dimensional residues (see [133]; cf. Chapter III for more details).

Remark 5.3 In a similar way one can define the current algebra for any compact Riemann surface (or complex curve) $\mathcal{C}$ of genus $\kappa > 0$. Let $H_\mathcal{C} =$

$H^0(\mathcal{C}, \Omega^1_{\mathcal{C}})$ be the vector space of holomorphic differentials on $\mathcal{C}$, which has the complex dimension $\dim(H_{\mathcal{C}}) = \kappa$. Define a complex κ-dimensional central extension of the Lie algebra $\mathfrak{g}^{\mathcal{C}}$ by the following 2-cocycle $\omega^{\mathcal{C}}$ with values in $H^*_{\mathcal{C}}$, dual to the space $H_{\mathcal{C}}$ of holomorphic differentials. Namely, set the value of this cocycle on any holomorphic 1-form α on $\mathcal{C}$ to be

$$\omega^{\mathcal{C}}(X, Y)(\alpha) := \int_{\mathcal{C}} \alpha \wedge \langle X, \bar{\partial} Y \rangle \,.$$

This central extension lifts to a central extension of the current group $G^{\mathcal{C}} = C^\infty(\mathcal{C}, G)$, this time by the Jacobian $J(\mathcal{C})$ of the complex curve $\mathcal{C}$:

$$e \to J(\mathcal{C}) \to \widehat{G}^{\mathcal{C}} \to G^{\mathcal{C}} \to e \,;$$

see [106]. If $\mathcal{C}$ is an elliptic curve Σ, its Jacobian is given by the curve Σ itself, and one recovers the central extension of the double loop group described above.

Remark 5.4 If M is of real dimension 2, but it is not equipped with a complex structure, or if M is higher-dimensional, then a priori, we do not have a preferred choice of a finite-dimensional cocycle. For any compact M the universal extension of the current algebra is by means of the space $\Omega^1(M)/d\Omega^0(M)$ of all 1-forms on M modulo exact 1-forms; see Remark 1.7 in Chapter II. For details, the interested reader is referred to [322, 287, 288], where one can find the study of this extension for a general manifold M, including the noncompact case and the case with boundary.

5.2 Coadjoint Orbits

In this section, we study the coadjoint representation of the elliptic Lie group $\widehat{G}^{\Sigma}$ for the elliptic curve Σ.

Definition 5.5 The (smooth part of the) dual space $(\widehat{\mathfrak{g}}^{\Sigma})^*$ of the Lie algebra $\widehat{\mathfrak{g}}^{\Sigma}$ is identified with the space $\mathfrak{g}^{\Sigma} \oplus \mathbb{C}$ via the pairing

$$\langle (A, a), (X, c) \rangle := \int_{\Sigma} \langle A, X \rangle dz \wedge d\bar{z} + a \cdot c \,,$$

where $\langle \, , \, \rangle$ is the Killing form on the finite-dimensional Lie algebra $\mathfrak{g}$, $(X, c) \in \widehat{\mathfrak{g}}^{\Sigma}$, and $(A, a) \in (\widehat{\mathfrak{g}}^{\Sigma})^* = \mathfrak{g}^{\Sigma} \oplus \mathbb{C}$.

Proposition 5.6 *In the coadjoint representation of the elliptic Lie group $\widehat{G}^{\Sigma}$ an element $g \in G^{\Sigma}$ acts on the dual space $(\widehat{\mathfrak{g}}^{\Sigma})^*$ as follows:*

$$g : (A, a) \mapsto (\mathrm{Ad}_g(A) + a(\bar{\partial} g)g^{-1}, a) \,.$$

In other words, elements of the dual space $(\widehat{\mathfrak{g}}^{\Sigma})^*$ can be thought of as $\bar{\partial}$-connections $\{-a\bar{\partial} + A(z, \bar{z})\}$, while the group coadjoint action is the gauge action on these connections, cf. Remark 1.13.

This proposition describes the coadjoint representation of the whole group $\widehat{G}^{\Sigma}$ completely, since the center acts trivially in the coadjoint representation.

PROOF. The proof repeats that for the corresponding statement for the affine Lie groups. $\qquad\square$

One of the main features of the centrally extended elliptic Lie algebras and Lie groups is a nice classification of their coadjoint orbits in terms of holomorphic principal G-bundles over the elliptic curve Σ. To study them we first note that the hyperplanes $a = $ const are invariant under the coadjoint action of the group G^{Σ} in the dual space $(\widehat{\mathfrak{g}}^{\Sigma})^*$. Fix a nonzero value of a and examine the orbits contained in the corresponding hyperplane $\mathcal{H}_a \subset (\widehat{\mathfrak{g}}^{\Sigma})^*$.

Theorem 5.7 ([106]) *Coadjoint orbits of the group G^{Σ} in the hyperplane $\mathcal{H}_a$ are in one-to-one correspondence with equivalence classes of holomorphic G-bundles over Σ.*

PROOF. Given an element $(A, a) \in (\widehat{\mathfrak{g}}^{\Sigma})^*$, we consider the operator $D = a\bar{\partial} - A$ and associate to it a holomorphic principal G-bundle over Σ as follows. Consider the partial differential equation

$$a\bar{\partial}\psi - A\psi = 0 \,.$$

(Here one can think of A as a $\mathfrak{g}$-valued $(0, 1)$-form, the operator $D = a\bar{\partial} - A$ as a partial $(0, 1)$-connection, while ψ is a horizontal section for this connection; cf. Section III.1.1.) One can find an open covering $U_i, i \in I$, of the curve Σ such that there exist local solutions $\psi_i : U_i \to G$ of this equation. Now define transition functions $\phi_{ij} : U_i \cap U_j \to G$ by $\varphi_{ij} = \psi_i^{-1}\psi_j$. It is easy to see that these functions are holomorphic and satisfy $\varphi_{ij} = \varphi_{ji}^{-1}$ on $U_i \cap U_j$ and $\varphi_{ij}\varphi_{jk}\varphi_{ki} = \mathrm{id}$ on $U_i \cap U_j \cap U_k$. Thus, they can be taken to be the gluing functions of a holomorphic principal G-bundle $P(A)$ on the curve Σ.

Conversely, for any holomorphic principal G-bundle P on Σ there exists an operator $D = a\bar{\partial} - A$ such that $P = P(A)$. Indeed, any principal G-bundle on Σ is topologically trivial, since G is simply connected. When a global trivialization is chosen, the local holomorphic trivializations over open sets U_i will be expressed by smooth functions $\psi_i : U_i \to G$ such that the transition functions $\varphi_{ij} = \psi_i^{-1}\psi_j$ are holomorphic on $U_i \cap U_j$. Then we have

$$\bar{\partial}\psi_i \cdot \psi_i^{-1} = \psi_j \cdot \bar{\partial}(\psi_j^{-1}\psi_i) \cdot \psi_i^{-1} - \psi_j \cdot \bar{\partial}\psi_j^{-1} = \bar{\partial}\psi_j \cdot \psi_j^{-1}$$

on $U_i \cap U_j$. Therefore there exists a $\mathfrak{g}$-valued 1-form A on Σ such that $A = a\bar{\partial}\psi_i \cdot \psi_i^{-1}$ on U_i for all $i \in I$. Set $D = a\bar{\partial} - A$. Then the bundle P is defined as the bundle $P(A)$.

Finally, two holomorphic principal G-bundles $P(A_1)$ and $P(A_2)$ are isomorphic if and only if there exists some $g \in G^\Sigma$ such that $A_2 = gA_1g^{-1} + a(\bar\partial g)g^{-1}$, i.e., if the differential operators $D_1 = a\bar\partial - A_1$ and $D_2 = a\bar\partial - A_2$ are related by a gauge transformation: $g \circ D_1 = D_2$. $\square$

Exercise 5.8 Define an elliptic analogue of the Virasoro algebra, i.e., a central extension of the Lie algebra $\mathrm{Vect}_{(0,1)}(\Sigma)$ of $(0,1)$-vector fields on the elliptic curve Σ and describe its coadjoint orbits. (Hint: cf. [106].)

In a similar way one can define an analogue of the Virasoro algebra for a complex curve $\mathcal{C}$ of genus $\kappa > 1$, where it is going to be a central extension of the Lie algebra $\mathrm{Vect}_{(0,1)}(\mathcal{C})$ of complex dimension κ; see [106]. This extension turns out to be universal [380]. One can also define the elliptic analogues of the Gelfand–Dickey brackets on the space of differential operators of higher order, $\{\sum_{k=0}^{n} u_k(z, \bar z)\bar\partial^k \mid u_k \in C^\infty(\Sigma, \mathbb{C})\}$; see [107].

Remark 5.9 For an arbitrary complex curve (i.e., a compact Riemann surface) $\mathcal{C}$, the orbits of the group $\widehat{G}^\mathcal{C}$ in the module of $(0,1)$- (or $\bar\partial$-) connections have a nice geometric interpretation similar to the one described above [106]. Namely, these orbits can be classified in terms of equivalence classes of holomorphic principal G-bundles on the complex curve $\mathcal{C}$.

However, for higher-genus curves, the space of such connections is not a hyperplane in the corresponding dual space of the Lie algebra $\widehat{\mathfrak{g}}^\mathcal{C}$. (The identification does not work directly, since the holomorphic differentials defining the 2-cocycle have zeros on $\mathcal{C}$.) The classification of coadjoint orbits of the corresponding group $\widehat{G}^\mathcal{C}$ for $\mathcal{C}$ of higher genus is still unknown.

5.3 Holomorphic Loop Groups and Monodromy

Recall that coadjoint orbits of a centrally extended loop group $\widehat{LG}$ can be classified in terms of conjugacy classes of the corresponding finite-dimensional Lie group G. Namely, the conjugacy class corresponding to a coadjoint orbit can be interpreted as the monodromy of a (necessarily flat) connection on the circle. As we shall see in this section, the coadjoint orbits of the centrally extended double loop group $\widehat{G}^\Sigma$ admit a similar description.

Definition 5.10 Let $HoLG$ denote the group of holomorphic maps from the cylinder $\mathbb{C}/\mathbb{Z}$ to a simply connected complex Lie group G with pointwise multiplication:

$$HoLG = \{g : \mathbb{C}/\mathbb{Z} \to G \mid g \text{ holomorphic}\}.$$

The group $HoLG$ is called the *holomorphic loop group* of G.

Fix some $\tau \in \mathbb{C}$. We define the *"τ-twisted" conjugacy classes* of the group $HoLG$ as the orbits of the $HoLG$-action on itself given by

$$g(z) : h(z) \mapsto h^g(z) := g(z)h(z)g(z+\tau)^{-1}. \qquad (5.53)$$

(These orbits depend only on the coset $\tau + \mathbb{Z}$, and below, τ is regarded as an element of $\mathbb{C}$ or $\mathbb{C}/\mathbb{Z}$ whenever this does not cause ambiguity.)

Remark 5.11 The τ-twisted conjugacy classes of the group $HoLG$ can be seen as "restricted conjugacy classes" of the bigger group $\widetilde{HoLG} = HoLG \rtimes \mathbb{C}/\mathbb{Z}$. More precisely, the cylinder $\mathbb{C}/\mathbb{Z}$ acts on the group $HoLG$ through translations of the argument (i.e., by "rotating" the loops, $\tau : f(z) \mapsto f(z+\tau)$), and we form the semidirect product $\widetilde{HoLG} = HoLG \rtimes \mathbb{C}/\mathbb{Z}$ associated to this action. The group $\widetilde{HoLG}$ acts on itself by conjugation.

Exercise 5.12 For (f,t) and $(g,\tau) \in \widetilde{HoLG}$ we have $(f,t)(g,\tau)(f,t)^{-1} = (h,\tau)$, where

$$h(z) = f(z)g(z+t)f(z+\tau)^{-1}$$

for $z \in \mathbb{C}/\mathbb{Z}$.

One can see that this action leaves the subset $\widetilde{HoLG} \times \tau$ with fixed τ invariant. The *restricted conjugacy classes* of the group $\widetilde{HoLG}$ are the $HoLG$-orbits in the "hyperplane" $HoLG \times \tau$, where $HoLG$ is viewed as a subgroup of the larger group $\widetilde{HoLG}$.

For what follows we assume that Im $\tau > 0$. The next theorem can be thought of as a direct analogue of the corresponding theorem for ordinary loop groups.

Proposition 5.13 *Let $\Sigma = \mathbb{C}/(\mathbb{Z} + \tau\mathbb{Z})$ be the elliptic curve with an elliptic parameter τ. For any fixed $a \neq 0$ there is a one-to-one correspondence between coadjoint orbits of the group $\widehat{G}^\Sigma$ in the hyperplane $\mathcal{H}_a = \{(A,a) \mid A \in \mathfrak{g}^\Sigma\} \subset (\widehat{\mathfrak{g}}^\Sigma)^*$ and τ-twisted conjugacy classes in the holomorphic loop group $HoLG$.*

PROOF.   Let A be a smooth function on the torus Σ. Consider the differential equation

$$a\bar{\partial}\psi - A\psi = 0 \qquad (5.54)$$

with respect to a G-valued function ψ on the cylinder $\mathbb{C}/\mathbb{Z}$.

Let $\psi_0(z)$ be a solution of this equation. Then $\psi_0(z+\tau)$ is also a solution, since the equation is periodic in τ. Therefore, $\nu_0(z) = \psi_0(z)^{-1}\psi_0(z+\tau)$ is also a G-valued function on the cylinder that is holomorphic, since $\bar{\partial}\nu_0/\partial\bar{z} = 0$. If we choose another solution of equation (5.54), say, ψ_1, it will have the form $\psi_1(z) = \psi_0(z)\mu(z)$, where μ is a holomorphic G-valued function on the same cylinder. Therefore, the function $\nu_1(z) = \psi_1(z)^{-1}\psi_1(z+\tau)$ can be expressed as follows: $\nu_1(z) = \mu(z)^{-1}\nu_0(z)\mu(z + \tau)$. This implies that the τ-twisted conjugacy class of the element (ν_0, τ) is independent of the choice of the solution ψ_0 of equation (5.54). Thus we have associated a twisted conjugacy class of $HoLG$ to any equation of the form (5.54).

Now we note that different equations of this form (5.54) associated to differential operators from one coadjoint orbit give the same twisted conjugacy class in $HoLG$. Indeed, fix some element $g \in G^\Sigma$ and, as before, let ψ_0 denote a solution of (5.54). Then the G-valued function $\phi = g \cdot \psi_0$ gives a solution of the equation

$$a\bar{\partial}\phi - (gAg^{-1} + a(\bar{\partial}g)g^{-1})\phi = 0\,.$$

Since g is periodic in τ, we obtain

$$\phi(z)^{-1}\phi(z + \tau) = \psi_0(z)^{-1}g(z)^{-1}g(z + \tau)\psi_0(z + \tau) = \psi_0(z)^{-1}\psi_0(z + \tau)\,.$$

This shows that the restricted conjugacy class of $HoLG$ associated to the element $(A, a) \in \mathfrak{g}^\Sigma \oplus \mathbb{C}$ with $a \neq 0$ is the same as the restricted conjugacy class associated to the element $(gAg^{-1} + a(\bar{\partial}g)g^{-1}, a)$, i.e., one and the same class corresponds to a coadjoint orbit of the group $\widehat{G^\Sigma}$.

It remains to show that every conjugacy class in $HoLG \times \tau$ comes from a certain equation (5.54). Given an element $g \in HoLG$, we take any smooth map $\psi : \mathbb{C}/\mathbb{Z} \to G$ satisfying $g(z) = \psi(z)^{-1}\psi(z + \tau)$ and set $A := a(\bar{\partial}\psi)\psi^{-1}$. The fact that $\psi(z)^{-1}\psi(z + \tau)$ is holomorphic implies that A is periodic in τ, since

$$0 = a\bar{\partial}(\psi(z)^{-1}\psi(z + \tau)) = \psi(z)^{-1}(-A(z) + A(z + \tau))\psi(z + \tau)\,.$$

Therefore, A is a smooth map from the elliptic curve Σ to the Lie algebra $\mathfrak{g}$, while the conjugacy class corresponding to the coadjoint orbit through $(A, a) \in \mathfrak{g}^\Sigma \oplus \mathbb{C}$ gives the τ-twisted conjugacy class through the element $g \in HoLG$. Thus we have established a one-to-one correspondence between orbits of the action of G^Σ in $\mathcal{H}_a$ and conjugacy classes of $\widehat{HoLG}$ in $HoLG \times \tau$. $\square$

Note that the τ-twisted conjugacy class of (ν_0, τ) plays the role of the conjugacy class of monodromy M for an ordinary differential equation $\frac{d}{d\theta}\psi - A\psi = 0$ in the orbit classification for affine algebras, cf. proof of Theorem 1.15.

Remark 5.14 Above we have identified (i) the coadjoint orbits in the elliptic Lie algebras with either (ii) equivalence classes of holomorphic principal G-bundles over the elliptic curve $\Sigma_\tau = \mathbb{C}/(\mathbb{Z} \oplus \tau\mathbb{Z})$, or ($iii$) restricted conjugacy classes of the group $HoLG$ of holomorphic loops lying inside the "hyperplane" $HoLG_\tau$. Below we describe a direct one-to-one correspondence between the bundles and the holomorphic loops (due to E. Looijenga, cf. [240]).

Obtain the torus Σ_τ from the annulus $\{z \in \mathbb{C}/\mathbb{Z} \mid 0 \leq \operatorname{Im} z \leq \operatorname{Im} \tau\}$ by gluing together the boundary components according to the rule $z \leftrightarrow z + \tau$. In order to define a holomorphic bundle on the corresponding elliptic curve, it is enough to present a G-valued holomorphic transition function $\varphi(z)$ in a neighborhood of the seam $\operatorname{Im} z = 0$. Therefore, one can naturally associate a holomorphic G-bundle to every element $(g, \tau) \in HoLG_\tau$ by setting $\varphi(z) = g(z)$. Observe that equation (5.53) expresses exactly the fact that the equivalence class of this bundle does not depend on the choice of the element inside the conjugacy class, and that different conjugacy classes give rise to inequivalent bundles. It remains to make sure that every holomorphic G-bundle over Σ_τ comes from a certain conjugacy class in $HoLG_\tau$. To see this, let us pick a bundle B over Σ_τ, and pull it back to the cylinder $\mathbb{C}/\mathbb{Z}$. The obtained bundle $\tilde{B}$ is holomorphically trivial (as is every holomorphic principal bundle on the cylinder with a connected structure group). Let us pick a global holomorphic section $\chi(z)$ of $\tilde{B}$. Then $\chi(z + \tau)$ is another holomorphic section. Therefore, $g(z) = \chi(z)^{-1}\chi(z + \tau)$ is a holomorphic function. By construction, the bundle B is associated to the conjugacy class of (g, τ).

Consider the space $\mathcal{S}$ of smooth maps $\psi : \mathbb{C}/\mathbb{Z} \to G$ such that $\psi(z)^{-1}\psi(z + \tau)$ is holomorphic. Such maps can be thought of as solutions to all differential equations $a\,\bar{\partial}\psi - A\psi = 0$ for some fixed $a \neq 0$, as we considered above. Now we can put all the considerations from this section into the following unifying picture:

$$\{\text{Maps } \psi : \mathbb{C}/\mathbb{Z} \to G \text{ such that } \psi(z)^{-1}\psi(z + \tau) \text{ is holomorphic}\}$$

The holomorphic loop group $HoLG$ acts on this set $\mathcal{S}$ of solutions via right multiplication and on itself by (twisted) conjugation. On the other hand, the double loop group G^Σ acts on the set of solutions by left multiplication and on the hyperplane $\mathcal{H}_a = \{(A, a) \mid A \in \mathfrak{g}^\Sigma\}$ by the coadjoint action. The map π_1 denotes factoring out the action of the group $HoLG$, and π_2 denotes factoring out the action of the double loop group G^Σ.

Remark 5.15 As we discussed before, the orbits of affine groups $\widehat{LG}$ (in a given affine hyperplane of the dual space) are labeled by conjugacy classes of the corresponding monodromy operator (cf. Section 1.2). In particular, their codimensions are given by codimensions of the conjugacy classes of the monodromy in the Lie group G, and hence these codimensions are bounded above by $\dim G$.

On the other hand, the codimension of orbits in elliptic Lie algebras can be arbitrarily large, even for $G = \mathrm{SL}(2, \mathbb{C})$. For instance, the conjugacy class of the element

$$\left(\begin{bmatrix} ae^{2\pi imz} & 0 \\ 0 & a^{-1}e^{-2\pi imz} \end{bmatrix}, \tau \right)$$

in $HoLG_\tau$ has codimension $2m + 2$ if $m > 0$; see [107].

Remark 5.16 The algebraic version of the above correspondence between all twisted conjugacy classes and isomorphism classes of holomorphic G-bundles was presented in [34]. It turns out that for an algebraic group G there is a natural bijection between "integral" twisted conjugacy classes in the group of formal loops in G and isomorphism classes of semistable holomorphic principal G-bundles on Σ.

5.4 Digression: Definition of the Calogero–Moser Systems

In this section, we take a step aside and describe a family of integrable systems, the so-called Calogero–Moser systems. We show how these systems can be obtained by Hamiltonian reduction from cotangent bundles of certain Lie algebras. It turns out that three different types of integrable potentials for such systems, rational, trigonometric, and elliptic ones, exactly correspond to the three different types of Lie algebras we discussed above: simple finite-dimensional, affine, and elliptic ones, respectively. This beautiful construction (due to [185, 148]) ties together these three classes of Lie algebras.

Consider a system of n interacting particles on the line that are governed by the Hamiltonian

$$H(q_1, \ldots, q_n, p_1, \ldots, p_n) = \frac{1}{2} \sum_{i=1}^{n} p_i^2 + \sum_{i>j} V(q_i - q_j), \qquad (5.55)$$

where the q_i denote the positions of the particles and the p_i denote their momenta. The potential V depends only on the distance $q_i - q_j$ between the particles, $q_i \in \mathbb{R}$.

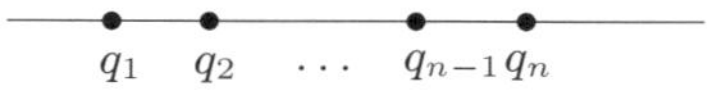

Fig. 5.2. n particles on the real line

Definition 5.17 The *Calogero–Moser systems* are Hamiltonian systems of type (5.55), where the potential is given by one of the functions

(a) $V(\xi) = 1/\xi^2$ (the rational case),

(b) $V(\xi) = \begin{cases} a^2/\sin^2(a\xi) \\ a^2/\sinh^2(a\xi) \end{cases}$ (the trigonometric/hyperbolic cases), and

(c) $V(\xi) = a^2 \wp(a\xi; \tau)$ (the elliptic case).

In the elliptic case, $\tau \in \mathbb{C}$ with $\mathrm{Im}\,\tau > 0$ is a fixed parameter that should be viewed as the modular parameter of the elliptic curve $\Sigma = \mathbb{C}/(\mathbb{Z} + \tau\mathbb{Z})$. The *Weierstrass $\wp$-function* $\wp(\,\cdot\,; \tau)$ with periods 1 and τ is defined by

$$\wp(\xi; \tau) = \frac{1}{\xi^2} + \sum_{\substack{(m_1, m_2) \in \mathbb{Z}^2 \\ (m_1, m_2) \neq (0,0)}} \left(\frac{1}{(\xi - m_1 - m_2\tau)^2} - \frac{1}{(m_1 + m_2\tau)^2} \right).$$

The Weierstrass $\wp$ function is a a meromorphic function on the elliptic curve $\Sigma = \mathbb{C}/(\mathbb{Z} + \tau\mathbb{Z})$ with a single pole of order 2 at $z = 0$.

Note that in the limit $a \to 0$, the trigonometric/hyperbolic Hamiltonian degenerates to the Hamiltonian of the rational Calogero–Moser system. Similarly, in the limit $\tau \to \mathbf{i}\infty$ with $\mathbf{i} = \sqrt{-1}$, the Hamiltonian of the elliptic Calogero–Moser system degenerates to the hyperbolic case, since the function $\wp(\xi; \tau)$ reduces to $\frac{\pi^2}{\sinh^2(\pi\xi)} - \frac{\pi^2}{3}$.

Remark 5.18 The Calogero–Moser systems can be defined for any simple root system [298]. In particular, the linear functions $(q_i - q_j)$ can be viewed as the roots of the root system A_{n-1}. How the latter can be replaced by another root system is briefly described in Appendix A.1.

Another direction for generalizations is defining the *Ruijsenaars–Schneider systems*, relativistic analogues of the Calogero–Moser systems. In the elliptic case the Ruijsenaars–Schneider Hamiltonian is

$$H(p, q) = \sum_{i=1}^{n} \cosh(\beta p_i) \prod_{j,\, j \neq i} \sqrt{1 - a^2 \wp(a(q_j - q_i); \tau)}.$$

Analogues of the rational and trigonometric Calogero–Moser systems can be obtained from this Hamiltonian by degenerations (see Appendix A.10.3).

The Calogero–Moser systems are known to be integrable and can be described within a group-theoretical framework. We start with the rational case.

Theorem 5.19 ([185]) *The rational Calogero–Moser system of n particles can be obtained by a Hamiltonian reduction of a system of free particles related to the algebra $\mathfrak{g} = \mathfrak{su}(n)$. In particular, it is completely integrable.*

PROOF.   For the simple Lie group $G = \mathrm{SU}(n)$ its Lie algebra $\mathfrak{g} = \mathfrak{su}(n)$ consists of traceless skew-Hermitian $n \times n$ matrices. Identify the dual space $\mathfrak{g}^*$ with $\mathfrak{g}$ itself using the nondegenerate bilinear form $\langle A, B \rangle = -\operatorname{tr}(AB)$. This identification gives an isomorphism $T^*\mathfrak{g} \cong \mathfrak{g} \oplus \mathfrak{g}$. The cotangent bundle of any manifold carries a natural symplectic structure. Using the identification of $\mathfrak{g}$ with $\mathfrak{g}^*$, the symplectic form ω on $T^*\mathfrak{g}$ is given by

$$\omega\left((A_1, B_1), (A_2, B_2)\right) = \langle A_1, B_2 \rangle - \langle A_2, B_1 \rangle \qquad (5.56)$$

for a pair of vectors $(A_1, B_1), (A_2, B_2)$ at any point $(P, Q) \in \mathfrak{g} \oplus \mathfrak{g} = T^*\mathfrak{g}$.

The group G acts on $\mathfrak{g}$ by conjugation. The induced action of G on the cotangent bundle $T^*\mathfrak{g} = \mathfrak{g} \oplus \mathfrak{g}$ is given by conjugation on each of the summands. In particular, the symplectic form ω is invariant under the G-action.

Exercise 5.20 Show that the map $\Phi : \mathfrak{g} \oplus \mathfrak{g} \to \mathfrak{g}$ given by

$$\Phi(Q, P) = [Q, P]$$

satisfies the moment map condition. That is, show that the map Φ is G-equivariant and $d\langle \Phi, X \rangle = \omega(\xi_X, .)$ for all $X \in \mathfrak{g}$, where ξ_X denotes the vector field on $T^*\mathfrak{g}$ generated by an element $X \in \mathfrak{g}$. The latter means that ξ_X at a point $(Q, P) \in \mathfrak{g} \oplus \mathfrak{g}$ is given by $([X, Q], [X, P])$.

To perform the Hamiltonian reduction for some element $\mu \in \mathfrak{g}^*$, we consider the manifold $\Phi^{-1}(\mu)/G_\mu$, where G_μ denotes the stabilizer subgroup of μ under the coadjoint representation. Let us fix $\mu := \mathbf{i}(-I + v \otimes v^*)$, where $v \in \mathbb{C}^n$ is the vector all of whose entries are equal to 1 in the standard basis, $*$ denotes the Hermitian conjugation, and I denotes the identity matrix (here and till the end of this section, we set $\mathbf{i} = \sqrt{-1}$ to distinguish the imaginary unit from the index i). Then μ is the matrix with 0's on the diagonal and $\mathbf{i}$'s in all other entries. The inverse image of the moment map $\Phi^{-1}(\mu)$ is the set of all $(Q, P) \in \mathfrak{g} \oplus \mathfrak{g}$ such that

$$[Q, P] = \mu.$$

Lemma 5.21 *Let (Q, P) be an element of $\Phi^{-1}(\mu)$. There is a simultaneous conjugation of Q and P by an element in G_μ such that Q becomes a diagonal matrix. Furthermore, for $Q = \operatorname{diag}(\mathbf{i}q_1, \ldots, \mathbf{i}q_n)$ one can assume the entries $q_j \in \mathbb{R}$ to be nonincreasing.*

PROOF.   Since every matrix in $\mathrm{SU}(n)$ is diagonalizable, one can choose some unitary matrix g such that $D = gQg^{-1}$ is diagonal. Set $E = gPg^{-1}$ and $w = gv$. Then $[D, E] = \mathbf{i}(I - w \otimes w^*)$ is a matrix with 0's on the diagonal (since D is diagonal). This implies that $w_j w_j^* = 1$, whence $w_j = e^{\mathbf{i}t_j}$ for some $t_j \in \mathbb{R}$. Hence the product of g and the diagonal matrix with entries $e^{-\mathbf{i}t_j}$ belongs to the stabilizer of μ. Finally, since multiplying v by a permutation

matrix does not change it (due to our choice of v), we can arrange the diagonal
entries iq_j of Q in such a way that the q_j are nonincreasing.   □

This lemma shows that the reduced space $\Phi^{-1}(\mu)/G_\mu$ is given by the
solutions of the equation

$$[Q, P] = \mu \tag{5.57}$$

for a fixed μ and unknown Q and P. Here $Q \in \mathfrak{su}(n)$ is a diagonal matrix,
$P \in \mathfrak{su}(n)$ is arbitrary, and μ is the matrix having zeros on the diagonal and
i's in all off-diagonal entries. Let ip_{jk} denote the entries of the matrix P, and
let Q be $\mathrm{diag}(iq_1, \ldots, iq_n)$. Then equation (5.57) translates to $p_{jk} = \frac{-i}{(q_j - q_k)}$
for $j \neq k$, while $p_{jj} \in \mathbb{R}$ are arbitrary.

Now consider a system of n particles freely moving over the vector space
of the Lie algebra $\mathfrak{g}$. The Hamiltonian of this system is given by its kinetic
energy:

$$H(Q, P) = -\frac{1}{2}\,\mathrm{tr}(P^2).$$

The function H is invariant under conjugation by elements of $SU(n)$. So it
descends to a function $\widetilde{H}$ on the quotient $\Phi^{-1}(\mu)/G_\mu$. The above consideration
shows that the reduced Hamiltonian is given by

$$\widetilde{H}(Q, P) = -\frac{1}{2}\,\mathrm{tr}(P^2) = \frac{1}{2}\sum_j p_{jj}^2 + \sum_{j>k} \frac{1}{(q_j - q_k)^2}.$$

This is the Hamiltonian of the rational Calogero–Moser system. First integrals
of this system are given by G-invariant functions $\mathrm{tr}(P^k)$ for $k = 2, \ldots, n$ on
the Lie algebra $\mathfrak{g} = \mathfrak{su}(n)$.   □

Remark 5.22 When constructing the rational Calogero–Moser system via
the Hamiltonian reduction, we had to solve the moment map equation $[Q, P] =$
μ, in which we diagonalized Q and solved for P. We could just as well have
diagonalized the matrix P first and then solved for Q. We would have obtained
the same Hamiltonian system with the p and q variables interchanged. This
shows that the rational Calogero–Moser system is *self-dual*.

However, in general, one can obtain a system in Q different from the one
in P. This change of variables is the passage to the *dual system*. For instance,
the trigonometric Calogero–Moser system turns out to be dual to the rational
Ruijsenaars–Schneider model, while the trigonometric Ruijsenaars–Schneider
model is also self-dual. For more details and facts about dualities in integrable
systems, see [123] and the references therein.

5.5 The Trigonometric Calogero–Moser System and Affine Lie Algebras

Now we turn to the trigonometric Calogero–Moser system.

Theorem 5.23 ([148, 185]) *The trigonometric Calogero–Moser system of n particles can be obtained by a Hamiltonian reduction of a system of free particles related to the affine algebra $\widehat{L\mathfrak{su}}(n)$. In particular, it is completely integrable.*

PROOF. Let G be the group $SU(n)$ with Lie algebra $\mathfrak{g} = \mathfrak{su}(n)$ and let LG denote the loop group of G, i.e., the group of smooth maps from $S^1 = \mathbb{R}/\mathbb{Z}$ to G. Furthermore, let us denote by $\mathfrak{a}^*$ the vector space of smooth maps from the circle to the dual $\mathfrak{g}^*$ of the Lie algebra $\mathfrak{g}$ enlarged by a "cocentral" direction:

$$\mathfrak{a}^* = C^\infty(S^1, \mathfrak{g}^*) \oplus \mathbb{R}.$$

The space $\mathfrak{a}^*$ should be viewed as the smooth part of the dual of the affine Lie algebra $\widehat{L\mathfrak{g}} = C^\infty(S^1, \mathfrak{g}) \oplus \mathbb{R}$ (hence the star * in the notation). As such, it carries a natural action of the loop group LG which comes from the coadjoint action of the centrally extended loop group $\widehat{LG}$. Namely, an element $g \in LG$ acts on the space $\mathfrak{a}^*$ via

$$g : (Q, a) \mapsto (gQg^{-1} + ag'g^{-1}, a),$$

where we have identified the Lie algebra $\mathfrak{g}$ with its dual $\mathfrak{g}^*$ via the Killing form.

Let $\mathfrak{a}^{**}$ denote the *distributional* (or *full*) *dual* of the space $\mathfrak{a}^*$. We present $\mathfrak{a}^{**}$ as a direct sum

$$\mathfrak{a}^{**} = \mathcal{F}(S^1, \mathfrak{g}) \oplus \mathbb{R},$$

where $\mathcal{F}(S^1, \mathfrak{g})$ denotes the space of $\mathfrak{g}$-valued distributions on S^1, and $\mathbb{R}$ denotes the "central" direction. Although the space $\mathfrak{a}^{**}$ contains the affine Lie algebra $\widehat{L\mathfrak{g}} = C^\infty(S^1, \mathfrak{g}) \oplus \mathbb{R}$ as a subspace, the Lie algebra structure of $\widehat{L\mathfrak{g}}$ does not extend to a Lie algebra structure on the vector space $\mathfrak{a}^{**}$. Nevertheless, the loop group LG acts naturally on the space $\mathfrak{a}^{**}$ via the dual to the coadjoint representation, and it contains the usual adjoint representation of the loop group LG on its Lie algebra as a subrepresentation. Explicitly, the action of LG on $\mathfrak{a}^{**}$ is given by

$$g : (P, c) \mapsto \left(gPg^{-1}, c - \int_0^1 \mathrm{tr}(Pg^{-1}g')\, d\theta\right).$$

Regard the direct sum $\mathfrak{a}^* \oplus \mathfrak{a}^{**}$ as the cotangent bundle of the vector space $\mathfrak{a}^*$. This allows one to endow $\mathfrak{a}^* \oplus \mathfrak{a}^{**}$ with an LG-action and a natural symplectic structure ω, which is invariant under the LG-action. (The symplectic structure can be written down explicitly by adapting equation (5.56) to the current situation.)

It turns out that the action of the loop group LG on the space $\mathfrak{a}^* \oplus \mathfrak{a}^{**}$ is Hamiltonian, i.e., it admits an LG-equivariant moment map $\Phi : \mathfrak{a}^* \oplus \mathfrak{a}^{**} \to (L\mathfrak{g})^*$. However, the moment map takes values not in the smooth part of the dual of the Lie algebra $L\mathfrak{g}$, but in the distributional dual, i.e., in the space $\mathcal{F}(S^1, \mathfrak{g}^*)$ of $\mathfrak{g}^*$-valued distributions on S^1. The analogue of Exercise 5.20 in this situation is the following

Proposition 5.24 *The action of the loop group LG on the space $\mathfrak{a}^* \oplus \mathfrak{a}^{**}$ is Hamiltonian with the moment map given by*

$$\Phi\left((Q, a), (P, c)\right) = [Q, P] - a\partial P.$$

Here, $[Q, P]$ denotes the pointwise commutator of Q and P, and the derivative $\partial P := dP/d\theta$ is understood in the distributional sense.

(Recall that if P has a finite jump at a point of S^1, its distributional derivative ∂P acquires a Dirac δ-type singularity at that point.)

Now perform the Hamiltonian reduction. Let us consider the element $\mu \otimes \delta_0 \in (L\mathfrak{g})^*$, where $\mu = \mathbf{i}(-I + v \otimes v^*)$ is the matrix with 0's on the diagonal and $\mathbf{i}$'s for all off-diagonal entries, and δ_0 is the Dirac δ-function centered at a point $\theta = 0 \in S^1$ (again, $\mathbf{i} = \sqrt{-1}$). Under the coadjoint action of the group LG on $(L\mathfrak{g})^*$, an element $g \in LG$ maps $\mu \otimes \delta_0$ to $g(0)\mu g(0)^{-1} \otimes \delta_0$. Note that the LG-orbit through $\mu \otimes \delta_0$ is *finite-dimensional*!

Lemma 5.25 *Let $((Q, a), (P, c))$ be an element in the inverse image $\Phi^{-1}(\mu \otimes \delta_0)$. Then, if $a \neq 0$, there is an element g in the stabilizer of $\mu \otimes \delta_0$ such that $gQg^{-1} + ag'g^{-1}$ is a diagonal matrix whose entries are constants $\mathbf{i}q_j$ with nonincreasing $q_j \in \mathbb{R}$.*

PROOF. The assertion that Q is conjugate to a diagonal matrix with nonincreasing diagonal entries follows from the classification of the coadjoint orbits of the smooth loop group LG in Section 1.2. The argument that g can be chosen in the stabilizer group $(LG)_{\mu \otimes \delta_0}$ is the same as in the proof of Lemma 5.21. $\square$

The moment map equation is given by

$$[Q, P] - a\partial P = \mu \otimes \delta_0,$$

where $Q \in \mathfrak{su}(n)$ is a diagonal matrix and $P \in (L\mathfrak{su}(n))^*$ is arbitrary. Let $\mathbf{i}q_j$ denote the (diagonal) entries of Q, and let $\mathbf{i}p_{jk}$ denote the entries of P. In these coordinates, the moment map equation reads as follows:

$$p_{jk}(\mathbf{i}q_j - \mathbf{i}q_k) - a\partial p_{jk} = \delta_0 \quad \text{for} \quad j \neq k, \qquad (5.58)$$

and

$$\partial p_{jj} = 0,$$

regarded as equations for distributions p_{jk} on $S^1 = \mathbb{R}/\mathbb{Z}$.

Lemma 5.26 *For a fixed $a \neq 0$, the solutions of equation (5.58) are given by*

$$p_{jk}(\theta) = \frac{e^{b_{jk}\theta}}{a(e^{b_{jk}} - 1)} \tag{5.59}$$

with $b_{jk} := \mathbf{i}(q_j - q_k)/a$ for $j \neq k$, and $p_{jj} \in \mathbb{R}$ are arbitrary constants.

PROOF.   Note that away from $\theta = 0$, equation (5.58) is an ordinary linear differential equation with constant coefficients. So the solution assumes the form (5.59). Regarded as a function on the circle, p_{jk} has a finite jump at $\theta = 0$; hence its derivative acquires a δ-type singularity. To prove the statement, we have to evaluate equation (5.58) on a test function $f : S^1 \to \mathbb{R}$. This gives

$$\langle (iq_j - iq_k)p_{jk} - a\partial p_{jk}, f \rangle = (iq_j - iq_k)\int_{S^1} p_{jk}(\theta)f(\theta)d\theta + a\int_{S^1} p_{jk}(\theta)f'(\theta)d\theta$$

$$= (iq_j - iq_k)\int_{S^1} p_{jk}(\theta)f(\theta)d\theta - a\int_{S^1} p'_{jk}(\theta)f(\theta)d\theta + af(0)(p_{jk}(1) - p_{jk}(0))$$

$$= f(0) =: \langle \delta_0, f \rangle,$$

where in the second step we used integration by parts and in the last step we used the fact that p_{jk} is of the form (5.59). In particular, we have $p_{jk}(1) - p_{jk}(0) = 1/a$.   □

Finally, the function $H : \mathfrak{a}^* \oplus \mathfrak{a}^{**} \to \mathbb{R}$ defined by

$$H((Q,a),(P,c)) = -\frac{1}{2}\int_{S^1} \mathrm{tr}(P^2)\,d\theta$$

is invariant under the action of the group LG on the space $\mathfrak{a}^* \oplus \mathfrak{a}^{**}$. Fixing a and restricting the function H to the solutions of the moment map equation, we get

$$H((Q,a),(P,c)) = \frac{1}{2}\sum_j p_{jj}^2 + \sum_{j>k} \frac{1}{4a^2 \sin^2(\frac{1}{2a}(q_j - q_k))}. \tag{5.60}$$

This is exactly the Hamiltonian function of the trigonometric Calogero–Moser system. First integrals of this system are given by (the reductions of) G-invariant functions on the Lie algebra $\mathfrak{g}$, integrated over the circle: $H_k(Q,P) = \int_{S^1} \mathrm{tr}(P^k)\,d\theta$, $k = 2, \ldots, n$.   □

5.6 The Elliptic Calogero–Moser System and Elliptic Lie Algebras

Finally, we show how the elliptic Calogero–Moser system naturally appears in the context of elliptic Lie algebras corresponding to $SL(n, \mathbb{C})$.

Theorem 5.27 ([148]) *The elliptic Calogero–Moser system of n particles with the modular parameter τ is obtained by a Hamiltonian reduction of a system of free particles related to the elliptic algebra $\widehat{\mathfrak{sl}(n, \mathbb{C})}^\Sigma$ on the elliptic curve Σ with the parameter τ. In particular, it is completely integrable.*

PROOF. Let $\Sigma = \mathbb{C}/(\mathbb{Z} \oplus \tau \mathbb{Z})$ be an elliptic curve and let $G = SL(n, \mathbb{C})$ denote the special linear group with Lie algebra $\mathfrak{g} = \mathfrak{sl}(n, \mathbb{C})$. We denote by $\mathfrak{b}^*$ the space of smooth maps from the curve Σ to the dual $\mathfrak{g}^*$ of the Lie algebra $\mathfrak{g}$ enlarged by one "cocentral" dimension:

$$\mathfrak{b}^* = C^\infty(\Sigma, \mathfrak{g}^*) \oplus \mathbb{C}.$$

The space $\mathfrak{b}^*$ can be thought of as the smooth part of the dual space of the elliptic Lie algebra $\widehat{\mathfrak{g}}^\Sigma$ (see Section 5.2). Hence, it carries the coadjoint action of the double loop group G^Σ, which is given by

$$g : (Q, a) \mapsto gQg^{-1} + a(\bar{\partial}g)g^{-1}$$

for an element $g \in G^\Sigma$. (As before, we identify the Lie algebra $\mathfrak{g}$ with its dual via the Killing form.)

Let $\mathfrak{b}^{**}$ denote the distributional dual of the space $\mathfrak{b}^*$, i.e., the space of $\mathfrak{g}$-valued distributions on the elliptic curve Σ enlarged by the "central" direction:

$$\mathfrak{b}^{**} = \mathcal{F}(\Sigma, \mathfrak{g}) \oplus \mathbb{C}.$$

The vector space $\mathfrak{b}^{**}$ carries the dual of the coadjoint representation of G^Σ, which contains the usual adjoint representation of the group G^Σ as a subspace.

As in the last section, the action of the group G^Σ on the cotangent bundle $T^*\mathfrak{b}^* = \mathfrak{b}^* \oplus \mathfrak{b}^{**}$ turns out to be Hamiltonian, with the moment map Φ taking values in the full dual of the current algebra $\mathfrak{g}^\Sigma$, i.e., in the space of $\mathfrak{g}^*$-valued distributions on the curve Σ:

Proposition 5.28 *The action of the current group G^Σ on the space $\mathfrak{b}^* \oplus \mathfrak{b}^{**}$ is Hamiltonian with the moment map given by*

$$\Phi((Q, a), (P, c)) = [Q, P] - a\bar{\partial}P.$$

Here, $[Q, P]$ denotes the pointwise commutator of Q and P, and the derivative $\bar{\partial}P := \bar{\partial}P/\partial\bar{z}$ is understood in the distributional sense.

Remark 5.29 Since we are considering the group $G = \mathrm{SL}(n, \mathbb{C})$, we have to deal with nondiagonalizable matrices: in contrast to the cases of finite-dimensional and affine Lie algebras based on the Lie algebra $\mathfrak{su}(n)$, it is not true anymore that in the coadjoint representation of the elliptic Lie algebra $\widehat{\mathfrak{g}}^{\Sigma}$, every element $(Q, a) \in \mathfrak{b}^*$ with $a \neq 0$ is conjugated to a constant diagonal matrix. However, recall that for fixed $a \neq 0$, the set of coadjoint orbits in the $(a = \mathrm{const})$-hyperplane in $\mathfrak{b}^*$ is in one-to-one correspondence with the set of equivalence classes of holomorphic $\mathrm{SL}(n, \mathbb{C})$-bundles over the elliptic curve Σ. We call a holomorphic $\mathrm{SL}(n, \mathbb{C})$-bundle *flat and unitary* if it admits a flat connection whose holonomy assumes values in the maximal compact subgroup $\mathrm{SU}(n) \subset \mathrm{SL}(n, \mathbb{C})$. It is a general fact that almost all holomorphic $\mathrm{SL}(n, \mathbb{C})$-bundles on the elliptic curve Σ are flat and unitary. To be more precise, if $\{B_t\}_{t \in \mathcal{T}}$ is a holomorphic family of holomorphic $\mathrm{SL}(n, \mathbb{C})$-bundles on Σ parametrized by a complex parameter space $\mathcal{T}$, then the subset $\mathcal{T}_0 \subset \mathcal{T}$ of those t for which B_t is flat and unitary is Zariski open in $\mathcal{T}$.

Lemma 5.30 *If the element* $(Q, a) \in \mathfrak{b}^*$ *corresponds to a flat and unitary bundle on the elliptic curve* Σ, *then there exists a gauge transformation* $g \in G^{\Sigma}$ *such that* $gQg^{-1} + a(\bar{\partial}g)g^{-1}$ *is a constant diagonal matrix.*

PROOF.   An $\mathrm{SL}(n, \mathbb{C})$-bundle ξ on the elliptic curve $\Sigma = \mathbb{C}/(\mathbb{Z} \oplus \tau\mathbb{Z})$ that is flat and unitary can be defined as a quotient $\xi = (\mathrm{SL}(n, \mathbb{C}) \times \mathbb{C})/(\mathbb{Z} \oplus \tau\mathbb{Z})$ as follows. Fix a flat connection with monodromies $t_1, t_2 \in \mathrm{SU}(n) \subset \mathrm{SL}(n, \mathbb{C})$. Such a connection exists by the assumption that the bundle ξ is flat and unitary. The group $\mathbb{Z} \oplus \tau\mathbb{Z}$ acts on $\mathrm{SL}(n, \mathbb{C}) \times \mathbb{C}$ via $(1, 0) : (h, z) \mapsto (t_1\, h, z+1)$, and $(0, \tau) : (h, z) \mapsto (t_2\, h, z + \tau)$. Since the group $\mathbb{Z} \oplus \tau\mathbb{Z}$ is abelian, the monodromies $t_1 =: \exp(H_1)$ and $t_2 =: \exp(H_2)$ have to commute. Hence they can be diagonalized simultaneously.

The bundle ξ is isomorphic to the bundle $\widetilde{\xi} = (\mathrm{SL}(n, \mathbb{C}) \times \mathbb{C}/\mathbb{Z})/\tau\mathbb{Z}$, where $\tau\mathbb{Z}$ acts via $\tau : (h, z) \mapsto (\exp(H_2 - \tau H_1)\, h, z + \tau)$. The isomorphism between the two bundles is delivered by the map $(h, z) \mapsto (\exp(-zH_1)\, h, z)$. Now, $\exp(H_2 - \tau H_1)$ can be viewed as an element of the holomorphic loop group $HoL(\mathrm{SL}(n, \mathbb{C}))$, and $\widetilde{\xi}$ is the holomorphic $\mathrm{SL}(n, \mathbb{C})$-bundle on the elliptic curve Σ that corresponds to the τ-twisted conjugacy class in $HoL(\mathrm{SL}(n, \mathbb{C}))$ containing the element $\exp(H_2 - \tau H_1)$ (see Remark 5.14). So the element $(Q, a) \in \mathfrak{b}^*$ is gauge-equivalent to $(H_2 - \tau H_1, a) \in \mathfrak{b}^*$, where $H_2 - \tau H_1$ is a constant diagonal matrix. $\square$

Since the set of all $(Q, a) \in \mathfrak{b}^*$ corresponding to flat and unitary bundles is open and invariant under the action of the group G^{Σ}, it makes sense to restrict the Poisson structure and the Hamiltonian reduction to this subset. To proceed with the Hamiltonian reduction, we fix the element $\alpha \otimes \delta_0 \in (\mathfrak{g}^{\Sigma})^*$, where α is the matrix with 0's on the diagonal and 1's everywhere else, and δ_0 is the Dirac δ-function centered at $0 \in \Sigma$.

Lemma 5.31 *Let $((Q, a), (P, c))$ be an element in the inverse image $\Phi^{-1}(\mu \otimes \delta_0)$. Then, if $a \neq 0$ and the element $(Q, a) \in \mathfrak{b}^*$ corresponds to a flat and unitary $\mathrm{SL}(n, \mathbb{C})$-bundle on the curve Σ, there exists an element g in the stabilizer of $\alpha \otimes \delta_0$ such that $gQg^{-1} + a(\bar{\partial}g)g^{-1}$ is a diagonal matrix whose entries are constants $q_j \in \mathbb{C}$.*

PROOF.   The proof of this lemma repeats that of Lemma 5.25, employing Remark 5.29 and Lemma 5.30.   □

Now the moment map equation is given by

$$[Q, P] - a\bar{\partial}P = \alpha \otimes \delta_0 \,,$$

where $Q \in \mathfrak{sl}(n, \mathbb{C})$ is a diagonal matrix and $P \in (\mathfrak{sl}(n, \mathbb{C})^{\Sigma})^*$ is arbitrary. Let q_j denote the diagonal entries of Q and let p_{jk} denote the entries of P. Then the moment map equation written in the entries of the matrices P and Q reads as follows:

$$p_{jk}(q_j - q_k) - a\bar{\partial}p_{jk} = \delta_0 \tag{5.61}$$

for any j and k with $j \neq k$, and where the equation is regarded in the sense of $\mathbb{C}$-valued distributions p_{jk} on the elliptic curve Σ. For p_{jj} the moment map equation becomes $a\bar{\partial}p_{jj} = 0$, which shows that p_{jj} are arbitrary complex constants. To solve equation (5.61), recall some facts about theta functions (see, e.g., [230]):

Definition / Proposition 5.32 *Fix some $\tau \in \mathbb{C}$ with $\mathrm{Im}\,\tau > 0$. The theta function*

$$\theta_{1,1}(z; \tau) = \sum_{n \in \mathbb{Z} + \frac{1}{2}} e^{2\pi i n(z + \frac{1}{2}) + \pi i \tau n^2}$$

is a holomorphic function in $z \in \mathbb{C}$ with a simple zero at $z = 0$. Furthermore, the function $\theta_{1,1}$ satisfies the identities

$$\theta_{1,1}(z + 1; \tau) = -\theta_{1,1}(z; \tau) \,,$$
$$\theta_{1,1}(z + \tau; \tau) = -e^{-2\pi i z - \pi i \tau}\theta_{1,1}(z; \tau) \,.$$

Using the theta function $\theta_{1,1}$, one can write down solutions for the moment map equation (5.61):

Lemma 5.33 *For $\tau = \tau_1 + i\tau_2$ we set $b_{jk} = \frac{\tau_2(q_k - q_j)}{\pi a}$. Then, for fixed $a \neq 0$, the solutions of the moment map equation (5.61) are given by*

$$p_{jk}(z) = e^{\pi b_{jk}(z - \bar{z})/\tau_2} \frac{\theta_{1,1}(z + b_{jk}; \tau)}{\theta_{1,1}(z; \tau)\theta_{1,1}(b_{jk}; \tau)} \tag{5.62}$$

for $j \neq k$, and $p_{jj} \in \mathbb{C}$ are arbitrary constants.

PROOF. Using the transformation properties of the theta function $\theta_{1,1}$, one checks that the functions p_{jk} are periodic with periods 1 and τ. Furthermore, the function

$$\psi_{jk}(z) = \frac{\theta_{1,1}(z + b_{jk}; \tau)}{\theta_{1,1}(z; \tau)\theta_{1,1}(b_{jk}; \tau)}$$

is meromorphic and has a first-order pole in $z = 0$. Therefore, its $\bar{\partial}$-derivative acquires a Dirac δ-type singularity at the point $z = 0$. Hence, evaluating equation (5.61) on a test function $f : \Sigma \to \mathbb{C}$ proves the assertion. $\square$

Finally, write down G^{Σ}-invariant Hamiltonians on the space $\mathfrak{b}^* \oplus \mathfrak{b}^{**}$. The quadratic one is given by

$$((Q, a), (P, c)) = \frac{1}{2} \int_{\Sigma} \operatorname{tr}(P^2) \, dz \wedge d\bar{z}. \tag{5.63}$$

Note that the functions $p_{jk}p_{kj} = \psi_{jk}(z)\psi_{kj}(z)$ are meromorphic and doubly periodic in z with periods 1 and τ. They have a second-order pole at $z = 0$ and zeros at $z = \pm b_{jk}$. This shows that one can express

$$p_{jk}p_{kj} = \lambda(\wp(b_{jk}) - \wp(z))$$

for some constant $\lambda \in \mathbb{C}$. So after rescaling, the Hamiltonian H descends to the Hamiltonian of the elliptic Calogero–Moser system on the symplectic quotient.

Similarly to the cases of rational and trigonometric Calogero–Moser systems, first integrals of this system can be obtained from G-invariant functions on the Lie algebra $\mathfrak{g}$ by integrating them over the elliptic curve:

$$H_k((Q, a), (P, c)) = \int_{\Sigma} \operatorname{tr}(P^k) \, dz \wedge d\bar{z}.$$

$\square$

In this way, three different types of integrable potentials of the Calogero–Moser systems turned out to be related to the "ladder" of the loop algebras: finite-dimensional (or "0-loop") algebras, loop algebras, and double loop algebras [148].

5.7 Bibliographical Notes

The elliptic Lie algebras were introduced in [106]. The equivalence of the bundles and the holomorphic loops discussed in Remark 5.14 is due to Looijenga; see [106]. The diagram in the same remark was suggested to us by P. Slodowy.

The correspondence between the set of equivalence classes of holomorphic G-bundles on the elliptic curve Σ and the set of τ-twisted conjugacy classes in the holomorphic loop group *HoLG* can be used to give a simple proof of Looijenga's theorem [240] that the (coarse) moduli space of semistable *G*-bundles on the elliptic curve Σ is a weighted projective space (see [135, 161, 271]). This construction also links the theory of loop groups to singularity theory [160].

It is also possible to define elliptic Lie algebras more abstractly, using the notion of elliptic root systems [334, 8, 315]. The coadjoint orbits of the corresponding Lie groups can be classified in terms of holomorphic bundles whose structure groups are nonsimply connected (and possibly even nonconnected), and in terms of twisted conjugacy classes of nonconnected loop groups [386].

The Krichever–Novikov-type algebras constitute another class of infinite-dimensional algebras related to holomorphic *G*-bundles over Riemann surfaces; see [219, 220, 353]. We discuss them and their relation to the affine and elliptic algebras in Appendix A.3.

The orbit classification for current Lie algebras on higher-dimensional manifolds (or even on surfaces, where the 2-cocycle does not rely on the choice of a complex structure) is a difficult problem. Note that having fixed a closed $(n-1)$-form defining the 2-cocycle of the current algebra, one can describe the orbits in terms of the corresponding one-dimensional foliation with a transverse measure on the manifold and a leafwise connection; see [63]. This brings in coadjoint invariants relying on the dynamical properties and geometry of this background foliation. In particular, the notion of asymptotic holonomy naturally appears in the context of coadjoint orbits for current groups on a two-dimensional torus.

The structure and representations of the current algebra on higher-dimensional manifolds with the infinite-dimensional extension are well studied in the case of toroidal Lie algebras (where M is an n-dimensional torus); see [40, 43, 276, 277].

There exists an extensive literature on the Calogero–Moser Hamiltonians, and here we merely mention several papers relevant to our discussion above. The rational Calogero–Moser systems were introduced by Calogero [66] and Moser [280], while the trigonometric potentials (the Sutherland model) were introduced in [361]. The hyperbolic and elliptic potentials were considered in [67, 68]. Olshanetsky and Perelomov [298] noted that one can replace the linear functions $(q_i - q_j)$ by the roots of an arbitrary root system and proved the integrability of the corresponding systems in many cases. The integrability for all root systems was proved in [50, 51]. The Ruijsenaars–Schneider systems were described in [333].

The construction of the rational Calogero–Moser system from Hamiltonian reduction from the cotangent bundle of the Lie algebra $\mathfrak{su}(n)$ appears in [185], which we follow in Section 5.4. In the sections on trigonometric and elliptic Calogero–Moser systems we follow the papers [148] and [290]. The latter paper also describes the integrable systems corresponding to the orbits with finitely many δ-functions on the curve. Dualities of various integrable systems are discussed in [123] and the references therein.

III

Applications of Groups: Topological and Holomorphic Gauge Theories

1 Holomorphic Bundles and Hitchin Systems

Here we recall some basic notions from the theory of holomorphic vector bundles. As an application we construct Hitchin systems, which are integrable systems related to vector bundles on Riemann surfaces.

1.1 Basics on Holomorphic Bundles

Let M be a complex manifold, let $E \to M$ be a complex vector bundle on M, and denote the set of C^∞ sections of the vector bundle E by $\Gamma(E)$.

Definition 1.1 A *holomorphic structure* on the complex vector bundle $E \to M$ is a covering of open sets $\{U_i\}_{i \in I}$ of the manifold M together with local trivializations $\varphi_i : E|_{U_i} \to U_i \times \mathbb{C}^n$ such that the transition functions

$$\varphi_i \circ \varphi_j^{-1} : (U_i \cap U_j) \times \mathbb{C}^n \to (U_i \cap U_j) \times \mathbb{C}^n \,,$$

which commute with the projections to the base, are holomorphic and $\mathbb{C}$-linear on fibers.

Remark 1.2 Since the transition functions $\varphi_i \circ \varphi_j^{-1}$ induce the identity on the first factor, they give rise to holomorphic maps $\varphi_{ij} : U_i \cap U_j \to \mathrm{GL}(n, \mathbb{C})$. It is easy to check that the maps φ_{ij} satisfy the cocycle condition:

$$\varphi_{ij}\varphi_{ji} = \mathrm{id} \quad \text{on } U_i \cap U_j \quad \text{and}$$
$$\varphi_{ij}\varphi_{jk}\varphi_{ki} = \mathrm{id} \quad \text{on } U_i \cap U_j \cap U_k \,.$$

On the other hand, given an open covering $\{U_i\}_{i \in I}$ of M together with a collection of holomorphic maps $\varphi_{ij} : U_i \cap U_j \to \mathrm{GL}(n, \mathbb{C})$ that satisfy the cocycle condition above, one can define a holomorphic vector bundle on M as follows. The total space E of the bundle is the union $\cup_{i \in I} U_i \times \mathbb{C}^n$ subject to the relation $(x, v) \sim (x, \varphi_{ij}(x)v)$ whenever $x \in U_i \cap U_j$. This is a well-defined equivalence relation, since the maps φ_{ij} satisfy the cocycle condition. The transition functions of this bundle are given by $(\mathrm{id}\,|_{U_i \cap U_j}, \varphi_{ij})$, which are holomorphic. Hence the bundle is indeed a holomorphic vector bundle on M.

Holomorphic bundles can also be defined by equipping the total space E with a structure of a complex manifold and requiring the trivializations $\varphi_i : E|_{U_i} \to U_i \times \mathbb{C}^n$ to be holomorphic maps of complex manifolds. Such trivializations are called holomorphic and in their terms the definition of *holomorphic sections* is immediate: A local section s of a holomorphic bundle E over $U \subset M$ is called holomorphic if the map $s : U \to E$ is a holomorphic map between two complex manifolds.

Remark 1.3 Recall that on a complex manifold M the complexified de Rham complex (Ω^*, d) splits into the double complex $(\Omega^{*,*}, \partial, \bar\partial)$, where $\partial : \Omega^{p,q} \to \Omega^{p+1\,q}$ and $\bar\partial : \Omega^{p,q} \to \Omega^{p,q+1}$ are the holomorphic and antiholomorphic differentials such that $\partial + \bar\partial = d$. In a local coordinate chart $(z_1, \ldots, z_n)$, the forms in $\Omega^{p,q}$ can be written as

$$\sum f_{i_1,\ldots,i_p,j_1,\ldots j_q}(z_1, \ldots, z_n, \bar z_1, \ldots, \bar z_n)\, dz_{i_1} \wedge \cdots \wedge dz_{i_p} \wedge d\bar z_{j_1} \wedge \cdots \wedge d\bar z_{j_q}.$$

Note that a function $f : U \to \mathbb{C}$ is *holomorphic* if $\bar\partial f = 0$.

For any complex vector bundle E on M one defines the set of E-valued (p, q)-forms as $\Omega^{p,q}(E) = \Gamma(E) \otimes \Omega^{p,q}$. Given a holomorphic structure on the bundle E, the operator $\bar\partial_E : \Omega^{p,q}(E) \to \Omega^{p,q+1}(E)$ is uniquely defined by the properties

1. $\bar\partial_E(fs) = (\bar\partial f)s + f(\bar\partial_E s)$ for all $f \in C^\infty(M)$ and $s \in \Gamma(E)$;
2. $\bar\partial_E(s)$ vanishes on an open $U \subset M$ if and only if the section s is holomorphic in U.

Indeed, in any local trivialization (U, φ_U) of E, the operator $\bar\partial_E$ can be taken to be the standard differential $\bar\partial$. To see that $\bar\partial_E$ is independent of the choice of local trivialization, consider a different trivialization (V, φ_V). On $U \cap V$ the trivializations are related by a holomorphic map $\varphi_{UV} : U \cap V \to \mathrm{GL}(n, \mathbb{C})$. The differential of a section s is given by $\bar\partial_E(\varphi_{UV} s) = (\bar\partial \varphi_{UV})s + \varphi_{UV} \bar\partial_E(s) = \varphi_{UV} \bar\partial_E(s)$, since φ_{UV} is holomorphic. This shows that the differential $\bar\partial_E$ is indeed well defined.

Thus a holomorphic structure on a complex vector bundle $E \to M$ gives rise to a differential operator $\bar\partial_E$ on the set of sections of E, which vanishes on the set of holomorphic sections of the bundle E. On the other hand, any connection d_A on a complex vector bundle E (i.e., any linear map $d_A : \Omega^0(E) \to \Omega^1(E)$ that satisfies the Leibniz rule $d_A(fs) = (df)s + f d_A(s)$ for all sections s and functions f) splits into *partial connections* $d_A = \partial_\beta + \bar\partial_\alpha$ with $\partial_\beta : \Omega^0(E) \to \Omega^{1,0}(E)$ and $\bar\partial_\alpha : \Omega^0(E) \to \Omega^{0,1}(E)$. Given a connection d_A on the bundle E, it is a natural question to ask whether $\bar\partial_\alpha$ is *integrable*, i.e., whether it comes from a holomorphic structure on E. To find a criterion for the integrability of a partial connection $\bar{\partial}_\alpha$, let us extend $\bar{\partial}_\alpha$ to a map $\Omega^{0,p}(E) \to \Omega^{0,p+1}(E)$. Then, in a local C^∞-trivialization (U, φ), we can write $\bar{\partial}_\alpha = \bar{\partial} + \alpha^\varphi$, where α^φ is a matrix of 1-forms on U. Set $\Phi_\alpha = \bar{\partial}_\alpha^2 \in \Omega^{0,2}(\text{End}\, E)$. In the local trivialization, we have $\Phi_\alpha^\varphi = \bar{\partial}\alpha^\varphi + \alpha^\varphi \wedge \alpha^\varphi$. Now we can state the integrability theorem for partial connections:

Theorem 1.4 *A partial connection $\bar{\partial}_\alpha$ on a complex vector bundle $E \to M$ is integrable if and only if $\bar{\partial}_\alpha^2 = 0$.*

Theorem 1.4 is essentially a bundle-valued version of the Nirenberg–Newlander theorem on the integrability of almost complex structures. The integrability condition $\bar{\partial}_\alpha^2 = 0$ ensures that for the equation $\bar{\partial}_\alpha s = 0$ there locally exists the maximal possible number of solutions linearly independent over the ring of holomorphic functions on M. For a proof of Theorem 1.4, see, e.g., [209].

Remark 1.5 Let d_A denote a connection in the vector bundle $E \to M$. The *curvature* F_A of the connection d_A is the $(\text{End}\, E)$-valued 2-form on M defined by $F_A = d_A^2 \in \Omega^2(\text{End}\, E)$. If E is a complex vector bundle, the curvature F_A of the connection d_A splits into $F_A = F_A^{2,0} + F_A^{1,1} + F_A^{0,2}$. Theorem 1.4 states that the $(0,1)$-part of the connection d_A on the complex vector bundle E defines a holomorphic structure on E if and only if the $(0,2)$-part of its curvature vanishes.

Definition 1.6 Fix a *Hermitian structure* on the vector bundle E over M, i.e. a Hermitian metric $(\ ,\ )$ on each fiber that varies smoothly along the manifold M. A connection d_A on the bundle E is called *unitary* if it is compatible with the Hermitian metric, i.e., if it satisfies

$$d(s,t) = (d_A s, t) + (s, d_A t)$$

for any two sections s, t of the bundle E.

Lemma 1.7 *Let $\bar{\partial}_\alpha$ be a partial connection on a complex Hermitian vector bundle E. Then there exists a unique unitary connection d_A on the bundle E such that the $(0,1)$-part of d_A is given by $\bar{\partial}_\alpha$.*

PROOF. Let A^φ be the matrix of 1-forms representing the connection d_A in a unitary trivialization (U, φ) of the bundle E. Then the unitarity condition on the connection d_A reads $A^\varphi = -(A^\varphi)^*$. This shows that given a partial connection $\bar{\partial}_\alpha$ with local connection matrix α^φ, one can define a matrix A^φ by $A^\varphi := \alpha^\varphi - (\alpha^\varphi)^*$. This defines the connection d_A globally, and its unitarity is clear from the construction. $\qquad\square$

Suppose that a complex Hermitian vector bundle E on the complex compact manifold M admits a holomorphic structure.

Theorem 1.8 *A unitary connection on a Hermitian complex vector bundle E over M is compatible with a holomorphic structure on E if and only if it has curvature of type $(1,1)$. In this case the connection is uniquely determined by the metric and the holomorphic structure.*

PROOF.   Indeed, define the holomorphic structure on E by an integrable partial $(0,1)$-connection. As we have seen above, this connection uniquely extends to a unitary connection d_A on the vector bundle E, which is now compatible with both Hermitian and holomorphic structures. Since the curvature F_A of a unitary connection is skew adjoint, we find that $F_A^{0,2} = (F_A^{2,0})^*$. On the other hand, the compatibility of the connection with the holomorphic structure forces $F_A^{0,2} = 0$, so that the curvature matrix F_A has to be a matrix of $(1,1)$-forms. $\square$

We now restrict our attention to bundles over Riemann surfaces. Recall that the *degree* $\deg(L)$ of a line bundle L over a compact Riemann surface Σ is defined to be the integral of its first Chern class: $\deg(L) = \int_\Sigma c_1(L) \in \mathbb{Z}$. For some facts on Chern classes and the degree, see, e.g., [150]. As an example, consider a line bundle L_P corresponding to a point $P \in \Sigma$. This bundle is glued from the trivial bundles over a local coordinate neighborhood U_0 of P and $U_1 = \Sigma \setminus \{P\}$. The bundle L_P is defined by means of the transition function $\varphi_{01} : U_0 \cap U_1 \to \mathbb{C}^* = GL(1, \mathbb{C})$ defined as $\varphi_{01}(z) = z$, where z in the left-hand side is a local coordinate around P. Then $\deg(L_P) = 1$.

The following proposition summarizes some important properties of the degree:

Proposition 1.9   *1. For any line bundles L and $\tilde{L}$ over Σ one has*

$$\deg(L \otimes \tilde{L}) = \deg(L) + \deg(\tilde{L}).$$

2. If $\deg(L) < 0$, then the bundle L has no nontrivial holomorphic sections.

For a vector bundle E its rank $\mathrm{rank}(E)$ is the dimension of its fibers, and the degree $\deg(E)$ of the bundle E is the degree of its *determinant bundle*, i.e., of the highest exterior power of the bundle E. Finally, let us recall the notion of a stable bundle.

Definition 1.10 A holomorphic vector bundle E on a Riemann surface is *stable* if for any proper subbundle $F \subset E$ one has

$$\frac{\deg(F)}{\mathrm{rank}(F)} < \frac{\deg(E)}{\mathrm{rank}(E)} .$$

The number $\mu(E) = \deg(E)/\mathrm{rank}(E)$ is called the *slope* of the vector bundle E. The bundle E is called *semistable* if the inequality above is not strict: $\mu(F) \leq \mu(E)$.

In the sequel, we are going to deal with the space of equivalence classes of holomorphic bundles. While in general, this is not a nice space, the situation becomes much better under the stability assumption on the bundles: the space of equivalence classes of stable bundles has the structure of a complex manifold. For instance, one of the consequences of the stability condition is the following:

Exercise 1.11 Let E be a stable vector bundle over a Riemann surface. Show that the only endomorphisms of E are scalars. (Hint: suppose that there is a nontrivial endomorphism, and apply the stability condition to the image and kernel subbundles; see details, e.g., in [239].)

Remark 1.12 The condition of stability of a holomorphic vector bundle E on a higher-dimensional Kähler manifold M^n is defined similarly. The degree of the bundle is now defined by $\deg(E) = \int_M \omega^{n-1} \wedge c_1(E)$, where ω is a Kähler class on M and $c_1(E)$ is the first Chern class of the bundle E. So the degree of the bundle E depends on the choice of the Kähler class ω on M. Furthermore, instead of considering just subbundles of the given vector bundle E, one considers coherent subsheaves of E.

1.2 Hitchin Systems

In this section we give a short introduction to the Hitchin systems. These are integrable systems on the cotangent bundle of the moduli space of stable vector bundle on a Riemann surface of genus $\varkappa > 1$ [163]. Similarly to the Calogero–Moser systems, which we have discussed in Section 5.4 of Chapter II, the Hitchin systems can be defined via symplectic reduction from the action of an infinite-dimensional Lie group on an affine space.

Throughout this section we fix a compact Riemann surface Σ of genus $\varkappa > 1$ and let $\mathcal{N}$ denote the space of equivalence classes of stable holomorphic vector bundles on Σ of a given topological type. That is, we fix a complex vector bundle E of rank n on the surface Σ and denote by $\mathcal{N}$ the set of equivalence classes of stable holomorphic structures on the bundle E. It is a classical theorem due to Narasimhan and Seshadri [286] that the set $\mathcal{N}$ is in fact a complex variety, the *moduli space of stable holomorphic structures on the bundle E*.

Definition / Proposition 1.13 *Let $\mathcal{A}^s$ denote the set of stable complex structures in the complex vector bundle E. This set can be identified with the set of partial connections $\bar\partial_\alpha$ in E. The partial connections on a complex curve are always integrable, and they give rise to stable holomorphic structures on the bundle E.*

The gauge transformation group $\mathcal{G}(E)$ *of the bundle E is the identity component of the group of smooth automorphisms of the bundle E. This group acts on the set $\mathcal{A}^s$ by gauge transformations, and the moduli space of stable*

holomorphic structures on the bundle E can be identified with the quotient
$\mathcal{N} = \mathcal{A}^s/\mathcal{G}(E)$.

Remark 1.14 For the bundle E of rank n over a curve of genus $\varkappa$ the moduli space $\mathcal{N}$ has complex dimension $n^2(\varkappa-1)+1$. This can be computed as follows. Let E_t be a family of holomorphic bundles depending on a parameter $t \in \mathbb{R}$ with $E_0 = E$ and let $\varphi_{ij}(t)$ denote the corresponding transition functions. Then the functions

$$\left(\varphi_{ij}^{-1}(t)\frac{\partial}{\partial t}\varphi_{ij}(t) \right)_{t=0}$$

on $U_i \cap U_j$ define a class in the sheaf cohomology group $H^1(\Sigma, \operatorname{End} E)$. Looking at all possible families E_t through a given holomorphic bundle E, one observes that the tangent space of $\mathcal{N}$ at E can be identified with the cohomology group $H^1(\Sigma, \operatorname{End} E)$. The dimension of $H^1(\Sigma, \operatorname{End} E)$ can be computed with the help of the Riemann–Roch theorem:

$$\dim H^0(\Sigma, \operatorname{End} E) - \dim H^1(\Sigma, \operatorname{End} E) = \deg(\operatorname{End} E) + \operatorname{rank}(\operatorname{End} E)(1 - \varkappa).$$

Note that $\deg(\operatorname{End} E) = \deg(E^* \otimes E) = 0$, since the nondegenerate bilinear form on $E^* \otimes E$ delivers a nonzero section of the determinant bundle $\det(E^* \otimes E)$. Moreover, the stability of E implies that $\dim H^0(\Sigma, \operatorname{End} E) = 1$, since this homology group consists of scalars only (see Exercise 1.11). So we obtain

$$\dim \mathcal{N} = n^2(\varkappa - 1) + 1.$$

Remark 1.15 The cotangent bundle $T^*\mathcal{N}$ admits a natural symplectic structure. We are going to describe $T^*\mathcal{N}$ as a symplectic quotient, from which it will be apparent how it becomes the phase space of an integrable system.

For this we consider the cotangent bundle $T^*\mathcal{A}^s$ to all stable complex structures on E. It can be identified with the set of pairs

$$T^*\mathcal{A}^s = \{(\bar{\partial}_\alpha, \phi)\},$$

where $\bar{\partial}_\alpha$ is a point in $\mathcal{A}^s$ and ϕ is an $(\operatorname{End} E)$-valued $(1,0)$-form on the surface Σ. Indeed, $\mathcal{A}^s$ is an open subset of the affine space of partial connections on the bundle E. Hence, its tangent space at any point is given by the space of $(\operatorname{End} E)$-valued $(0,1)$-forms on Σ. The pairing between an $(\operatorname{End} E)$-valued $(0,1)$-form α and an $(\operatorname{End} E)$-valued $(1,0)$-form ϕ is given by

$$\langle \alpha, \phi \rangle = \int_\Sigma \operatorname{tr}(\alpha \wedge \phi). \tag{1.1}$$

The form ϕ is called a *Higgs field,* and the pair $(\bar{\partial}_\alpha, \phi) \in T^*\mathcal{A}^s$ defines a *Higgs bundle.*

There exists a natural symplectic structure ω on the cotangent bundle $T^*\mathcal{A}^s$. The evaluation of the 2-form ω on two tangent vectors $(\delta_1\bar{\partial}_\alpha, \delta_1\phi)$ and $(\delta_2\bar{\partial}_\alpha, \delta_2\phi)$ at a point $(\bar{\partial}_\alpha, \phi) \in T^*\mathcal{A}^s$ is given by

$$\omega_{(\bar{\partial}_\alpha, \phi)}\left((\delta_1\bar{\partial}_\alpha, \delta_1\phi), (\delta_2\bar{\partial}_\alpha, \delta_2\phi)\right) = \int_\Sigma \mathrm{tr}(\delta_1\bar{\partial}_\alpha \wedge \delta_2\phi) - \int_\Sigma \mathrm{tr}(\delta_2\bar{\partial}_\alpha \wedge \delta_1\phi)\,.$$

The gauge transformation group $\mathcal{G}(E)$ of the complex vector bundle E acts on the space $T^*\mathcal{A}^s$ by conjugation: $g : \phi \mapsto g\phi g^{-1}$ and $\bar{\partial}_\alpha \mapsto g\bar{\partial}_\alpha g^{-1}$. The symplectic form ω is invariant under this action.

Exercise 1.16 The (smooth) dual of the Lie algebra of the gauge transformation group $\mathcal{G}(E)$ is naturally identified with the space of $(\mathrm{End}\,E)$-valued 2-forms on the Riemann surface Σ. Show that the $\mathcal{G}(E)$-action on the space $T^*\mathcal{A}^s$ is Hamiltonian with moment map

$$\Phi : T^*\mathcal{A}^s \to \Omega^2(\Sigma, \mathrm{End}\,E)$$

given by

$$\Phi(\bar{\partial}_\alpha, \phi) = [\bar{\partial}_\alpha, \phi] := \bar{\partial}\phi + [\alpha, \phi]\,.$$

Exercise 1.17 Prove that the symplectic quotient $\Phi^{-1}(0)/\mathcal{G}(E)$ is naturally identified with the cotangent bundle $T^*\mathcal{N}$. (Hint: show that if $[\bar{\partial}_\alpha, \phi] = 0$, then $\phi \in T^*_\alpha\mathcal{A}$ vanishes on the tangent space to the $\mathcal{G}(E)$-orbit through $\bar{\partial}_\alpha \in \mathcal{A}^s$, where the pairing is given by formula (1.1).)

Exercise 1.17 shows that the cotangent bundle $T^*\mathcal{N}$ can be described as a symplectic quotient. Our next goal is to define an integrable system on this quotient by finding a set of Poisson commuting Hamiltonians on $T^*\mathcal{N}$ which are functionally independent and whose number is equal to $\dim(\mathcal{N})$. Let K denote the canonical bundle on the Riemann surface Σ, and let $T = K^{-1}$ be the holomorphic tangent bundle of Σ. Furthermore, fix some $j \in \mathbb{N}$ and consider a basis $\nu_{j,k}$ of $H^1(\Sigma, K \otimes T^j)$. The elements $\nu_{j,k} \in H^1(\Sigma, K \otimes T^j)$ should be thought of as independent holomorphic $(1 - j, 1)$-differentials. The latter means that in some local coordinate z on the complex curve Σ, such differentials look like $f(z)dz^{(1-j)}d\bar{z}$ with a holomorphic function f.

Exercise 1.18 Use the Riemann–Roch theorem as in Remark 1.14 to show that the dimension of $H^1(\Sigma, K \otimes T^j)$ is equal to $(2j - 1)(\varkappa - 1)$ for $j > 1$ and to $\varkappa$ for $j = 1$. (Hint: the degree of the canonical bundle K is $(2\varkappa - 2)$.)

The $(1 - j, 1)$-differentials can be integrated against a $(j, 0)$-differential on the complex curve. For any $(\mathrm{End}\,E)$-valued $(1, 0)$-form ϕ on Σ, the trace $\mathrm{tr}(\phi^j)$ of ϕ^j is a $(j, 0)$-differential.

Hence, one can define functions $H_{j,k} : T^*\mathcal{A}^s \to \mathbb{C}$ by

$$H_{j,k}(\bar{\partial}_\alpha, \phi) = \int_\Sigma \nu_{j,k}\,\mathrm{tr}(\phi^j)\,,$$

where $j = 1, \ldots, n$, $k = 1, \ldots, \varkappa$ for $j = 1$ and $k = 1, \ldots, (2j - 1)(\varkappa - 1)$ for $j > 1$ (recall that $n = \mathrm{rank}(E)$). The functions $H_{j,k}$ depend only on ϕ, so they Poisson commute on $T^*\mathcal{A}^s$.

Exercise 1.19 Show that the functions $H_{j,k}$ are gauge invariant. Therefore they descend to the symplectic quotient and Poisson commute there.

The total number of the functions $H_{j,k}$ is

$$\varkappa + \sum_{j=2}^{n}(2j - 1)(\varkappa - 1) = n^2(\varkappa - 1) + 1 \,,$$

which is exactly half the (complex) dimension of $T^*\mathcal{N}$. Evidently, the functions $H_{j,k}$ are functionally independent. Hence they form an integrable system on $T^*\mathcal{N}$. This integrable system is called the *Hitchin system* associated to the Riemann surface Σ and the complex vector bundle E.

1.3 Bibliographical Notes

Holomorphic vector bundles on a Riemann surface were classified by Grothendieck [152] for genus $\varkappa = 0$, Atiyah [25] in the case $\varkappa = 1$, and Narasimhan and Seshadri [286] in the case $\varkappa > 1$. Their results were generalized to the case of principal bundles with any reductive algebraic structure group over $\mathbb{C}$ in [325]. For more details on holomorphic bundles and connections we refer to [84, 171, 209, 297]. More facts on the moduli space of stable vector bundles on Riemann surfaces can be found in [239, 135, 136], and for those on complex surfaces, see [170].

The Hitchin systems were introduced in [163] (see also [165]). There is a close connection between Hitchin systems and the Calogero–Moser systems that goes beyond the similarity in their construction. In [103, 289] it is shown that a spin generalization of the elliptic Calogero–Moser system, the elliptic version of the Gaudin model, and some other systems can be treated as degenerations of Hitchin systems. The geometry and quantization of such systems are widely discussed in the literature; see, e.g., [169, 104].

2 Poisson Structures on Moduli Spaces

In this section we give a comparative description of the Poisson structures on the moduli spaces of flat connections on *real* surfaces and holomorphic Poisson structures on the moduli spaces of holomorphic bundles on *complex* surfaces, following [194]. Their relation is similar to that of the loop and double loop groups described in Chapter II. It can be thought of as a "geometric complexification" (cf. [14]), that is, a natural extension of the correspondence between de Rham and Dolbeault complexes as follows:

$$d \leftrightarrow \bar{\partial}$$
$$\text{de Rham complex} \leftrightarrow \text{Dolbeault complex}$$
$$\text{locally constant functions or sections} \leftrightarrow \text{local holomorphic functions or sections}$$
$$\text{flat connections} \leftrightarrow \text{holomorphic bundles}$$

The key ingredient of this "complexification" is the Cauchy–Stokes formula, a complex analogue of the Stokes formula, which we describe below.

While the affine and elliptic Lie groups manifest this correspondence in dimension one (respectively, real and complex), in two dimensions it is provided by the moduli spaces of flat connections and those of (stable) holomorphic bundles on surfaces. A somewhat similar picture exists in any dimension (cf. the discussion of the topological and holomorphic Chern–Simons action functional in 3D in Section 3, or of the 4D Yang–Mills functional in [85]).

2.1 Moduli Spaces of Flat Connections on Riemann Surfaces

Here we describe Poisson structures on the spaces of flat connections on *real* surfaces with boundary, which constitute the "real side" in the above-mentioned correspondence.

In the real case, G stands for a simply connected simple *compact* Lie group. On its Lie algebra $\mathfrak{g}$ we fix a nondegenerate invariant (Killing) bilinear form, denoted by $\langle X, Y \rangle = \text{tr}(XY)$. Let Σ be a compact Riemann surface, possibly with boundary $\Gamma = \partial \Sigma$ consisting of several components, $\Gamma = \cup_1^k \Gamma_j$; see Figure 2.1.

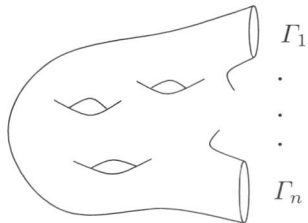

Fig. 2.1. A Riemann surface with boundary components $\Gamma_1, \ldots, \Gamma_n$.

Consider a principal G-bundle E over Σ. Note that this bundle is necessarily trivial, since the group G is simply connected. We denote by $\mathcal{A}^\Sigma$ the affine space of all smooth connections on the bundle E. It is convenient to fix a trivialization of E and identify $\mathcal{A}^\Sigma$ with the vector space $\Omega^1(\Sigma, \mathfrak{g})$ of smooth $\mathfrak{g}$-valued 1-forms on the surface Σ:

$$\mathcal{A}^\Sigma = \{d + A \mid A \in \Omega^1(\Sigma, \mathfrak{g})\}.$$

Thus the tangent space at any point of $\mathcal{A}^\Sigma$ is also identified with the vector space $\Omega^1(\Sigma, \mathfrak{g})$.

Definition 2.1 The space $\mathcal{A}^\Sigma$ can be equipped with a natural *symplectic structure*

$$\omega(\delta_1 A, \delta_2 A) = \int_\Sigma \operatorname{tr}(\delta_1 A \wedge \delta_2 A), \qquad (2.2)$$

where $\delta_1 A$ and $\delta_2 A$ denote tangent vectors of $\mathcal{A}^\Sigma$ at the point $d + A$, which are $\mathfrak{g}$-valued 1-forms on Σ. Here $\operatorname{tr}(\cdot \wedge \cdot)$ denotes the wedge product on Σ and symmetric pairing in $\mathfrak{g}$, thus producing a (real-valued) 2-form on Σ.

Since the principal G-bundle E on the surface Σ is topologically trivial, the gauge group of the bundle E is isomorphic to the current group $G^\Sigma = C^\infty(\Sigma, G)$, while its Lie algebra is given by the current algebra $\mathfrak{g}^\Sigma = C^\infty(\Sigma, \mathfrak{g})$. The action of the group G^Σ on the affine space $\mathcal{A}^\Sigma$ of connections is the usual action by gauge transformations:

$$g : d + A \mapsto d + gAg^{-1} - (dg)g^{-1},$$

where $g \in G^\Sigma$ is a smooth G-valued function on the surface.

Exercise 2.2 Prove that the symplectic structure ω is invariant with respect to gauge transformations.

The infinitesimal gauge transformations forming the Lie algebra $\mathfrak{g}^\Sigma$ are generated on the symplectic manifold $\mathcal{A}^\Sigma$ by certain Hamiltonian functions.

Proposition 2.3 (see, e.g., [125]) *An infinitesimal gauge transformation* $\epsilon \in \mathfrak{g}^\Sigma$ *is generated by the Hamiltonian function*

$$H_\epsilon(A) = \int_\Sigma \operatorname{tr}\left(\epsilon(dA + A \wedge A)\right) - \int_{\partial\Sigma} \operatorname{tr}(\epsilon A). \qquad (2.3)$$

PROOF. The Hamiltonian vector field X corresponding to any function H on $\mathcal{A}^\Sigma$ is defined by its action on functions $f(A)$ via

$$L_X f = \{H, f\} = \int_\Sigma \operatorname{tr}\left(\frac{\delta H}{\delta A} \wedge \frac{\delta f}{\delta A}\right),$$

where the latter expression is the Poisson bracket corresponding to the symplectic structure on $\mathcal{A}^\Sigma$ given by Formula (2.2). In order to determine the Hamiltonian function H_ϵ, it suffices to consider the coordinate function $f(A) = A$:

$$L_X A = \{H, A\} = \frac{\delta H}{\delta A}.$$

Then for the above Hamiltonian H_ϵ from equation (2.3), we obtain

$$L_{X_\epsilon} A = \frac{\delta H_\epsilon}{\delta A} = \nabla_A \epsilon,$$

where $\nabla_A \epsilon := d\epsilon - [\epsilon, A]$. That is, $L_{X_\epsilon} A$ is the infinitesimal gauge transformation of A by ϵ. Indeed, for $F(A) = dA + A \wedge A$ we have

$$\delta F(A) = \delta(dA + A \wedge A) = d\delta A + \delta A \wedge A + A \wedge \delta A = \delta dA + [A, \delta A],$$

by keeping only the terms linear in δA. This gives

$$\begin{aligned}
\delta H_\epsilon &= H_\epsilon(A + \delta A) - H_\epsilon(A) = \int_\Sigma \operatorname{tr}(\epsilon \delta F) - \int_{\partial \Sigma} \operatorname{tr}(\epsilon \delta A) \\
&= \int_\Sigma \operatorname{tr}\left(\epsilon(\delta dA + [A, \delta A])\right) - \int_{\partial \Sigma} \operatorname{tr}(\epsilon \delta A) \\
&= \int_\Sigma \operatorname{tr}\left(([\epsilon, A] - d\epsilon) \wedge \delta A\right) = \int_\Sigma \operatorname{tr}\left(\delta A \wedge \nabla_A \epsilon\right).
\end{aligned}$$

In the second-to-last equality we have used the Stokes formula. $\qquad \square$

Remark 2.4 In general, the Hamiltonian function H_ϵ corresponding to a vector field X_ϵ is defined only up to an additive constant. Thus, the Poisson bracket of two Hamiltonian functions reproduces the commutation relation in the gauge algebra $\mathfrak{g}^\Sigma$ only up to a 2-cocycle:

$$\{H_{\epsilon_1}, H_{\epsilon_2}\} = H_{[\epsilon_1, \epsilon_2]} + c(\epsilon_1, \epsilon_2).$$

Proposition 2.5 *For the Hamiltonian functions H_ϵ considered above, the cocycle c is given by*

$$c(\epsilon_1, \epsilon_2) = \int_{\partial \Sigma} \operatorname{tr}(\epsilon_1 d\epsilon_2).$$

PROOF. Let us calculate $\{H_{\epsilon_1}, H_{\epsilon_2}\}$. Using the expression $\delta H_\epsilon / \delta A = d\epsilon - [\epsilon, A]$, we get

$$\{H_{\epsilon_1}, H_{\epsilon_2}\} = \int_{\Sigma} \mathrm{tr}\left((d\epsilon_1 - [\epsilon_1, A]) \wedge (d\epsilon_2 - [\epsilon_2, A])\right)$$

$$= \int_{\Sigma} \mathrm{tr}(d\epsilon_1 \wedge d\epsilon_2) - \int_{\Sigma} \mathrm{tr}(d\epsilon_1 \wedge [\epsilon_2, A]) - \int_{\Sigma} \mathrm{tr}([\epsilon_1, A] \wedge d\epsilon_2)$$

$$+ \int_{\Sigma} \mathrm{tr}([\epsilon_1, A] \wedge [\epsilon_2, A])$$

$$= \int_{\partial\Sigma} \mathrm{tr}(\epsilon_1 d\epsilon_2) + \int_{\Sigma} \mathrm{tr}\left([\epsilon_1, \epsilon_2](dA + A \wedge A)\right) - \int_{\partial\Sigma} \mathrm{tr}([\epsilon_1, \epsilon_2]A)$$

$$= \int_{\partial\Sigma} \mathrm{tr}(\epsilon_1 d\epsilon_2) + H_{[\epsilon_1, \epsilon_2]}.$$

We have used the Stokes formula and the invariance of the Killing form several times.   □

Remark 2.6 The cocycle c is an obstruction to the existence of a moment map for the gauge group action on the space $\mathcal{A}^{\Sigma}$. Whenever the cocycle c is nontrivial, in order to define the corresponding moment map one has to consider the action of the centrally extended group $\widehat{G}^{\Sigma}$, which corresponds to the extension of the Lie algebra $\mathfrak{g}^{\Sigma}$ by this cocycle.

Exercise 2.7 Prove nontriviality of the cocycle c from Proposition 2.5. (Hint: restrict it to one boundary component and use the nontriviality of the cocycle defining the universal central extension of the loop algebra $L\mathfrak{g}$; see Section II.1.1.)

Corollary 2.8 *The gauge action of G^{Σ} on $\mathcal{A}^{\Sigma}$ extends to a Hamiltonian action of the centrally extended group $\widehat{G}^{\Sigma}$.*

Our next goal is to describe explicitly the corresponding moment map of the action of the extended current group $\widehat{G}^{\Sigma}$ on the space $\mathcal{A}^{\Sigma}$. The Lie algebra $\widehat{\mathfrak{g}}^{\Sigma}$ of this group is the Lie algebra of gauge transformations centrally extended by the cocycle c. The infinite-dimensional space $\widehat{\mathfrak{g}}^{\Sigma}$ is the space of pairs (ϵ, a), where ϵ is a $\mathfrak{g}$-valued function on the surface Σ and a is a real number:

$$\widehat{\mathfrak{g}}^{\Sigma} = \{(\epsilon, a) \mid \epsilon \in \mathfrak{g}^{\Sigma} = C^{\infty}(\Sigma, \mathfrak{g}), \ a \in \mathbb{R}\}.$$

Definition 2.9 The smooth part $(\widehat{\mathfrak{g}}^{\Sigma})^*$ of the *dual of the Lie algebra* $\widehat{\mathfrak{g}}^{\Sigma}$ is the space of triples

$$(\widehat{\mathfrak{g}}^{\Sigma})^* = \{(F, C, \lambda) \mid F \in \Omega^2(\Sigma, \mathfrak{g}), \ C \in \Omega^1(\partial\Sigma, \mathfrak{g}), \ \lambda \in \mathbb{R}\},$$

where F is a $\mathfrak{g}$-valued 2-form on Σ, C is a $\mathfrak{g}$-valued 1-form on the boundary of Σ, and λ is a real number.

The nondegenerate pairing between the spaces $\widehat{\mathfrak{g}}^{\Sigma}$ and $(\widehat{\mathfrak{g}}^{\Sigma})^*$ is given by

$$\langle (F, C, \lambda), (\epsilon, a) \rangle := \int_{\Sigma} \mathrm{tr}(\epsilon F) - \int_{\partial\Sigma} \mathrm{tr}(\epsilon C) + a\lambda.$$

Let us consider the action of the extended group $\widehat{G}^{\Sigma}$ on $\mathcal{A}^{\Sigma}$ generated by the gauge action of G^{Σ}. We note that the center of $\widehat{G}^{\Sigma}$ acts trivially. Now we can sharpen Corollary 2.8 as follows.

Proposition 2.10 ([125]) *The centrally extended group $\widehat{G}^{\Sigma}$ of gauge transformations acts on $\mathcal{A}^{\Sigma}$ in a Hamiltonian way. The moment map for the action of the corresponding gauge algebra $\widehat{\mathfrak{g}}^{\Sigma}$ is the mapping $\Phi : \mathcal{A}^{\Sigma} \to (\widehat{\mathfrak{g}}^{\Sigma})^*$, which is given by taking the curvature $F(A) = dA + A \wedge A$ of the connection and by restricting the connection form to the boundary:*

$$\Phi : A \mapsto (dA + A \wedge A, \quad A|_{\partial\Sigma}, \quad 1) \ .$$

PROOF.   The Hamiltonian functions H_ϵ corresponding to infinitesimal gauge transformations ϵ form the centrally extended algebra $\widehat{\mathfrak{g}}^{\Sigma}$; see Proposition 2.5. According to Definition I.5.4 of the moment map, it remains to check that the Hamiltonian H_ϵ is equal to the pairing of the element $\epsilon \in \mathfrak{g}^{\Sigma}$ with $\Phi(A)$:

$$H_\epsilon(A) = \langle \Phi(A), \epsilon \rangle \ .$$

(Here we take only the nonextended part of $\Phi(A)$, i.e., we omit its third component.) But this is exactly the explicit form of H_ϵ given in Proposition 2.3. $\square$

Definition 2.11 For any submanifold $\Gamma \subset \Sigma$ let G_{Γ}^{Σ} be the *group of gauge transformations on Σ "based on Γ"* :

$$G_{\Gamma}^{\Sigma} = \{ g \in C^{\infty}(\Sigma, G) \mid g|_{\Gamma} = \mathrm{id} \in G \} \ .$$

Denote the corresponding Lie algebra by $\mathfrak{g}_{\Gamma}^{\Sigma}$. In this section we always set Γ to be the boundary of the surface Σ.

A slight modification of Proposition 2.10 gives the following corollary.

Corollary 2.12 *For $\Gamma = \partial\Sigma$ the group G_{Γ}^{Σ} acts on $\mathcal{A}^{\Sigma}$ in a Hamiltonian way. The moment map Φ_{Γ} for the action of the corresponding Lie algebra $\mathfrak{g}_{\Gamma}^{\Sigma}$ is the map $\mathcal{A}^{\Sigma} \to (\mathfrak{g}_{\Gamma}^{\Sigma})^*$ given by the curvature:*

$$\Phi_{\Gamma} : \ A \mapsto dA + A \wedge A \ .$$

Note that the group G_{Γ}^{Σ} is not centrally extended, but we can still think of it as a (normal) subgroup $G_{\Gamma}^{\Sigma} \subset \widehat{G}^{\Sigma}$ (cf. the quotient construction of the centrally extended group $\widehat{LG}$ in Section II.1.3, where the surface Σ is the unit disk).

Now we consider the symplectic quotient $\mathcal{M}_{\Sigma,\Gamma} = \mathcal{A}^{\Sigma} /\!/ G_{\Gamma}^{\Sigma}$ of the space of connections $\mathcal{A}^{\Sigma}$ with respect to the group G_{Γ}^{Σ} of gauge transformations equal to the identity on the boundary $\Gamma = \partial\Sigma$. This yields the space

$$\mathcal{M}_{\Sigma,\Gamma} = \Phi_\Gamma^{-1}(0)/G_\Gamma^\Sigma = \{d + A \in \mathcal{A}^\Sigma \mid F(A) = 0\}/G_\Gamma^\Sigma$$

of flat connections on Σ modulo based gauge transformations from G_Γ^Σ.

Remark 2.13 Since $\mathcal{M}_{\Sigma,\Gamma}$ is defined via Hamiltonian reduction, it inherits a symplectic structure from the space $\mathcal{A}^\Sigma$. However, the quotient can have singularities. We shall be concerned only with the nonsingular part of the moduli space $\mathcal{M}_{\Sigma,\Gamma}$.

If Σ has no boundary, we obtain that the moduli space $\mathcal{M}_\Sigma$ of flat connections on Σ is a (finite-dimensional) symplectic manifold; see [28]. A point in this manifold can be described by holonomies around the handles of the surface modulo conjugation, i.e., by the equivalence class of a G-representation of the fundamental group of the surface:

$$\mathcal{M}_\Sigma = \mathrm{Rep}\,(\pi_1(\Sigma) \to G)/G\,.$$

If the boundary $\Gamma = \partial\Sigma$ of the Riemann surface Σ is nonempty, the moduli space $\mathcal{M}_{\Sigma,\Gamma}$ is infinite-dimensional. It can be mapped to certain familiar Poisson manifolds. Consider the restriction of a connection $A \in \mathcal{A}^\Sigma$ to any boundary component Γ_i of the surface Σ, which is a connection on the curve Γ_i. This restriction can be viewed as an element in the dual space $(\widehat{\mathfrak{g}}^{\Gamma_i})^*$ of the affine Lie algebra $\widehat{\mathfrak{g}}^{\Gamma_i}$ corresponding to the boundary component Γ_i. In turn, the space $(\widehat{\mathfrak{g}}^{\Gamma_i})^*$ is naturally equipped with the linear Lie–Poisson structure as the dual of a Lie algebra.

To summarize, we obtain the restriction map of two Poisson spaces

$$\mathcal{A}^\Sigma \to \prod_i (\widehat{\mathfrak{g}}^{\Gamma_i})^*\,, \qquad (2.4)$$

where the product ranges over the set of boundary components of Σ. Furthermore, this mapping descends to that of the moduli

$$\mathcal{M}_{\Sigma,\Gamma} \to \prod_i (\widehat{\mathfrak{g}}^{\Gamma_i})^*\,,$$

since all connections are flat on the boundary and the group G_Γ^Σ acts trivially on Γ. The relation of the corresponding Poisson structures on $\mathcal{M}_{\Sigma,\Gamma}$ (where the structure is, in fact, symplectic) and on $(\widehat{\mathfrak{g}}^{\Gamma_i})^*$ is given by the following proposition.

Proposition 2.14 *The mapping from the space $\mathcal{M}_{\Sigma,\Gamma}$ to the coadjoint representation space $(\widehat{\mathfrak{g}}^{\Gamma_i})^*$ sending a flat connection on the surface Σ to its restriction to a boundary component Γ_i is a Poisson mapping.*

PROOF.   This mapping is essentially the moment map for the action of gauge transformations on the boundary.   □

Consider the quotient of the space $\mathcal{M}_{\Sigma,\Gamma}$ by the whole group $\widehat{G}^{\Sigma}$ of centrally extended gauge transformations. The latter group acts on $\mathcal{M}_{\Sigma,\Gamma}$, since gauge transformations equal to the identity on the boundary form a normal subgroup G_{Γ}^{Σ} in the group $\widehat{G}^{\Sigma}$ of all gauge transformations. The quotient space

$$\mathcal{M}_{\Sigma} = \{d + A \in \mathcal{A}^{\Sigma} \mid dA + A \wedge A = 0\}/\widehat{G}^{\Sigma}$$

is a finite-dimensional Poisson manifold (with singularities). As before, we shall be interested only in the nonsingular part of it.

Exercise 2.15 Show that the *quotient of any symplectic* manifold by a symplectic action of a group is always a *Poisson* manifold. For a Hamiltonian action the symplectic leaves in this Poisson manifold are "labeled" by coadjoint orbits in the image of the corresponding moment map. (Hint: this is a direct generalization of the theorem on the Marsden–Weinstein reduction discussed in Chapter I.)

We can use the above restriction map $\mathcal{M}_{\Sigma,\Gamma} \to \prod_i (\widehat{\mathfrak{g}}^{\Gamma_i})^*$ to describe the symplectic leaves of the Poisson manifold $\mathcal{M}_{\Sigma}$.

Proposition 2.16 ([125]) *The space $\mathcal{M}_{\Sigma}$ of flat G-connections modulo gauge transformations on a surface Σ with holes inherits a Poisson structure from the space $\mathcal{A}^{\Sigma}$ of all (smooth) G-connections. The symplectic leaves of this structure are parametrized by the conjugacy classes of holonomies around the holes (that is, a symplectic leaf is singled out by fixing the conjugacy class of the holonomy around each boundary component).*

PROOF. The symplectic leaves of $\mathcal{M}_{\Sigma}$ are in one-to-one correspondence with the coadjoint orbits of the (centrally extended) affine Lie algebra $\widehat{\mathfrak{g}}^{\Gamma}$ on a circle (or the direct product of several copies of the affine algebras if the boundary Γ of the surface Σ consists of several components). As we have seen in Section II.1.2, these coadjoint orbits are parametrized by the conjugacy classes of holonomies around the circle. □

Remark 2.17 The above proposition should not be understood in the sense that the conjugacy classes of holonomies around the holes can be taken arbitrarily: the holonomies of a flat connection on the surface obey certain relations coming from the fundamental group $\pi_1(\Sigma)$. For example, if Σ is a sphere with n holes then the product of all n holonomies has to be id $\in G$ (provided one has chosen the same base point and a convenient orientation for all n loops encircling the holes).

Remark 2.18 A beautiful purely finite-dimensional description of the Poisson structure on the space $\mathcal{M}_\Sigma$ was obtained in [125]. Consider a triangulation of the surface Σ. Let the graph L be the 1-skeleton of the triangulation; see Figure 2.2. The space of flat connections on the surface Σ is replaced by the set $\mathcal{A}^{\Sigma}$ of *graph connections* on L. By definition, the set $\mathcal{A}^L = \prod_{\text{edges of } L} G$

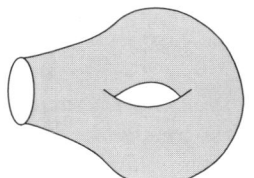 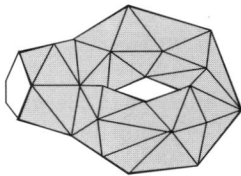

Fig. 2.2. A surface Σ is "replaced" by the 1-skeleton of its triangulation.

is taken to be the product of G over all edges of the graph L and should be viewed as the "holonomies along the (oriented) edges." The gauge group $G^L = \prod_{\text{vertices of } L} G$ acts by conjugation at the vertices. The moduli space $\mathcal{M}_\Sigma$ of flat connections is diffeomorphic to the quotient $\mathcal{M}_L = \mathcal{A}^L/G^L$.

It turns out that one can define Poisson structures on both $\mathcal{A}^L$ and G^L such that the *action* of the group G^L on the space $\mathcal{A}^L$ is *Poisson* (i.e., the map $G^L \times \mathcal{A}^L \to \mathcal{A}^L$ is a Poisson map with respect to the sum of the two Poisson structures on $G^L \times \mathcal{A}^L$). In particular, the quotient $\mathcal{M}_L = \mathcal{A}^L/G^L$ is a Poisson manifold, and one obtains an explicit description of the Poisson structure on $\mathcal{M}_\Sigma$; see [125]. (A bit more precisely, a flat graph connection satisfies the condition that the monodromies around all the faces of the triangulation are equal to id $\in G$. This restriction has to be taken into account when defining $\mathcal{A}^L$. Note that for a surface Σ with at least one hole, one can always choose a graph with all faces empty. Observe also that for a graph coming from a triangulation, the orientation of Σ induces a cyclic order of the ends of edges incident to each vertex.) Although the Poisson structure on $\mathcal{A}^L$ and G^L depends on some additional choices, the Poisson structure on moduli $\mathcal{M}_L$, after the reduction, does not.

2.2 Poincaré Residue and the Cauchy–Stokes Formula

In the next section we are going to develop the symplectic geometry related to holomorphic bundles on complex surfaces in a way analogous the symplectic geometry of flat connections on real surfaces. For this we need a complex analogue of the Stokes formula, which turns out to be a simple multidimensional generalization of the Cauchy residue formula.

Let X be a compact complex n-dimensional manifold and γ a meromorphic n-form on X with poles on a smooth complex hypersurface $Y \subset X$. Here and below we consider forms with logarithmic singularities only, i.e., forms with first-order poles. Let $z : U \to \mathbb{C}$ be a function defining Y in a neighborhood $U \subset X$ of some point $p \in Y$. Then locally in U the n-form γ can be decomposed into the sum

$$\gamma = \frac{dz}{z} \wedge \alpha + \beta \, ,$$

where α and β are, respectively, holomorphic $(n-1)$-form and n-form in U.

Exercise 2.19 Show that the restriction $\alpha|_Y$ is a well-defined holomorphic $(n-1)$-form on Y, that is, $\alpha|_Y$ is independent of the choice of z. (Hint: use local coordinates.)

Definition 2.20 The holomorphic $(n-1)$-form $\alpha|_Y$ on Y is called the *Poincaré residue* of the meromorphic form γ and is denoted by $\mathrm{res}_Y \gamma$.

Note that for the case of top-degree meromorphic forms having singularities on smooth divisors, which we consider here, a logarithmic singularity is the same as a first-order pole. In the situation of a nonsmooth divisor of poles one should keep the formulation "γ with logarithmic singularities."

Theorem 2.21 (Cauchy–Stokes formula, [151, 133]) *Let X, Y, and γ be as above, and let u be a smooth $(n-1)$-form on X. Then the form $\gamma \wedge du$ is L^1-integrable on X and*

$$\int_X \gamma \wedge du = 2\pi i \int_Y \mathrm{res}_Y \, \gamma \wedge u \, .$$

PROOF. This formula is proved by applying the Stokes formula to reduce the integral to the tubular neighborhood of Y, and then by using the standard Cauchy formula in the transversal direction to Y, whence the name.

Let z be a coordinate function locally defining Y. To see that $\gamma \wedge du$ is integrable, observe that $\gamma \wedge du$ grows like $1/|z|$ as $z \to 0$. Since the real codimension of Y in X equals 2, the $2n$-form $\gamma \wedge du$ is indeed integrable on X.

To calculate the integral, choose some tubular neighborhood U_ϵ of radius ϵ for the manifold Y (see Figure 2.3). Then by definition of the improper integral we have

$$\int_X \gamma \wedge du := \lim_{\epsilon \to 0} \int_{X \setminus U_\epsilon} \gamma \wedge du = -\lim_{\epsilon \to 0} \int_{\partial U_\epsilon} \gamma \wedge u = \lim_{\epsilon \to 0} \int_Y \int_{|z|=\epsilon} \gamma \wedge u \, ,$$

where we have used the Stokes theorem to reduce the integral over $X \setminus U_\epsilon$ to an integral over its boundary. Note that the boundary $\partial(X \setminus U_\epsilon)$ is oriented by an outer normal, which is an inner normal for the tubular neighborhood:

$\partial(X \setminus U_\epsilon) = -\partial U_\epsilon$. This orientation implies the positive (counterclockwise) orientation of the contour $|z| = \epsilon$.

Finally, by decomposing $\gamma = \frac{dz}{z} \wedge \alpha + \beta$ and using the Cauchy integral formula in the z-direction we obtain

$$\lim_{\epsilon \to 0} \int_Y \int_{|z|=\epsilon} \gamma \wedge u = 2\pi i \int_Y \alpha \wedge u = 2\pi i \int_Y \mathrm{res}_Y \gamma \wedge u,$$

while the term involving $\beta \wedge u$ vanishes as $\epsilon \to 0$. $\square$

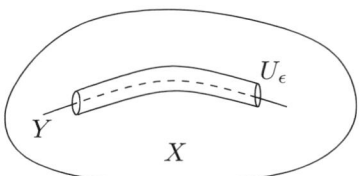

Fig. 2.3. A tubular neighborhood U_ϵ of the polar divisor $Y \subset X$.

Another way to rewrite the Cauchy–Stokes formula is to use $\bar{\partial}u$ instead of du:

$$\int_X \gamma \wedge \bar{\partial}u = 2\pi i \int_Y \mathrm{res}_Y \gamma \wedge u,$$

since it is only the $(0, n-1)$-part of the form u that is essential here.

Remark 2.22 Theorem 2.21 can be extended to the case in which the meromorphic n-form γ has first-order poles on a normal crossing divisor $Y = \cup_i Y_i \subset X$. (A normal crossing divisor means that the smooth irreducible components Y_i of Y, each one appearing with multiplicity one, intersect generically.) Analogously to Definition 2.20, one can define the residue $\mathrm{res}_{Y_i} \gamma$ on each component of Y. The resulting $(n-1)$-forms $\mathrm{res}_{Y_i} \gamma$ are meromorphic on Y_i and have first-order poles at the intersections $Y_{i,j} = Y_i \cap Y_j$. One can consider the repeated Poincaré residue of γ at $Y_{i,j}$. Let $z_i = 0$ and $z_j = 0$ be local equations in X of the components Y_i and Y_j respectively. Then representing γ as $\gamma = \frac{dz_i}{z_i} \wedge \frac{dz_j}{z_j} \wedge \alpha + \beta$, one finds that

$$\mathrm{res}_{i,j} \gamma := \mathrm{res}_{Y_{i,j}}(\mathrm{res}_{Y_i} \gamma) = \mathrm{res}_{z_j=0}\left(\mathrm{res}_{z_i=0} \frac{dz_i}{z_i} \wedge \frac{dz_j}{z_j} \wedge \alpha\right) = \alpha|_{Y_{i,j}}.$$

Note that the sign of the repeated residue depends on the order in which the latter was taken:

$$\mathrm{res}_{i,j} \gamma = -\mathrm{res}_{j,i} \gamma.$$

We shall denote by $\mathrm{res}\,\gamma$ the collection of $(n-1)$-forms $\mathrm{res}_{Y_i} \gamma$, the residues of γ at the components Y_i of the normal crossing divisor $Y = \mathrm{div}_\infty \gamma$.

Remark 2.23 We also note that for the Cauchy–Stokes formula above, the *polar divisor* plays the role of the *boundary* in the usual Stokes formula. In the same spirit, the meromorphic *n-form* can be seen as a "complex analogue" of an *orientation* of a real manifold. We will encounter similar analogies between notions from differential topology and complex analysis in subsequent sections.

2.3 Moduli Spaces of Holomorphic Bundles

Let S be a compact *complex* surface ($\dim_{\mathbb{C}} S = 2$) and let G be a simply connected *complex* reductive group ($G \subset \mathrm{GL}(n, \mathbb{C})$). In this section, we describe a Poisson structure on the moduli space of (stable) holomorphic vector bundles on S with the structure group G.[13] This will be done by analogy with our considerations of the moduli space of flat connections on a Riemann surface in Section 2.1. The basic idea for this analogy is to replace the "real" notions of orientation and boundary of a Riemann surface by their "complex analogues": a meromorphic 2-form on the surface (thought of as a "complex orientation" of S) and its divisor of poles (playing the role of the "boundary" of S); see Remark 2.23.

On the complex surface S we fix some meromorphic 2-form β that has only poles of the first order and whose polar divisor $P \subset S$ is a smooth curve in S. Let us assume additionally that β has *no zeros*. (The latter corresponds to the *smoothness* of a real surface: we will see below that *zeros* of the 2-form β on a complex manifold would correspond to *singularities* of a real manifold). If it happens that β has neither zeros nor poles (i.e., S is "smooth, oriented, without boundary") this means that we are dealing with either a K3 or an abelian surface S. If the 2-form β is meromorphic, then S is a *complex Poisson surface*, since the bivector field dual to β defines a holomorphic Poisson structure on S.

For the meromorphic 2-form β without zeros, its polar divisor P is an anticanonical divisor on S, and it has to be an elliptic curve or the union of several nonintersecting elliptic curves. (Indeed, in this case, $\mathrm{res}_P \beta$ defines a nonvanishing holomorphic 1-form on P.) These elliptic curves can be seen as the analogues of the circles constituting the boundary components of a smooth real surface. An example of such is $S = \mathbb{CP}^2$ with a smooth cubic P as an anticanonical divisor. (As a matter of fact, many Fano surfaces admit such a form β, i.e., fall into this class. The considerations below can also be extended with minimal changes to the case of a nonsmooth divisor P, in particular, to P consisting of several components intersecting transversally. Example: for $S = \mathbb{CP}^2$ with $\beta = dx \wedge dy/xy$, the divisor P is a union of three $\mathbb{CP}^1$'s.)

Exercise 2.24 Prove that if a meromorphic form β is nonvanishing, so is its residue $\mathrm{res}_P \beta$. (Hint: use local coordinates.)

---

[13] As before, when speaking of a moduli space we shall always mean its nonsingular part. Here under the moduli space we understand a local universal family near a smooth point.

Now let E be a smooth G-bundle over the surface S that can be endowed with a holomorphic structure, and let $\operatorname{End} E$ be the corresponding bundle of endomorphisms with fiber $\mathfrak{g} = \operatorname{Lie}(G)$. Furthermore, let $\mathcal{A}^S$ denote the infinite-dimensional affine space of smooth $(0,1)$-connections (or $\bar{\partial}$-connections) in E. Finally, let us choose a reference holomorphic structure, i.e., a $(0,1)$-connection $\bar{\partial}_0$ such that $\bar{\partial}_0^2 = 0$. Then the space $\mathcal{A}^S$ can be identified with the vector space $\Omega^{0,1}(S, \operatorname{End} E)$ of $(\operatorname{End} E)$-valued $(0,1)$-forms on S, i.e.

$$\mathcal{A}^S = \{\bar{\partial}_0 + A \mid A \in \Omega^{0,1}(S, \operatorname{End} E)\}\,.$$

In what follows, instead of $\bar{\partial}_0$, we shall write simply $\bar{\partial}$, keeping in mind that this corresponds to a reference holomorphic structure in E when it applies to sections of E or associated bundles.

Definition 2.25 The affine space $\mathcal{A}^S$ possesses a natural *holomorphic symplectic structure* given by

$$\omega_{\mathbb{C}}(\delta_1 A, \delta_2 A) = \int_S \beta \wedge \operatorname{tr}(\delta_1 A \wedge \delta_2 A)\,,$$

where β is the meromorphic 2-form of S (i.e., the "complex orientation" on S), $\delta_1 A$ and $\delta_2 A$ are two tangent vectors to the affine space $\mathcal{A}^S$ at the point $\bar{\partial} + A$, and $\operatorname{tr}(\cdot \wedge \cdot)$ stands for the wedge product on S and the nondegenerate pairing on $\mathfrak{g}$.

Recall that a *holomorphic symplectic structure* on a complex manifold is defined by a holomorphic closed 2-form on it that is nondegenerate over $\mathbb{C}$. (Note, for instance, that the 2-form $dz \wedge dw$ defines a holomorphic symplectic structure in $\mathbb{C}^2$, but it is not symplectic in $\mathbb{R}^4 \simeq \mathbb{C}^2$ as a 2-form over $\mathbb{R}$, since it is of rank 2.) In the infinite-dimensional context, the nondegeneracy is always understood as the absence of a tangent vector skew-orthogonal to all others; see Definition I.4.4.

Following this definition, we can appropriately "complexify" the constructions for moduli of flat connections from Section 2.1 step by step. Abusing notation, let us denote by G^S the group of gauge transformations in E, i.e, the group of automorphisms of the smooth bundle E.

Exercise 2.26 Prove that the symplectic structure $\omega_{\mathbb{C}}$ is invariant with respect to the gauge transformations

$$A \mapsto gAg^{-1} - \bar{\partial}gg^{-1}\,,$$

for any $g \in G^S$.

The infinitesimal gauge transformations form the Lie algebra $\mathfrak{g}^S = \Gamma(S, \operatorname{End} E)$, where Γ denotes the space of C^∞-sections. As before, they are generated by certain Hamiltonian functions on the symplectic manifold $\mathcal{A}^S$.

Proposition 2.27 *An infinitesimal gauge transformation ϵ is generated by the Hamiltonian function*

$$H_\epsilon(A) = \int_S \beta \wedge \operatorname{tr}(\epsilon(\bar{\partial}A + A \wedge A)) - 2\pi i \int_P \operatorname{res}_P \beta \wedge \operatorname{tr}(\epsilon A) \,.$$

PROOF. The only modification in comparison with the proof of Proposition 2.3 is that now we are going to use the Cauchy–Stokes formula instead of the usual Stokes formula. Indeed, now we have

$$\begin{aligned}
\delta H_\epsilon(A) &= \int_S \beta \wedge \operatorname{tr}\left(\epsilon\,\delta F(A)\right) - 2\pi i \int_P \operatorname{res}_P \beta \wedge \operatorname{tr}(\epsilon\,\delta A) \\
&= \int_S \beta \wedge \operatorname{tr}(\epsilon\,\overline{\nabla}_A\,\delta A) - 2\pi i \int_P \operatorname{res}_P \beta \wedge \operatorname{tr}(\epsilon\,\delta A) \\
&= \int_S \beta \wedge \operatorname{tr}\left(\delta A \wedge \overline{\nabla}_A\,\epsilon\right) \,.
\end{aligned}$$

Here $F(A) := \bar{\partial}A + A \wedge A$ is the curvature, $\overline{\nabla}_A\,\epsilon := \bar{\partial}\epsilon - [\epsilon, A]$ is the infinitesimal gauge transformation of a $(0,1)$-connection $\bar{\partial} + A$ by the element $\epsilon \in \mathfrak{g}^S$, and $\overline{\nabla}_A\,\delta A = \bar{\partial}\,\delta A - [\delta A, A]$.   □

Similarly, the commutation relations for these Hamiltonians with respect to the holomorphic Poisson structure induced by ω become centrally extended:

$$\{H_{\epsilon_1}, H_{\epsilon_2}\} = H_{[\epsilon_1, \epsilon_2]} + c(\epsilon_1, \epsilon_2)\,,$$

where c is the following 2-cocycle on the Lie algebra $\mathfrak{g}^S$:

Proposition 2.28 *The value of the cocycle c on a pair of infinitesimal gauge transformations $\epsilon_1, \epsilon_2 \in \mathfrak{g}^S$ is given by*

$$c(\epsilon_1, \epsilon_2) = 2\pi i \int_P \operatorname{res}_P \beta \wedge \operatorname{tr}(\epsilon_1\,\bar{\partial}\epsilon_2)\,.$$

Remark 2.29 Recall that a smooth divisor P must be an elliptic curve (or a union of such), while the residue $\operatorname{res}_P \beta$ is a holomorphic 1-form on P. In Section II.5 we have used the same cocycle to construct a central extension $\widehat{\mathfrak{g}}^P$ of the double loop group $\mathfrak{g}^P$. This suggests that the centrally extended (elliptic) Lie algebra $\widehat{\mathfrak{g}}^P$ should play a similar role in the theory of holomorphic bundles on complex surfaces to that played by the centrally extended loop algebras (the affine Lie algebras) in the theory of flat connections on Riemann surfaces.

Let $\widehat{\mathfrak{g}}^{S,\beta}$ denote the Lie algebra of gauge transformations of the bundle E over the surface S, which is centrally extended by the cocycle c from Proposition 2.28, and let $\widehat{G}^{S,\beta}$ be the corresponding group. (One can consider any of the possible central extensions of the group G^S: it is only the Lie algebra that matters here.) The infinite-dimensional space $\widehat{\mathfrak{g}}^{S,\beta}$ is the space of pairs (ϵ, a), where ϵ is a $\mathfrak{g}$-valued function on the surface S and a is a complex number.

Definition 2.30 The (smooth) dual $(\widehat{\mathfrak{g}}^{S,\beta})^*$ to the algebra $\widehat{\mathfrak{g}}^{S,\beta}$ is the space of triples (F, C, λ), where F is an $(\operatorname{End} E)$-valued $(0,2)$-form on S, C is an $(\operatorname{End} E)$-valued $(0,1)$-form on the "boundary" P of the surface S (i.e., on the polar divisor of the 2-form β), and λ is a complex number. The nondegenerate pairing $\langle\,,\,\rangle$ between the spaces $\widehat{\mathfrak{g}}^{S,\beta}$ and $(\widehat{\mathfrak{g}}^{S,\beta})^*$ is the following:

$$\langle (F, C, \lambda), (\epsilon, a) \rangle = \int_S \beta \wedge \operatorname{tr}(\epsilon F) - 2\pi i \int_P \operatorname{res}_P \beta \wedge \operatorname{tr}(\epsilon C) + a\lambda \,.$$

Let us consider the action of $\widehat{G}^{S,\beta}$ on the affine space $\mathcal{A}^S$ generated by the action of G^S. This is the action by gauge transformations, while the center of $\widehat{G}^{S,\beta}$ acts trivially. The following result immediately follows from Proposition 2.27, similarly to the "real" case.

Proposition 2.31 ([194]) *The centrally extended group $\widehat{G}^{S,\beta}$ of gauge transformations acts on the affine space $\mathcal{A}^S$ in a Hamiltonian way. The moment map for the action of the corresponding gauge algebra $\widehat{\mathfrak{g}}^{S,\beta}$ is the mapping $\mathcal{A}^S \to (\widehat{\mathfrak{g}}^{S,\beta})^*$ given by the $(0,2)$-curvature and by the restriction of the $\bar{\partial}$-connection to the "boundary":*

$$A \mapsto (\bar{\partial} A + A \wedge A, \; A|_P, \; 1) \,.$$

As before, let us consider the group G_P^S of gauge transformations on S based at the polar divisor P,

$$G_P^S = \{ g \in G^S \mid g|_P = \operatorname{id} \} \,,$$

and denote by $\mathfrak{g}_P^S$ the corresponding Lie algebra. The latter can be viewed as a subalgebra of the centrally extended Lie algebra $\widehat{\mathfrak{g}}^{S,\beta}$, since the cocycle c is trivial on $\mathfrak{g}_P^S$. Accordingly, the group G_P^S can be viewed as a subgroup of the centrally extended gauge group $\widehat{G}^{S,\beta}$, and we can restrict the action of the large group to this subgroup. A slight modification of the last proposition gives the following corollary.

Corollary 2.32 *The group G_P^S of based gauge transformations acts on the affine space $\mathcal{A}^S$ in a Hamiltonian way. The moment map $\Phi_P : \mathcal{A}^S \to (\mathfrak{g}_P^S)^*$ for this action is given by the $(0,2)$-curvature:*

$$\Phi_P : \; A \mapsto \bar{\partial} A + A \wedge A \,.$$

First consider the (holomorphic) Hamiltonian reduction

$$\mathcal{M}_{S,P} = \mathcal{A}^S /\!/ G_P^S := \Phi_P^{-1}(0)/G_P^S$$

of the space of $\bar{\partial}$-connections $\mathcal{A}^S$ with respect to the group G^S_P. The result will be the space of *integrable $\bar{\partial}$-connections* in the bundle E on S modulo gauge transformations from G^S_P. Indeed, the set $\Phi_{\bar{P}}^{-1}(0)$ consists of $(0, 1)$-connections whose $(0, 2)$-curvature vanishes. Such connections are in one-to-one correspondence with *holomorphic structures* in the complex bundle E.

Thus the holomorphic Hamiltonian reduction leads us to the consideration of the *space of all holomorphic structures in the bundle E modulo gauge equivalences that act as the identity over P*. The corresponding quotient space $\mathcal{M}_{S,P}$, which we consider only locally, near some of its smooth points, is, by construction, an (infinite-dimensional) symplectic manifold, since it comes from the Hamiltonian reduction of a symplectic manifold.

Remark 2.33 Before we treat the general case with a nonempty P, let us look at the particular situation in which the meromorphic form β has no poles, i.e., it is actually a holomorphic 2-form without zeros. (As we mentioned, this can happen if the surface S is K3 or abelian.) If $P = \emptyset$, the group $G^S_P = G^S$ is the group of all gauge transformations on S, while (a nonsingular part of) the quotient $\mathcal{M}_{S,P}$ recovers the (finite-dimensional) moduli space of (stable) holomorphic G-bundles over S. This way we obtain a "visualization" of the following theorem of Mukai:

Theorem 2.34 ([283]) *Let S be a K3 or abelian surface. Then the moduli space of stable holomorphic G-bundles over S admits a holomorphic symplectic structure.*

Returning to the case of a nonempty divisor $P \subset S$, consider now the space of holomorphic structures in a smooth bundle over a complex one-dimensional manifold by taking P as such a manifold and $E|_P$ as the bundle over P:

$$\mathcal{C} := \{\bar{\partial} + C \,|\, C \in \Omega^{0,1}(P, \operatorname{End} E|_P)\} \,.$$

Here $\bar{\partial}$ is understood as the restriction to P of our reference holomorphic structure. The space $\mathcal{C}$ of holomorphic structures in a bundle on an elliptic curve (or a sum of such spaces if P consists of several disjoint components) is in fact an affine subspace in a vector space dual to the elliptic Lie algebra $\hat{\mathfrak{g}}^{P,\alpha}$. The latter Lie algebra is defined as the central extension of $\mathfrak{g}^P = \Gamma(P, \operatorname{End} E|_P)$ by the cocycle

$$c_\alpha(\epsilon_1, \epsilon_2) = 2\pi i \int_P \alpha \wedge \operatorname{tr}(\epsilon_1 \bar{\partial}\epsilon_2) \,;$$

cf. Remark 2.29. We set $\alpha := \operatorname{res}_P \beta$, in which case we obtain the elliptic Lie algebra $\hat{\mathfrak{g}}^{P,\alpha} = \hat{\mathfrak{g}}^{S,\beta}/\mathfrak{g}^S_P$ as the corresponding quotient. Consider the linear Lie–Poisson structure on the dual $(\hat{\mathfrak{g}}^{P,\alpha})^*$ of the elliptic algebra and, consequently, on its affine subspace $\mathcal{C}$. Recall that the symplectic leaves of this structure (or coadjoint orbits of the elliptic Lie group) correspond to isomorphism classes of holomorphic bundles on P; see Section II.5.2. We have obtained the restriction map

$$\mathcal{A}^S \to (\widehat{\mathfrak{g}}^{P,\alpha})^*$$

(respectively, the map to the product of the corresponding dual spaces if P consists of several connected components), which is Poisson and which factors through the symplectic quotient to produce the map $\mathcal{M}_{S,P} \to (\widehat{\mathfrak{g}}^{P,\alpha})^*$.

Proposition 2.35 *The mapping from the space $\mathcal{M}_{S,P}$ to the coadjoint representation space $(\widehat{\mathfrak{g}}^{P,\alpha})^*$, sending an integrable $\bar{\partial}$-connection on the surface S to its restriction to the divisor P, is a Poisson mapping.*

PROOF. This mapping is essentially the moment map for the action of gauge transformations on the polar divisor P.

In the second step, we consider the quotient of the space $\mathcal{M}_{S,P}$ by the whole group $\widehat{G}^{S,\beta}$ of centrally extended gauge transformations. The latter group acts on $\mathcal{M}_{S,P}$, since gauge transformations that are equal to the identity on P form a normal subgroup G_P^S in $\widehat{G}^{S,\beta}$.

The quotient space

$$\mathcal{M}_S := \mathcal{M}_{S,P}/\widehat{G}^{S,\beta} = \{\bar{\partial} + A \in \mathcal{A}^S \mid \bar{\partial}A + A \wedge A = 0\}/\widehat{G}^{S,\beta}$$

represents the set of isomorphism classes of holomorphic bundles on S (corresponding to a given underlying topological bundle E). Then, by construction, the local smooth moduli space $\mathcal{M}_S$ of holomorphic bundles on S is a finite-dimensional Poisson manifold. Its symplectic leaves are described in terms of coadjoint orbits in $(\widehat{\mathfrak{g}}^{P,\alpha})^*$ as follows.

Proposition 2.36 (*i*) [**55, 371**] *The local moduli space $\mathcal{M}_S$ of holomorphic G-bundles on the Poisson surface S possesses a (holomorphic) Poisson structure.*

(*ii*) [**194**] *The symplectic leaves of this structure are parametrized by the moduli of their restrictions to the polar divisor $P \subset S$ of the nonvanishing meromorphic 2-form β on S. (That is, a symplectic leaf is singled out by fixing the isomorphism class of the restriction to the elliptic curve P, or the isomorphism classes of restrictions to each curve if P consists of several such curves.)*

Remark 2.37 The above proposition should not be understood in the sense that the isomorphism classes of bundles on P can be taken arbitrarily. Rather they have to satisfy certain conditions that come from the fact that they arise as restrictions of bundles defined over S.

The discussion above applies with minor modifications to the case in which the polar divisor P consists of several components intersecting transversally. In the latter case, the corresponding group $\widehat{G}^{P,\alpha}$ is the extended current group over the (reducible) complex curve P.

2.4 Bibliographical Notes

The classical reference for the moduli space of flat connections on a Riemann surface is the paper by Atiyah and Bott [28]. The case of a surface with boundary was studied in [83, 125]. In our exposition we follow the papers [125] and [194] for the real and complex sides of the story.

The finite-dimensional description [125] of the moduli space $\mathcal{M}_\Sigma$ with the help of connections on graphs had various extensions and generalizations (see, e.g., [10, 184]), and it is related to recent progress in such diverse areas as [6, 71, 105, 121, 122]. For a beautiful introduction and survey of the surrounding area we refer the reader to [29].

For other finite-dimensional constructions of the symplectic moduli space see, e.g., [173, 183]. In [6] this space was studied by introducing the notion of a group-valued moment map, which proved to be a powerful tool in symplectic geometry. In Appendix A.10 we discuss torus actions and integrable systems on such moduli spaces, after [175, 120].

The Cauchy–Stokes formula can be found, in particular, in [151]. The approach using this formula can also be viewed as parallel to the geometric complexification suggested by Arnold in [14].

The symplectic structure on the moduli space of stable sheaves on a K3 surface or an abelian surface was found by Mukai [283] by algebro-geometric methods, while the holomorphic Poisson structures were discussed in [55, 371]. The description of the corresponding holomorphic symplectic leaves was obtained in [194].

3 Around the Chern–Simons Functional

In this section we move from surfaces to threefolds, along with connections and bundles on them. By studying the topological Chern–Simons action functional on the space of connections on a three-dimensional manifold with boundary, we recover the definition of the symplectic structure on the moduli space of flat connections on a compact Riemann surface. Similarly, the holomorphic Chern–Simons functional on $\bar\partial$-connections over three-dimensional Fano manifolds is related to the holomorphic symplectic structure on the moduli spaces of stable bundles over $K3$ or abelian surfaces.

Furthermore, the corresponding path integrals for these Chern–Simons functionals in the abelian case can be used to define the Gauss linking number of oriented curves in three-dimensional space and its holomorphic analogue, the polar linking number of holomorphic curves.

3.1 A Reminder on the Lagrangian Formalism

A motion of a particle on a manifold can be described by the least action principle. Consider an action functional

$$S[q] = \int_{t_0}^{t_1} L(q(t), \dot{q}(t), t)\, dt$$

defined on the space $\mathcal{C}[t_0, t_1]$ of smooth maps $q : [t_0, t_1] \to M$ of the interval $[t_0, t_1]$ to the manifold M. Here L is a (time-dependent) Lagrangian function, $L : TM \times \mathbb{R} \to \mathbb{R}$, which we assume to depend only on t, q, and its first derivative $\dot{q} := dq/dt$.

For a path variation δq one can find the corresponding *variation of the action functional*, i.e., the linear-in-δq term of the difference $S[q + \delta q] - S[q]$:

$$\delta S[q] = \int_{t_0}^{t_1} E\, \delta q\, dt + p\, \delta q|_{t_0}^{t_1}\ ,$$

where

$$E := \frac{\partial L(q, \dot{q}, t)}{\partial q} - \frac{d}{dt} \frac{\partial L(q, \dot{q}, t)}{\partial \dot{q}}$$

and

$$p := \frac{\partial L(q, \dot{q}, t)}{\partial \dot{q}}\ .$$

(Here and below we assume the summation over the coordinates $q = (q^1, \ldots, q^d)$: $p\delta q := \sum_j p_j \delta q^j$, $p_j := \partial L(q, \dot{q}, t)/\partial \dot{q}^j$, etc.)

Exercise 3.1 Prove the variation formula. (Hint: use integration by parts.)

This way the variation δS can be regarded as a 1-form on the infinite-dimensional space $\mathcal{C}[t_0, t_1]$ of "virtual trajectories" of the particle.

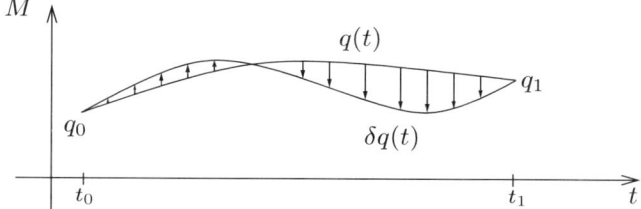

Fig. 3.1. A small variation of the path $q(t)$ with fixed endpoints.

Definition 3.2 The *least action principle* states that the actual trajectories of the particle are the critical points of this action functional: $\delta S[q] = 0$.

By confining ourselves to variations with fixed ends, $\delta q(t_0) = \delta q(t_1) = 0$, we come to a necessary condition on the extremals. Namely, actual particle trajectories satisfy the *Euler–Lagrange equation* $E = 0$, i.e.,

$$\frac{\partial L(q, \dot{q}, t)}{\partial q} - \frac{d}{dt} \frac{\partial L(q, \dot{q}, t)}{\partial \dot{q}} = 0 \,.$$

Denote by $\mathcal{E}[t_0, t_1]$ the space of all solutions to the Euler–Lagrange equation, i.e., the space of such trajectories.

Exercise 3.3 A free particle of mass m moving in the space $\mathbb{R}^d$ with a potential energy $V : \mathbb{R}^d \to \mathbb{R}$ has the Lagrangian $L(q, \dot{q}, t) = m|\dot{q}|^2/2 - V(q)$, the difference of its kinetic and potential energies. Prove that the Euler–Lagrange equation for this L gives the *Newton equation* of motion:

$$m\ddot{q} = -\mathrm{grad}\, V(q) \,.$$

Now we restrict the variation 1-form δS to the space of extremals $\mathcal{E}[t_0, t_1]$, which is singled out by the Euler–Lagrange equation. On this space of "trajectories with free ends" we obtain

$$\delta S = p\, \delta q|_{t_0}^{t_1} = \sigma_1 - \sigma_0 \,, \tag{3.5}$$

where $\sigma_i := p\, \delta q|_{t_i}$, $i = 0, 1$ are the corresponding 1-forms on $\mathcal{C}[t_0, t_1]$. One can regard the above as a relation between these three 1-forms: σ_0, σ_1, and δS, which holds for their restrictions to the space of extremals $\mathcal{E}[t_0, t_1]$.

Now, by applying the exterior differential δ (on the infinite-dimensional manifold $\mathcal{C}[t_0, t_1]$) to both sides of the relation (3.5) above and using $\delta^2 = 0$, we obtain $\delta\sigma_0 = \delta\sigma_1$, which holds on $\mathcal{E}[t_0, t_1]$. This means that the space $\mathcal{E}[t_0, t_1]$ turns out to be naturally equipped with a closed 2-form ω defined by

$$\omega := \delta\sigma_0 = \delta\sigma_1 \,.$$

Definition 3.4 A manifold N equipped with a closed 2-form ω (not necessarily nondegenerate) is called *presymplectic*.

Consider the distribution of null-spaces of this 2-form in N.

Exercise 3.5 (*i*) Assuming that this distribution has constant rank, prove that it is integrable, i.e., it is tangent to a foliation in N.

(*ii*) Assuming that this null-foliation is a fibration $\pi : N \to N'$, prove that the base of this fibration carries a natural symplectic structure, i.e., (N', ω') is a symplectic manifold such that $\pi^* \omega' = \omega$.

The above discussion shows that whenever the space of extremals $\mathcal{E}[t_0, t_1]$ is a manifold, it is in fact a *presymplectic* manifold. However, the 2-form ω is often degenerate. The *phase space* $\mathcal{P}$ of the particle can be described as the corresponding *symplectic* manifold. (Here we implicitly assume that various regularity conditions are satisfied to guarantee that both $\mathcal{E}[t_0, t_1]$ and the phase space are smooth manifolds.)

Exercise 3.6 Check that for the above example of a particle motion in $\mathbb{R}^d$ this definition of the phase space $\mathcal{P}$ coincides with $T^* \mathbb{R}^d$ equipped with the natural symplectic structure.

Remark 3.7 [393, 79, 341] The discussed Lagrangian formalism can be generalized to infinite-dimensional target manifolds M or to higher-dimensional domains instead of the interval $[t_0, t_1]$. These are the objects that a field theory deals with. Consider, for example, a local action functional

$$S[\varphi] = \int_N L(\varphi(x), \partial\varphi(x)) \, d^n x$$

describing a field theory on an n-dimensional manifold N with boundary ∂N. Here $x = (x_1, \ldots, x_n)$ are local coordinates on N, φ is a map from N to a target manifold M or a section of some bundle on N, $\partial\varphi$ are the first derivatives of φ, while the Lagrangian L can depend on additional structures on N. As in the one-dimensional situation described above, one can pose a variational problem $\delta S[\varphi] = 0$, which leads to the Euler–Lagrange equations.

Suppose first that $N = I \times \Sigma$, where I is an interval and a manifold Σ has dimension $n - 1$. One can consider $t \in I$ as the time variable and identify the field theory with an infinite-dimensional classical mechanics, where the space of maps $\varphi : \Sigma \to M$ plays the role of the target. In particular, one has a presymplectic manifold of extremals $\mathcal{E}_N$ and the symplectic phase space $\mathcal{P}$ associated to N (or, rather, to Σ).

Alternatively, one can associate the phase spaces $\mathcal{P}_0$ and $\mathcal{P}_1$ to the corresponding boundary components $\partial N = \Sigma_1 - \Sigma_0$ of N, and equip the total phase space $\mathcal{P}_0 \times \mathcal{P}_1$ with the product symplectic structure. There is a natural projection α_N of the space $\mathcal{E}_N$ of extremals into the product $\mathcal{P}_0 \times \mathcal{P}_1$, since it "tautologically" projects to each factor: one describes the extremals via different boundary components, taking the orientation of the latter into account. Then the relation $0 = \delta^2 S = \delta\sigma_1 - \delta\sigma_0$ that held on $\mathcal{E}_N$ now reads that the image $\alpha_N(\mathcal{E}_N)$ of $\mathcal{E}_N$ is an *isotropic submanifold* in the symplectic manifold $\mathcal{P}_0 \times \mathcal{P}_1$.

Definition 3.8 A submanifold of a symplectic manifold is *isotropic* if the restriction of the symplectic form to this submanifold is zero.

Exercise 3.9 Let $f : (N_1, \omega_1) \to (N_2, \omega_2)$ be a diffeomorphism between two symplectic manifolds. Prove that f is a symplectic map, i.e., $\omega_1 = f^*\omega_2$, if and only if the graph of f is an isotropic submanifold in the symplectic manifold $(N_1 \times N_2, \omega_1 \ominus \omega_2)$.

(An isotropic submanifold of maximal possible dimension, which is equal to half the dimension of the symplectic manifold, is called a *Lagrangian submanifold*; cf. Section I.4.5. This is the case for the graph of f.)

One can see that the image $\alpha_N(\mathcal{E}_N)$ is indeed isotropic in $\mathcal{P}_0 \times \mathcal{P}_1$, since the 2-form $\delta\sigma_1 - \delta\sigma_0$ is exactly the restriction of the product symplectic structure of $\mathcal{P}_0 \times \mathcal{P}_1$ (with different orientations of the boundary components) to this image.

The latter formulation of the presymplectic/isotropic properties of the space of extremals $\mathcal{E}_N$ extends naturally to the general case of a manifold N with boundary consisting of several components $\Sigma_1, \ldots, \Sigma_k$. Associate the phase space $\mathcal{P}_j$ to each component Σ_j, thinking of a neighborhood of Σ_j in N as a product $I \times \Sigma$. One has the relations $\delta S = \sigma_1 + \cdots + \sigma_k$ and $\delta\sigma_1 + \cdots + \delta\sigma_k = 0$ on the space of extremals $\mathcal{E}_N$, where σ_j stands for the contribution of the corresponding boundary component. The latter shows that the image $\alpha_N(\mathcal{E}_N)$ under the natural map $\alpha_N : \mathcal{E}_N \to \mathcal{P}_1 \times \cdots \times \mathcal{P}_k$ is *isotropic* with respect to the product symplectic structure on the phase space $\mathcal{P}_1 \times \cdots \times \mathcal{P}_k$. We refer to [341, 79] for more details.

Remark 3.10 The philosophy of holomorphic orientation (see Sections 2.2 and 2.3) can be applied to field-theoretic notions in the following way. Suppose we have an action functional

$$S[\varphi] = \int_M L(\varphi, \partial\varphi) \, d^n x$$

on *smooth* fields φ (e.g., functions, connections, etc.) on a *real* (oriented) manifold M, and this functional is defined by an n-form $L \, d^n x$, which depends on the fields and their derivatives.

Then one can suggest the following complex analogue $S_{\mathbb{C}}$ of the action functional S for a *complex* n-dimensional manifold X equipped with a "polar orientation," i.e., with a holomorphic or meromorphic n-form μ:

$$S_{\mathbb{C}}[\varphi] := \int_X \mu \wedge L(\varphi, \bar{\partial}\varphi) \, d^n \bar{x} \, .$$

Here $\wp$ stands for *smooth* fields on a complex manifold X. Now the $(0, n)$-form $L\, d^n \bar{x}$ is integrated against the holomorphic orientation μ over X.

Furthermore, the interrelation between the extremals of the real functional $\mathcal{S}[\varphi]$ (on smooth fields) on the real manifold M and the boundary values of those fields on ∂M is replaced by the analogous interrelation for the complex functional $\mathcal{S}_{\mathbb{C}}[\varphi]$ (still on smooth fields) on a complex manifold X (equipped with an n-form μ) and on the polar divisor $Y := \mathrm{div}_{\infty}\mu \subset X$ (equipped with the residue $(n - 1)$-form $\nu := \mathrm{res}\,\mu$).

The above discussion will allow us to see in the next two sections how the symplectic structures on the moduli of flat connections and holomorphic bundles on surfaces arise naturally from the Lagrangian formalism related to the topological and holomorphic Chern–Simons functionals.

3.2 The Topological Chern–Simons Action Functional

Let N be a real compact oriented three-dimensional manifold with boundary $\partial N = \Sigma$. As usual in the "real case," we take G to be a compact simply

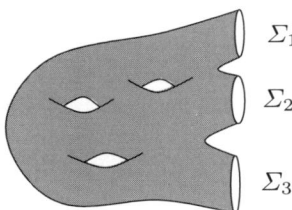

Fig. 3.2. Three-dimensional manifold N with boundary $\partial N = \Sigma_1 \cup \Sigma_2 \cup \Sigma_3$.

connected simple Lie group with the corresponding Lie algebra $\mathfrak{g}$. Denote the nondegenerate invariant (Killing) bilinear form on $\mathfrak{g}$ by $\mathrm{tr}(XY) := \langle X, Y \rangle$. Fix a trivial G-bundle E over N and let $\mathcal{A}$ denote the space of connections in the bundle E. Upon fixing a reference flat connection, we think of $\mathcal{A}$ as the space $\Omega^1(N, \mathfrak{g})$.

Definition 3.11 The topological *Chern–Simons action functional* is the following real-valued function on the space of connections $\mathcal{A}$:

$$\mathrm{CS}(A) := \int_N \mathrm{tr}(A \wedge dA) + \frac{2}{3}\int_N \mathrm{tr}(A \wedge A \wedge A)\,,$$

where a connection $A \in \mathcal{A}$ is understood as a $\mathfrak{g}$-valued 1-form on N.

Proposition 3.12 *The set of extremals, i.e., solutions of the Euler–Lagrange equation, for the Chern–Simons functional* CS *is the space of flat connections in the G-bundle E over the manifold N.*

PROOF.   For a small variation δA of a connection $A \in \mathcal{A}$ the corresponding variation of the functional is

$$
\begin{aligned}
\delta\,\mathrm{CS} &= \int_N \mathrm{tr}(\delta A \wedge dA) + \int_N \mathrm{tr}(A \wedge d\delta A) + 2 \int_N \mathrm{tr}(\delta A \wedge A \wedge A) \\
&= \int_N d\,\mathrm{tr}(A \wedge \delta A) + 2 \int_N \mathrm{tr}\,(\delta A \wedge (dA + A \wedge A)) \\
&= \int_{\partial N} \mathrm{tr}(A \wedge \delta A) + 2 \int_N \mathrm{tr}\,(\delta A \wedge (dA + A \wedge A))\,,
\end{aligned}
$$

where at the last step we used the Stokes formula.

By imposing the boundary condition $\delta A|_{\partial N} = 0$ on variations δA, we obtain the Euler–Lagrange equation

$$
dA + A \wedge A = 0\,,
$$

i.e., the equation of vanishing curvature $F(A) = 0$ on N. Hence the space of solutions of this equation is exactly the space of flat connections on the real threefold N. $\square$

The first term in the above calculation of $\delta\,\mathrm{CS}$ gives the boundary contribution, the 1-form $\sum \sigma_j$ on the extremals, where the summation is taken over the boundary components of ∂N. Take $N = I \times \Sigma$ to be a finite cylinder over a closed two-dimensional surface Σ. Then the presymplectic structure on the space of flat connections on N, i.e., on the extremals for our action functional, is $\omega = \delta\sigma$ for

$$
\sigma := \int_\Sigma \mathrm{tr}(a \wedge \delta a)\,,
$$

where $a := A|_\Sigma$ denotes the restriction of a flat connection A from the manifold N to either of its boundary components Σ. (Here we omit the index $j = 0, 1$ for σ_j, since $\omega = \delta\sigma_0 = \delta\sigma_1$.)

Exercise 3.13 Verify that the 2-form $\omega = \delta\sigma$ is degenerate on the space of flat connections on the surface Σ exactly along the gauge equivalence classes of the connections $\{a\}$. (Hint: the 2-form $\delta\sigma = \int_\Sigma \mathrm{tr}(\delta a \wedge \delta a)$ is the restriction of the canonical 2-form ω from the set of all connections to the subset of flat connections on Σ; cf. Definition 2.1.)

Thus the moduli space of flat connections $\mathcal{M}^\Sigma$ on the surface Σ appears as the natural symplectic (or phase) space for this presymplectic space of flat connections on Σ, and we obtain yet another definition of the symplectic structure on $\mathcal{M}^\Sigma$ from Section 2.1.

Corollary 3.14 *The moduli space $\mathcal{M}^\Sigma$ of flat connections on a surface Σ is naturally symplectic as the phase space for extremals of the Chern–Simons action functional for connections on the threefold $N = I \times \Sigma$.*

Remark 3.15 To see why this action functional is called *topological* we now check the invariance property of the Chern–Simons action with respect to gauge transformations of the connections. Let M be a compact three-dimensional manifold *without boundary* and suppose that A and $\widetilde{A}$ are connections in a G-bundle over M that are sent to each other by a gauge transformation g:

$$\widetilde{A} = gAg^{-1} - dgg^{-1} \,.$$

Then the Chern–Simons actions for them are related as follows:

$$\mathrm{CS}(\widetilde{A}) = \mathrm{CS}(A) + \frac{1}{3} \int_M \mathrm{tr}\left(g^{-1}dg \wedge g^{-1}dg \wedge g^{-1}dg\right) \,.$$

Recall that the 3-form $\frac{1}{24\pi^2} \mathrm{tr}(g^{-1}dg)^{\wedge 3}$ is the pullback under the map $g : M \to G$ of an integral closed 3-form η on the compact simply connected simple Lie group G (see Proposition 2.16 in Appendix A.2; cf. Section II.1.3). Thus the integral of this form depends only on topological properties of the map g and can be expressed as

$$\frac{1}{24\pi^2} \int_M \mathrm{tr}(g^{-1}dg)^{\wedge 3} = \int_M g^* \eta \,,$$

which is an integer, since the 3-form η generates $H^3(G, \mathbb{Z})$. The latter implies that the exponential $\exp\left(\frac{i}{4\pi} \mathrm{CS}(A)\right)$ is gauge invariant:

$$\frac{i}{4\pi} \mathrm{CS}(\widetilde{A}) - \frac{i}{4\pi} \mathrm{CS}(A) = 2\pi i \cdot \frac{1}{24\pi^2} \int_M \mathrm{tr}(g^{-1}dg)^{\wedge 3} \in 2\pi i \cdot \mathbb{Z} \,.$$

Remark 3.16 An interesting integer-valued invariant for a homology 3-sphere M was introduced by Casson and is closely related to the gauge-theoretic constructions above [70]. Roughly speaking, the *Casson invariant* $\mathrm{Cas}(M)$ is defined as the algebraic number of the conjugacy classes of irreducible SU(2)-representations of the fundamental group $\pi_1(M)$. In other words, it counts the number of irreducible flat SU(2)-connections on M modulo conjugation. The homology restriction on the threefold M is related to the fact that if $H_1(M) \neq 0$, then the moduli space of flat connections on M might not be zero-dimensional, and in particular, it would not consist of a finite number of points. The reason for restricting to SU(2) is clarified in the following exercise.

Exercise 3.17 Show that the only reducible representation $\rho : \pi_1(M) \to$ SU(2) is the trivial one. (Hint: Reducible representations of $\pi_1(M)$ in SU(2) are necessarily abelian and hence factor through the homology $H_1(M)$. This homology group is trivial for a homology 3-sphere.)

Now consider a Heegaard splitting of M into two handlebodies $M = M_1 \cup_\Sigma M_2$ glued together along their common boundary, an embedded surface $\Sigma \subset M$. Consider the moduli space $\mathcal{M}^\Sigma$ of flat connections in the trivial

SU(2)-bundle on the surface Σ. Define two submanifolds L_1 and L_2 of the symplectic manifold $\mathcal{M}^\Sigma$ as those (equivalence classes of) flat connections on the surface Σ that extend to M_1 and M_2 respectively. One can show that these submanifolds are Lagrangian. Their intersection points $L_1 \cap L_2$ correspond to flat connections extendable to the whole of M. Thus the Casson invariant is defined as the intersection number of these submanifolds,

$$\mathrm{Cas}(M) = \#(L_1, L_2),$$

where we assume that the submanifolds intersect transversally, and exclude the intersection corresponding to the trivial representation; see details, for example, in [364].

3.3 The Holomorphic Chern–Simons Action Functional

A complex three-dimensional manifold X equipped with a nowhere vanishing meromorphic 3-form μ can be regarded as a complex analogue of a real oriented manifold with boundary, following the general philosophy that we adopted in Sections 2.2 and 2.3. Accordingly, one can complexify the Lagrangian formalism to this situation. Here we define a holomorphic analogue of the Chern–Simons action functional for (X, μ) and relate it to Mukai's holomorphic symplectic structures on moduli of holomorphic bundles over complex surfaces, following [195, 85].

Let $G_\mathbb{C}$ be a complex simple and simply connected Lie group and $E_\mathbb{C}$ a complex $G_\mathbb{C}$-bundle over the manifold X. As before, let us denote by $\mathcal{A}_\mathbb{C}^X$ the space of $(0, 1)$-connections in the bundle $E_\mathbb{C}$.

Definition 3.18 The *holomorphic Chern–Simons action functional* $\mathrm{CS}_\mathbb{C}$: $\mathcal{A}_\mathbb{C}^X \to \mathbb{C}$ is defined via

$$\mathrm{CS}_\mathbb{C}(A) := \int_X \mu \wedge \left(\langle A \wedge \bar{\partial} A \rangle + \frac{2}{3} \langle A \wedge A \wedge A \rangle \right)$$

for any $(0, 1)$-connection $A \in \mathcal{A}_\mathbb{C}^X$ thought of as a $\mathfrak{g}_\mathbb{C}$-valued $(0, 1)$-form on X. As usual, we assume that the 3-form μ has only first-order poles, and hence the integral above is well defined.

Proposition 3.19 *The extremals of the holomorphic Chern–Simons functional are holomorphic structures in the complex bundle $E_\mathbb{C}$.*

PROOF. Indeed, in the same way as in the real case and by using the Cauchy–Stokes formula we come to the Euler–Lagrange equation

$$\bar{\partial} A + A \wedge A = 0$$

in the holomorphic setting. Its solutions are $(0,1)$-connections A with vanishing $(0,2)$-curvature, $F^{0,2}(A) = 0$, and each such connection defines the corresponding holomorphic structure in the complex bundle $E_{\mathbb{C}}$. □

Consider now the "boundary term" of the variation $\delta \, \mathrm{CS}_{\mathbb{C}}$, which now descends to the polar divisor of the meromorphic 3-form μ. Denote this polar divisor by $Y := \mathrm{div}_{\infty}\mu \subset X$. Note that the residue $\nu := \mathrm{res}_Y \, \mu$ is a nonvanishing 2-form on the divisor Y, since μ itself is nonvanishing (see Exercise 2.19). In particular, the canonical bundle of Y has to be trivial, so that Y is either a $K3$ surface or a complex torus.

To define the presymplectic structure in the real case we considered a cylinder $M = I \times \Sigma$ over a Riemann surface Σ. Here we look at the complex analogue of such a cylinder. Namely, let $X = \mathbb{CP}^1 \times Y$ be the product of $\mathbb{CP}^1$ and a $K3$ surface or abelian surface Y. Suppose that Y is endowed with a holomorphic (necessarily nonvanishing) 2-form ν, and consider the meromorphic 3-form $\mu = (dz/z) \wedge \nu$ on X, where dz/z is a 1-form on the complex line $\mathbb{CP}^1$. One can see that $\nu = \mathrm{res}_{z=0} \, \mu = -\mathrm{res}_{z=\infty} \, \mu$.

Now the variation of the holomorphic Chern–Simons functional satisfies the relation $\delta \, \mathrm{CS}_{\mathbb{C}} = \sigma_{0,\mathbb{C}} + \sigma_{\infty,\mathbb{C}}$ on the space of extremals, which are the integrable $(0,1)$-connections on X, i.e., the connections with vanishing $(0,2)$-curvature. Here $\sigma_{0,\mathbb{C}}$ and $\sigma_{\infty,\mathbb{C}}$ stand for the contributions of the corresponding components $z = 0$ and $z = \infty$ of the polar divisor of μ.

This allows us to introduce the *holomorphic presymplectic* structure $\omega_{\mathbb{C}} = \delta\sigma_{\mathbb{C}}$ on the "boundary values" of the extremals, i.e., on the space of integrable connections on the surface Y. Explicitly, the holomorphic 1-form $\sigma_{\mathbb{C}}$ is

$$\sigma_{\mathbb{C}} := \int_Y \nu \wedge \mathrm{tr}(a \wedge \delta a) \,,$$

where $a := A|_{z=0}$ is the restriction of a $(0,1)$-connection A in $E_{\mathbb{C}}$ from the threefold X to the surface Y (understood as one component $\{z = 0\} \times Y \subset X$ of the polar divisor of μ), δa is the corresponding variation of a, and $\nu = \mathrm{res}_{z=0} \, \mu$ is a holomorphic 2-form on Y.

One can show that, similarly to the real case, the presymplectic structure $\omega_{\mathbb{C}}$ is degenerate along the orbits of the action of the complex group of gauge transformations $G_{\mathbb{C}}^Y$ on integrable $(0,1)$-connections (i.e., holomorphic structures) in the bundle $E_{\mathbb{C}}$ over Y. After taking the quotient with respect to the group action, we obtain a nondegenerate holomorphic symplectic structure on the moduli space of (stable) holomorphic bundles on the $K3$ or abelian surface Y. (Here, as usual, we are concerned with the moduli space only locally around a smooth point.) Thus the holomorphic Lagrangian formalism gives an alternative approach to Mukai's result discussed before:

Theorem 3.20 ([283]) *There exists a holomorphic symplectic structure $\omega_{\mathbb{C}}$ on the moduli space $\mathcal{M}_Y$ of stable holomorphic $G_{\mathbb{C}}$-bundles over a $K3$ or abelian surface Y.*

Remark 3.21 It turns out that there exists a holomorphic analogue of the Casson invariant for a Calabi–Yau manifold X; see [85, 366]. Instead of a Heegaard splitting of a real manifold, one considers a degeneration of this CY manifold to an intersection of two Fano manifolds. The divisor of intersection is a $K3$ or abelian surface, and one counts in a special way the holomorphic bundles over Y extendable to both of these two Fano manifolds.

We also note that the holomorphic Chern–Simons action functional has more complicated transformation properties with respect to gauge transformations. After a "large" gauge transformation, the value of the functional differs by a multiple of the integrals $\int_X \mu \wedge g^*\eta$. The latter can be viewed as the integrals of the meromorphic 3-form μ over the three-cycles in X that are Poincaré dual to the 3-form $g^*\eta$ for various maps $g : X \to G_{\mathbb{C}}$. The values of these integrals can form a lattice or even a dense set in $\mathbb{C}$; hence considering the exponential similar to $\exp\left(\frac{i}{4\pi}CS(A)\right)$ does not allow one to extract a gauge-invariant quantity in the holomorphic setting.

3.4 A Reminder on Linking Numbers

Let M be a simply connected oriented manifold and let γ_1 and γ_2 be two nonintersecting oriented closed curves in M. Pick an oriented surface $D_1 \subset M$ (a Seifert surface for the curve γ_1) such that the curve γ_1 is the oriented boundary of the surface D_1 and such that D_1 and γ_2 intersect transversally.

Definition 3.22 The *linking number* $\mathrm{lk}(\gamma_1, \gamma_2)$ of the curves γ_1 and γ_2 is the intersection number of the surface D_1 and the curve γ_2, i.e., the number of intersections of the curve γ_2 with the surface D_1 counted with orientation (see Figure 3.3):

$$\mathrm{lk}(\gamma_1, \gamma_2) = \#(D_1, \gamma_2).$$

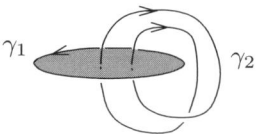

Fig. 3.3. Linking of two oriented curves.

The sign at each intersection point is obtained by forming there a frame from the orientation frames for D_1 and γ_2, and comparing it with the orientation of the ambient manifold M.

Proposition 3.23 *The linking number* $\mathrm{lk}(\gamma_1, \gamma_2)$ *is*
(i) independent of the choice of a Seifert surface D_1,

(*ii*) *symmetric in γ_1 and γ_2,*

(*iii*) *invariant with respect to isotopy of the curves, provided they do not intersect each other,*

(*iv*) *well defined in any (not necessarily simply connected) oriented three-dimensional manifold M, provided that both curves γ_1 and γ_2 are homologous to 0 in M.*

Note that if the manifold M is not simply connected and only one of the curves is homologous to 0 in M, but the other is not, the linking number might not be well defined. For instance, take $M = \mathbb{T}^3$ and two curves, one of which is homologous to 0, while the other is a generator in $H_1(\mathbb{T}^3, \mathbb{Z})$. Then by taking different Seifert surfaces for the first curve one obtains either 0 or 1 for their linking number; see Figure 3.4.

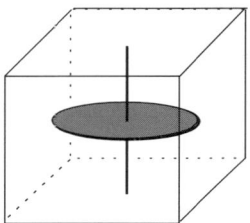 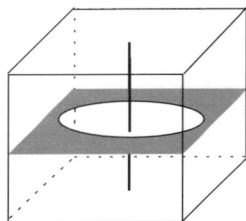

Fig. 3.4. Two Seifert surfaces for the horizontal circle in the cube-torus $\mathbb{T}^3 = \mathbb{R}^3/\mathbb{Z}^3$, one "inside" and one "outside," give linking numbers ± 1 and 0, respectively, for the intersection with the "vertical" cycle.

Exercise 3.24 Prove the above proposition. Furthermore, show also that the linking number is actually invariant when the curve γ_1 changes to a curve (or a collection of curves) $\widetilde{\gamma}_1$ homologous to γ_1 in the complement $M \setminus \gamma_2$ (see Figure 3.5).

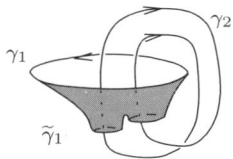

Fig. 3.5. The homologous curves γ_1 and $\widetilde{\gamma}_1$ have the same linking number with the curve γ_2.

Needless to say, the linking number easily generalizes to manifolds of any dimension n, provided that the linking submanifolds are homologous to zero and have "linking dimensions": the sum of their dimensions equals $n - 1$.

Remark 3.25 There exists the Gauss integral formula for the linking number of two curves γ_1 and γ_2 in $\mathbb{R}^3$. Recall it here in a somewhat "symbolic form," which we need further.

Let $\Delta \subset M \times M$ denote the diagonal in $M \times M$ and let δ stand for its Poincaré dual current, a closed 3-form supported on the diagonal, $[\delta] \in H^3(M \times M, \mathbb{R})$. Then we can write

$$\mathrm{lk}(\gamma_1, \gamma_2) = \#(\Delta, D_1 \times \gamma_2) = \int_{x \in D_1} \int_{y \in \gamma_2} \delta(x, y), \qquad (3.6)$$

where $\#(\Delta, D_1 \times \gamma_2)$ is the intersection number of Δ and $D_1 \times \gamma_2 \subset M \times M$. One can split the 3-form δ into the homogeneous components

$$\delta = \delta^{3,0} + \delta^{2,1} + \delta^{1,2} + \delta^{0,3},$$

where $\delta^{i,j}$ denotes the component that is an i-form on the first factor of $M \times M$, and a j-form on the second factor. Note that in equation (3.6) we had to integrate only over the component $\delta^{2,1}$, since all the other integrals vanish. This component is an exact 2-form in x on D_1, which allows us to apply the Stokes formula:

$$\mathrm{lk}(\gamma_1, \gamma_2) = \int_{x \in D_1} \int_{y \in \gamma_2} \delta^{2,1}(x, y) = \int_{x \in \gamma_1} \int_{y \in \gamma_2} d_x^{-1} \delta^{2,1}(x, y).$$

For $\mathbb{R}^3$ the $(1,1)$-form $d_x^{-1} \delta^{2,1}(x, y)$ on the torus $\gamma_1 \times \gamma_2$ assumes the standard *Gauss form*

$$\frac{1}{4\pi} \cdot \frac{(\overrightarrow{x - y}, \overrightarrow{dx}, \overrightarrow{dy})}{\|\overrightarrow{x - y}\|^3},$$

where $(\cdot, \cdot, \cdot)$ is the mixed product of three vectors in $\mathbb{R}^3$.

Remark 3.26 In what follows we need a bit of calculus of such δ-type forms. Let δ_γ be the Dirac δ-type 2-form supported on a closed oriented curve γ in a simply connected threefold M. (Alternatively, the curve γ can be regarded as a de Rham current, a linear functional on 1-forms on M, whose value is the integral of the 1-form over γ.) The integral of this 2-form δ_γ over a two-dimensional surface counts the intersection number of this surface with the curve γ. Then by using the decomposition of the diagonal 3-form δ into the homogeneous components, we can express

$$\delta_\gamma(x) = \int_{y \in \gamma} \delta^{2,1}(x, y),$$

where we denote the coordinates on the first and the second factors of $M \times M$ by x and y respectively. Choose a surface $D \subset M$ whose boundary is $\gamma = \partial D$. Similarly, we can define the δ-type 1-form supported on the surface D by

$$\delta_D(x) = \int_{y \in D} \delta^{1,2}(x, y) \,.$$

The relation $\partial D = \gamma$ is equivalent to the relation between the corresponding δ-forms: $d_x \delta_D(x) = \delta_\gamma(x)$, due to the Stokes theorem, or more explicitly,

$$\delta_\gamma(x) = \int_{y \in D} d_x(\delta^{1,2})(x, y) \,,$$

where d_x denotes the exterior derivative applied to the x-coordinates only. Finally, if γ_1 and γ_2 are two nonintersecting curves, we have

$$\mathrm{lk}(\gamma_1, \gamma_2) = \int_{x \in D_1} \delta_{\gamma_2}(x) = \int_M \delta_{D_1}(x) \wedge \delta_{\gamma_2}(x) = \int_M \delta_{D_1} \wedge d\delta_{D_2} \,, \qquad (3.7)$$

where $\partial D_2 = \gamma_2$. The latter form suggests a common nature of the linking number and the $A \wedge dA$-part of the Chern–Simons functional, which we are going to study below.

3.5 The Abelian Chern–Simons Path Integral and Linking Numbers

We start with a reminder on finite-dimensional Gaussian integrals. Let (x, Qx) be a symmetric negative-definite form in the Euclidean $\mathbb{R}^n$. The classical Gauss integral

$$\int_{\mathbb{R}} \exp(-qx^2/2) dx = \sqrt{2\pi/q}$$

has the multidimensional analogue

$$\int_{\mathbb{R}^n} e^{\frac{1}{2}(x, Qx)} \, d^n x = \left(\frac{(2\pi)^n}{\det(-Q)} \right)^{\frac{1}{2}} \,.$$

Now fix a vector $J \in \mathbb{R}^n$ and consider the integral

$$Z_Q(J) := \int_{\mathbb{R}^n} e^{\frac{1}{2}(x, Qx) + (x, J)} \, d^n x = \int_{\mathbb{R}^n} e^{S_J(x)} \, d^n x$$

corresponding to the shift $S_J(x) := \frac{1}{2}(x, Qx) + (x, J)$ of the quadratic form by a linear term. (The initial integral is $Z_Q(0)$.) This integral can easily be solved by completing the square. Indeed, let x_0 be a solution of the equation $Qx_0 + J = 0$, i.e., $x_0 = -Q^{-1}J$. Then by introducing a shifted variable $\tilde{x} = x - x_0$ and using the translation invariance of the measure $d^n x$, we obtain

$$Z_Q(J) = \int_{\mathbb{R}^n} e^{S_J(\widetilde{x}+x_0)} \, d^n x$$

$$= \int_{\mathbb{R}^n} \exp\left\{\frac{1}{2}(\widetilde{x}+x_0, Q(\widetilde{x}+x_0)) + (\widetilde{x}+x_0, J)\right\} d^n x$$

$$= \int_{\mathbb{R}^n} \exp\left\{\frac{1}{2}(\widetilde{x}, Q\widetilde{x}) + \frac{1}{2}(x_0, Qx_0) + (x_0, J)\right\} d^n \widetilde{x}$$

$$= e^{S_J(x_0)} \int_{\mathbb{R}^n} e^{\frac{1}{2}(\widetilde{x}, Q\widetilde{x})} \, d^n \widetilde{x} = e^{\frac{1}{2}(x_0, J)} Z_Q(0).$$

Thus, we have

$$\frac{Z_Q(J)}{Z_Q(0)} = e^{S_J(x_0)} = e^{\frac{1}{2}(x_0, J)} = e^{-\frac{1}{2}(Q^{-1}J, J)}. \tag{3.8}$$

Remark 3.27 When the space $\mathbb{R}^n$ is replaced by some infinite-dimensional vector space, the integrals defining $Z_Q(0)$ and $Z_Q(J)$ usually do not make sense. However, one can "calculate" their ratio, which often turns out to be well defined. Note that the second of the equivalent expressions for the ratio $Z_Q(J)/Z_Q(0)$ in formula (3.8) has the form $\exp(\frac{1}{2}(x_0, J)) = \exp(-\frac{1}{2}(x_0, Qx_0))$, which allows us to avoid looking for the inverse Q^{-1} of the corresponding operator in the infinite-dimensional space.

Consider an application of this idea to the abelian Chern–Simons path integral. Let $\mathcal{A}$ be the space of connections in a U(1)-bundle over a real three-dimensional simply connected manifold M without boundary. We can think of such connections as real-valued 1-forms on M. Denote by CS : $\mathcal{A} \to \mathbb{R}$ the Chern–Simons action functional on $\mathcal{A} = \Omega^1(M, \mathbb{R})$, which now becomes a quadratic form

$$\mathrm{CS}(A) = \int_M A \wedge dA,$$

since the group U(1) is abelian and the cubic term $A \wedge A \wedge A$ vanishes. Note that the kernel of this quadratic form is the space of exact 1-forms $d\Omega^0 \subset \Omega^1(M, \mathbb{R})$.

Fix some linear functional J on $\Omega^1(M, \mathbb{R})$, i.e., a de Rham current on this space, and define

$$S_J(A) := \frac{1}{2} \int_M A \wedge dA + \int_M A \wedge J.$$

for $A \in \Omega^1(M, \mathbb{R})$. We also impose the condition $dJ = 0$, so that the linear term $\int_M A \wedge J$ is well defined on the quotient $\Omega^1(M)/d\Omega^0(M)$. Now make the following "formal" definition.

Definition 3.28 The *abelian Chern–Simons path integral* is the expression

$$Z_{\mathrm{CS}}(J) := \int_{\Omega^1/d\Omega^0} e^{S_J(A)} \, DA \,,$$

where DA stands for a translation-invariant measure on the infinite-dimensional space $\Omega^1(M)/d\Omega^0(M)$.

Rather than trying to define the measure and the path integral precisely, we are going to see what the above formal manipulations with Gaussian integrals give us in this situation, where, in a sense, the operator Q is replaced by the outer derivative d. By formula (3.8) for the ratio $Z_Q(J)/Z_Q(0)$ we obtain

$$\frac{Z_{\mathrm{CS}}(J)}{Z_{\mathrm{CS}}(0)} = e^{S_J(A_0)} = e^{\frac{1}{2}\int_M A_0 \wedge J} \,,$$

where A_0 is a solution of the equation $dA_0 + J = 0$. (Recall that J is a closed current on a simply connected M, and hence it is exact, i.e., this equation formally has a solution.)

Now we would like to specify the functional J on 1-forms $A \in \Omega^1(M, \mathbb{R})$ to be the integral of the form over a collection of curves in the simply connected manifold M. Let γ_i, $i = 1, \ldots, k$, be closed oriented nonintersecting curves in the manifold M. We set $J = \sum_i q_i \delta_{\gamma_i}$, where δ_{γ_i} is the δ-type 2-form on M supported on the curve γ_i, while q_i are real parameters. By applying the calculus of δ-forms (see Remark 3.26) we obtain that the ratio $Z_{\mathrm{CS}}(J)/Z_{\mathrm{CS}}(0)$ assumes the following explicit form:

$$\frac{Z_{\mathrm{CS}}(J)}{Z_{\mathrm{CS}}(0)} = \exp\left\{\frac{1}{2}\int_M A_0 \wedge J\right\} = \exp\left\{\frac{1}{2}\int_M A_0 \wedge \sum_i q_i \int_{y\in\gamma_i} \delta^{2,1}(x,y)\right\}$$

$$= \exp\left\{\frac{1}{2}\int_M A_0 \wedge \sum_i q_i \int_{y\in D_i} d_x\delta^{1,2}(x,y)\right\}$$

$$= \exp\left\{\frac{1}{2}\int_M -dA_0 \wedge \sum_i q_i \int_{y\in D_i} \delta^{1,2}(x,y)\right\}$$

$$= \exp\left\{\frac{1}{2}\int_M \left(\sum_j q_j \int_{z\in\gamma_j} \delta^{2,1}(x,z)\right) \wedge \left(\sum_i q_i \int_{y\in D_i} \delta^{1,2}(x,y)\right)\right\}$$

$$= \exp\left\{\frac{1}{2}\sum_{i,j} q_i q_j \int_M \delta_{\gamma_j}(x) \wedge \delta_{D_i}(x)\right\} = \exp\left\{\frac{1}{2}\sum_{i,j} q_i q_j \, lk(\gamma_j, \gamma_i)\right\}$$

Here we have used the Stokes theorem, as well as the definition of A_0 as a solution of the equation $dA_0 + J = 0$.

Corollary 3.29 ([340, 318]) *For the functional J defined as the integral of 1-forms over a collection of curves in a threefold, the ratio $Z_{\mathrm{CS}}(J)/Z_{\mathrm{CS}}(0)$ counts the pairwise linking numbers of these curves.*

Note also that above, in the latter sum, we had to assume that $i \neq j$, so that the linking number was defined. The case of self-linking is much more subtle. It leads to divergences of the path integral and requires some additional specifications, such as framing, for its normalization; see [54]. The value $Z_{\mathrm{CS}}(0)$ in the case without any curve corresponds to the Ray–Singer torsion of the manifold M [340].

The topological Chern–Simons path integral has a holomorphic analogue.

Definition 3.30 (cf. [390]) For a three-dimensional Calabi–Yau manifold X with a holomorphic 3-form μ the *holomorphic abelian Chern–Simons path integral* is the expression

$$Z_{\mathbb{C}S}(J) := \int_{\Omega^{0,1}/\bar{\partial}\Omega^{0,0}} e^{S_{\mathbb{C}J}(A)} \, DA \,,$$

where

$$S_{\mathbb{C}J}(A) := \frac{1}{2} \int_X \mu \wedge A \wedge \bar{\partial} A + \langle \mathbb{C}J, A \rangle$$

is the quadratic form shifted by the linear functional $\mathbb{C}J$ on the space of $(0,1)$-connections $A \in \Omega^{0,1}(X, \mathbb{C})$.

Remark 3.31 For a complex curve $C \subset X$ equipped with a holomorphic 1-form α define the linear functional on $(0,1)$-connections A by assigning $\langle \mathbb{C}J_C, A \rangle := \int_C \alpha \wedge A$. Similarly to the topological case, if such a functional $\mathbb{C}J$ corresponds to a collection of complex curves, the holomorphic abelian Chern–Simons path integral can be described in terms of the *polar linking number*, a holomorphic analogue of the Gauss linking number, which we define in Section 4.3. The relation of this functional with the holomorphic analogue of linking was established in [134, 195, 366].

The abelian theory is a particular case of the general Chern–Simons path integral. In the topological case we consider a link $L = \cup_i \gamma_i$ in a compact real threefold M. Let $\mathcal{A}$ be the affine space of all connections in the (trivial) G-bundle over M for a compact simply connected simple Lie group G. We identify $\mathcal{A}$ with the space $\Omega^1(M, \mathfrak{g})$ of 1-forms on M with values in the Lie algebra $\mathfrak{g}$ of G. Finally, let $G^M = C^\infty(M, G)$ be the group of gauge transformations in the bundle.

Definition 3.32 The *nonabelian Chern–Simons path integral* for a link $L \subset M$ is the following function of a parameter k:

$$Z_{\mathrm{CS}}(L; k) = \int_{\mathcal{A}/G^M} \left\{ \exp\left\{ ik \int_M \mathrm{tr}\left(A \wedge dA + \frac{2}{3} A \wedge A \wedge A \right) \right\} \right.$$

$$\left. \times \prod_{\gamma_i \subset L} \mathrm{tr}\left(P \exp \int_{\gamma_i} A \right) \right\} DA,$$

where $P\exp$ is the path-ordered exponential integral of a nonabelian connection A over γ_i, and DA is an appropriate measure on the moduli space of the connections $\mathcal{A}/G^M$.

Remark 3.33 Witten showed in [389] that for $M = S^3$ and $G = \mathrm{SU}(2)$ this path integral leads to the Jones polynomial for the link L. Other link or knot invariants can be obtained by changing the group. Note that they are always Vassiliev-type invariants of finite order [35, 36]. There are various ways to give $Z_{\mathrm{CS}}(L; k)$ and the corresponding link invariants rigorous definitions (see, e.g., the combinatorial [327] or probabilistic [5] approaches).

The extension of these results to a holomorphic version of the nonabelian Chern–Simons path integral is an intriguing open problem. The more complicated gauge transformation property of the holomorphic Chern–Simons action functional already makes the first step, writing out the corresponding path integral for an arbitrary collection of complex curves in a Calabi–Yau threefold, a serious problem; see some discussion in [391, 134, 366].

3.6 Bibliographical Notes

The Chern–Simons functional was introduced in [72]. For the relation of the abelian Chern–Simons functional to linking numbers we refer to [340, 318]. The appearance of the Jones polynomial and other knot invariants from the Chern–Simons functional was discovered by Witten [389]; see more details in [35, 210]. An excellent account of the relation between this functional to knot theory is contained in the book by Atiyah [27]. The relation between the Chern–Simons functional and the Vassiliev knot invariants is described in [36, 210].

The holomorphic Chern–Simons functional was introduced in [390] and studied in a number of papers [85, 134, 195, 196, 367]. For a higher-dimensional version of the Chern–Simons functional and its relation to linking numbers of several submanifolds see [124].

The classical Lagrangian formalism can be found, for example, in [18]. The formalism of the Lagrangian field theory was described in [393]; see also the presentations in [79, 341] for more details and examples. For preliminaries on linking numbers one can look at any book on differential topology, e.g., [162]. The question of when the space of extremals (more precisely, geodesics on a manifold) is a smooth manifold by itself is addressed, for example, in [38, 39].

4 Polar Homology

In the preceding sections we have encountered several analogies between notions from differential topology and complex algebraic geometry, which we list in the following "complexification table":

Real	Complex
real n-dimensional manifold	complex n-dimensional manifold
orientation of the manifold	nonvanishing meromorphic n-form
manifold's boundary	form's divisor of poles
orientation of the boundary	residue of the form
manifold's singularity or infinity	form's divisor of zeros
flat connection	holomorphic bundle
affine Lie algebra	elliptic Lie algebra
topological Chern–Simons functional	holomorphic Chern–Simons functional
Stokes formula	Cauchy–Stokes formula
singular homology	polar homology

The last analogy of this table is the subject of the present section. We begin with an informal introduction to polar homology, which is followed by the precise definition [195, 196]. Then we treat polar analogues of the intersection and linking numbers, and finally, we briefly introduce polar homology of affine curves.

4.1 Introduction to Polar Homology

In this section we discuss the naturality of the correspondence between the notions of an orientation of a real manifold and a meromorphic form on a complex manifold. As we show below, this correspondence can be thought of as an extension of the analogy between de Rham and Dolbeault cochains ($d \leftrightarrow \bar{\partial}$) to an analogy at the level of the corresponding *chain* complexes.

Consider a compact complex (ambient) manifold X. Let $W \subset X$ be a k-dimensional submanifold equipped with a holomorphic k-form ω. We are going to regard the top-degree holomorphic form ω on a complex submanifold as the submanifold's "holomorphic orientation."

More generally, assume that the form ω is allowed to have first-order poles on a smooth hypersurface $V \subset W$. A pair (W, ω), which consists of a k-dimensional submanifold W equipped with such a *meromorphic* top-degree form ω, will be thought of as an analogue of a compact oriented submanifold *with boundary*.

In the polar homology theory the pairs (W, ω) will play the role of k-chains. The corresponding boundary operator ∂ assumes the form $\partial(W, \omega) = 2\pi i (V, \operatorname{res} \omega)$, where V is the polar set of the k-form ω, while $\operatorname{res} \omega$ is the Poincaré residue of ω, which is a $(k-1)$-form on V. Note that in the situation under consideration, when the polar set V of the form ω is a smooth $(k-1)$-dimensional submanifold in a smooth k-dimensional W, the induced "orientation" on V is given by a regular $(k-1)$-form $\operatorname{res} \omega$. This means that

$\partial\,(V, \operatorname{res}\omega) = 0$, or the boundary of a boundary is zero. The latter is the source of the identity $\partial^2 = 0$, which allows one to define *polar homology* groups $\operatorname{HP}_k(X) = \ker \partial/\operatorname{im} \partial$.

Example 4.1 To illustrate the identity $\partial^2 = 0$ consider the example $\omega = dx \wedge dy/xy$ in $\mathbb{C}^2$. Then we have

$$\operatorname{res}|_{y=0} \operatorname{res}|_{x=0}\, \omega = \operatorname{res}|_{y=0} dy/y = 1$$
$$= -\operatorname{res}|_{x=0}\,(-dx/x) = -\operatorname{res}|_{x=0} \operatorname{res}|_{y=0}\, \omega\,.$$

Thus, the iterated residues differ by the sign corresponding to the order in which they are taken. Hence, the total second residue on the polar divisor of the form ω equals 0.

Note that the example of the polar divisor $\{xy = 0\}$ for the form $\omega = dx \wedge dy/xy$ in $\mathbb{C}^2$ should be viewed as a complexification of a vertex of a polygon in $\mathbb{R}^2$. Indeed, the cancellation of the repeated residues on different components of the divisor is mimicking the calculation of the boundary of a boundary of a polygon: every vertex of the polygon appears twice with different signs as a boundary point of two sides (see Figure 4.1).

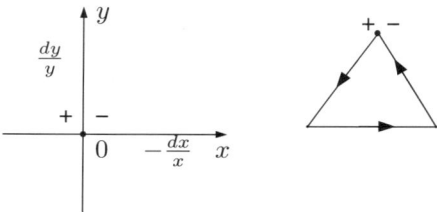

Fig. 4.1. The relation $\partial^2 = 0$ in polar homology is a complex analogue of the relation $\partial^2 = 0$ in singular homology: in the boundary of the boundary every vertex of a polygon appears twice with opposite signs.

Example 4.2 Let us find the polar homology groups $\operatorname{HP}_k(Z)$ of a complex projective curve Z. In this (and in any) case, all the 0-chains are cycles. Let (P, a) and (Q, b) be two 0-cycles, where P, Q are points on Z and $a, b \in \mathbb{C}$ (see Figure 4.2). These two 0-cycles are polar homologically equivalent if and only if $a = b$. Indeed, $a = b$ is necessary and sufficient for the existence of a meromorphic 1-form α on Z such that $\operatorname{div}_\infty \alpha = P + Q$ and $\operatorname{res}_P \alpha = 2\pi i\, a$, $\operatorname{res}_Q \alpha = -2\pi i\, b$. (The sum of all residues of a meromorphic differential on a projective curve is zero by the Cauchy residue theorem.) Then, in terms of the polar chain complex (to be formally defined in the next section) one can write that $(P, a) - (Q, a) = \partial\,(Z, \alpha)$. Thus, $\operatorname{HP}_0(Z) = \mathbb{C}$.

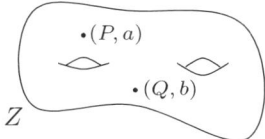

Fig. 4.2. A complex curve Z with two polar 0-chains.

Polar 1-cycles correspond to all possible holomorphic 1-forms on Z. On the other hand, there are no 1-boundaries, since there are no polar 2-chains in Z. Hence $\mathrm{HP}_1(Z) = \mathbb{C}^\varkappa$, where $\varkappa$ is the genus of the curve Z. The absence of polar 2-chains in Z also gives $\mathrm{HP}_k(Z) = 0$ for $k \geq 2$.

Remark 4.3 Polar homology is an analogue of singular homology with coefficients in $\mathbb{R}$. Consider real analogues of complex curves, i.e., real manifolds whose singular homology groups have the same dimensions over $\mathbb{R}$ as the polar homology groups of complex curves over $\mathbb{C}$. For a curve Z of genus $\varkappa$ its real analogue is a wedge of $\varkappa$ circles if $\varkappa \geq 1$ and a closed interval (i.e., homotopic to a point) if $\varkappa = 0$. These manifolds can be drawn as graphs as in Figure 4.3. We would like to emphasize that $\mathbb{CP}^1$ is a complex analogue of a real interval, while an elliptic curve $\mathcal{E}$ is a complex counterpart of a circle in this precise sense. We exploited this analogy in the previous chapters.

We also note that the number of trivalent points (i.e., points of nonsmoothness) for such graphs is equal to $2\varkappa - 2$. This is exactly the number of zeros of a holomorphic differential on the curve Z of genus $\varkappa$. We obtain yet another line in the complexification dictionary we started with: the *divisor of zeros* of a holomorphic (or meromorphic) volume form on a complex manifold correspond to *nonsmoothness points* of real manifolds.

Remark 4.4 There is a natural pairing between polar chains and smooth differential forms on a manifold: For a polar k-chain (W, ω) and any $(0, k)$-form u this pairing is given by the integral

$$\langle (W, \omega), u \rangle = \int_W \omega \wedge u .$$

In other words, the polar k-chain (W, ω) defines a de Rham current on X of degree $(n, n - k)$, where $n = \dim X$.

This pairing descends to (co)homology classes by virtue of the Cauchy–Stokes formula. Indeed, recall that for a meromorphic k-form ω on W having first-order poles on a smooth hypersurface $V \subset W$, the Cauchy–Stokes formula (see Theorem 2.21) states that

$$\int_W \omega \wedge \bar{\partial} v = 2\pi i \int_V \operatorname{res} \omega \wedge v ,$$

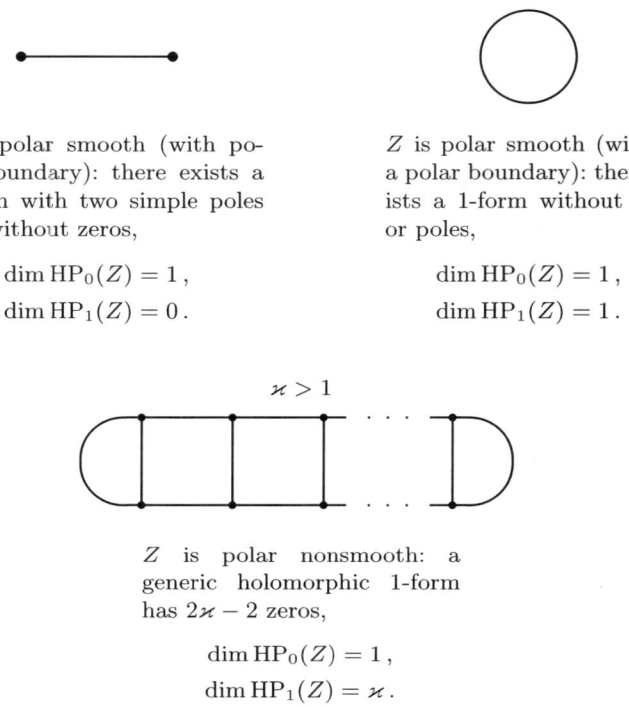

$\varkappa = 0$

$\varkappa = 1$

Z is polar smooth (with polar boundary): there exists a 1-form with two simple poles and without zeros,

$$\dim \mathrm{HP}_0(Z) = 1\,,$$
$$\dim \mathrm{HP}_1(Z) = 0\,.$$

Z is polar smooth (without a polar boundary): there exists a 1-form without zeros or poles,

$$\dim \mathrm{HP}_0(Z) = 1\,,$$
$$\dim \mathrm{HP}_1(Z) = 1\,.$$

$\varkappa > 1$

Z is polar nonsmooth: a generic holomorphic 1-form has $2\varkappa - 2$ zeros,

$$\dim \mathrm{HP}_0(Z) = 1\,,$$
$$\dim \mathrm{HP}_1(Z) = \varkappa\,.$$

Fig. 4.3. The real analogue of a smooth projective curve Z of genus $\varkappa$.

where v is any smooth $(0, k-1)$-form on X. Using the definition of the polar boundary operator, the latter can be rewritten as

$$\langle (W, \omega)\,, \bar{\partial} v \rangle = \langle\, \partial(W, \omega)\,, v \rangle\,.$$

In this way, the Cauchy–Stokes formula defines the pairing between the polar homology groups of a complex manifold X and the Dolbeault cohomology groups $H_{\bar{\partial}}^{0,k}(X)$.

Remark 4.5 One can define polar (k, p)-chains as pairs (A, α), where $A \subset X$ is a k-dimensional subvariety and α is a meromorphic p-form on A. The relation $\partial^2 = 0$ still holds in this more general setup, and one obtains polar homology groups $\mathrm{HP}_{k,p}$. Above, we have been considering the groups $\mathrm{HP}_k = \mathrm{HP}_{k,k}$. Another interesting case is the case $p = 0$, in which we consider pairs (A, a), where $a \in \mathbb{C}$. Two pairs (A_1, a_1) and (A_2, a_2) are equivalent if there exists a 1-form α on X that has opposite residues on A_1 and A_2.

We also note that the intuitive definition of polar homology above has to be made more precise, since one cannot restrict to the case of smooth complex submanifolds only: the divisor of poles is already not necessarily smooth. Instead of considering submanifolds, we will define polar chains as smooth varieties along with their maps to our manifold. This is similar to the definition of singular homology in topology, where one considers maps of abstract simplices into a manifold. We give a formal definition of the polar homology groups in the next section. Below we recall the construction of the pushforward of a differential form, which we need for that definition.

Definition 4.6 For a finite covering $f : X \to Y$ and a function u on X one can define its *pushforward*, or the *trace*, $f_* u$ as a function on Y whose value at a point is calculated by summing over the preimages taken with multiplicities. The operation f_* can be generalized to p-forms and to the maps f that are only generically finite.

Suppose that $f : X \to Y$ is a proper, surjective holomorphic mapping where both X and Y are smooth complex manifolds of the same dimension n. The *pushforward map* is a mapping

$$f_* : \Gamma(X, \Omega_X^p) \to \Gamma(Y, \Omega_Y^p) .$$

The pushforward map is also defined for meromorphic forms, f_* : $\Gamma(X, \mathcal{M}^p) \to \Gamma(Y, \mathcal{M}^p)$.

Its construction is as follows. First note that f is generically finite, i.e., there is an analytic hypersurface $D \subset Y$ such that f is a finite unramified covering off this hypersurface D. Hence, for a sufficiently small open neighborhood U of any point in $Y^* := Y \smallsetminus D$, the inverse image $f^{-1}(U) = U_1 \sqcup \cdots \sqcup U_d$ is a disjoint union of d open sets U_j, such that $f|_{U_j}$ is an isomorphism with the inverse $s_j : U \to U_j$. Given a form ω on X, one defines its pushforward

$$f_* \omega := s_1^* \omega + \cdots + s_d^* \omega$$

in U, and therefore in Y^*. One can check that the form $f_* \omega$ extends across the smooth points of D and hence to the whole of the manifold Y, since the remaining part of D has codimension at least two. The resulting form $f_* \omega$ is holomorphic (respectively, meromorphic) on Y provided the form ω was holomorphic (respectively, meromorphic) on X; see details, e.g., in [149].

Furthermore, the operations of pushforward and residue commute:

Proposition 4.7 *Let $f : X \to Y$ be a proper surjective holomorphic map between complex manifolds of the same dimension. Let ω be a meromorphic form on X with only first-order poles on a smooth hypersurface $V \subset X$. Suppose that $f(V)$ is a smooth hypersurface in Y. Then $f_* \omega$ has first-order poles on $f(V)$, and*

$$\operatorname{res} f_* \omega = \widetilde{f}_* \operatorname{res} \omega ,$$

where $\widetilde{f} : V \to f(V)$ is the restriction to V of the map f.

Example 4.8 To visualize how $f_*\omega$ extends across the smooth points of D, consider the following one-dimensional example. Let $f\colon \mathbb{C} \to \mathbb{C}$ be defined as $f\colon x \mapsto y = x^m$. Then any holomorphic 1-form $\omega = \psi(x)dx$ has a holomorphic pushforward 1-form $f_*\omega = \varphi(y)dy$. Indeed, by definition of the pushforward, for any $y \neq 0$, we obtain

$$\varphi(y)dy := \sum_{j=1}^{m} \psi(x_j)\, dx = \frac{1}{m} \sum_{j=1}^{m} \frac{x_j \psi(x_j)}{y} dy \ ,$$

where x_j are all mth roots of y, and in the last equality we used the relation $dy/y = m\, dx/x$ for $y = x^m$. This form $\varphi(y)dy$ is well defined for $y \neq 0$ and we need to check that $\varphi(y)$ can be extended to $y = 0$. Expanding ψ into a power series

$$\psi(x) = \sum_{l \geq 0} a_l x^l \ ,$$

we obtain

$$\varphi(y) = \frac{1}{my} \sum_{j=1}^{m} x_j \psi(x_j) = \frac{1}{my} \sum_{l \geq 0} a_l \sum_{j=1}^{m} x_j^{l+1}$$

$$= \frac{1}{my} \sum_{k \geq 1} a_{km-1}(my^k) = \sum_{k \geq 1} a_{km-1} y^{k-1} \ ,$$

where we have used that $\sum_{j=1}^{m} x_j^n = 0$, unless $n = km$. This power expansion in y proves that $\varphi(y)$ is holomorphic at 0, and hence over $\mathbb{C}$.

4.2 Polar Homology of Projective Varieties

In this section we deal with complex projective varieties, i.e., closed subvarieties of a complex projective space. For a smooth variety X, we denote by Ω_X^p the sheaf of holomorphic p-forms on X.

The space of polar k-chains for a complex projective variety X of dimension n is defined as a $\mathbb{C}$-vector space with certain generators and relations.

Definition 4.9 The space of *polar k-chains* $\mathcal{C}_k(X)$ is a vector space over $\mathbb{C}$ defined as the quotient $\mathcal{C}_k(X) = \hat{\mathcal{C}}_k(X)/\mathcal{R}_k$, where the vector space $\hat{\mathcal{C}}_k(X)$ is freely generated by the triples (A, f, α) described in (i), (ii), (iii) and $\mathcal{R}_k$ is defined as relations (R1), (R2), (R3) imposed on the triples:

   (i) A is a smooth complex projective variety, $\dim A = k$;
   (ii) $f\colon A \to X$ is a holomorphic map of projective varieties;
   (iii) α is a rational k-form on A with first-order poles on $V \subset A$,
       where V is a normal crossing divisor in A, i.e., $\alpha \in \Gamma(A, \Omega_A^k(V))$.

The relations are:

(R1) $\lambda(A, f, \alpha) = (A, f, \lambda\alpha)$ for $\lambda \in \mathbb{C}$.

(R2) $\sum_i (A_i, f_i, \alpha_i) = 0$ provided that $\sum_i f_{i*}\alpha_i \equiv 0$, where $\dim f_i(A_i) = k$ for all i and the pushforwards $f_{i*}\alpha_i$ are considered on the smooth part of $\cup_i f_i(A_i)$.

(R3) $(A, f, \alpha) = 0$ if $\dim f(A) < k$.

Remark 4.10 The relation (R3) implies that $\mathcal{C}_k(X) = 0$ for $k > \dim X$. Also by definition, $\mathcal{C}_k(X) = 0$ for $k < 0$.

The relation (R2) in particular represents additivity with respect to α, that is,

$$(A, f, \alpha_1) + (A, f, \alpha_2) = (A, f, \alpha_1 + \alpha_2).$$

Here we make no distinction between a triple and its equivalence class. In particular, if the polar divisor $\mathrm{div}_\infty(\alpha_1 + \alpha_2)$ is not normal crossing, one can replace A by an appropriate blowup, by the Hironaka theorem, where the pullback of $\alpha_1 + \alpha_2$ already has a normal crossing polar divisor.

In this way, the relation (R2) allows us to deal with polar chains as pairs replacing a triple (A, f, α) by a pair $(\hat{A}, \hat{\alpha})$, where $\hat{A} = f(A) \subset X$, $\hat{\alpha}$ is defined only on the smooth part of $\hat{A}$ and $\hat{\alpha} = f_*\alpha$ there. Due to the relation (R2), such a pair $(\hat{A}, \hat{\alpha})$ carries precisely the same information as (A, f, α). (Note, however, that such pairs cannot be arbitrary. In fact, by the Hironaka theorem on resolution of singularities, any subvariety $\hat{A} \subset X$ can be the image of some regular A, but the form $\hat{\alpha}$ on the smooth part of $\hat{A}$ cannot be arbitrary.)

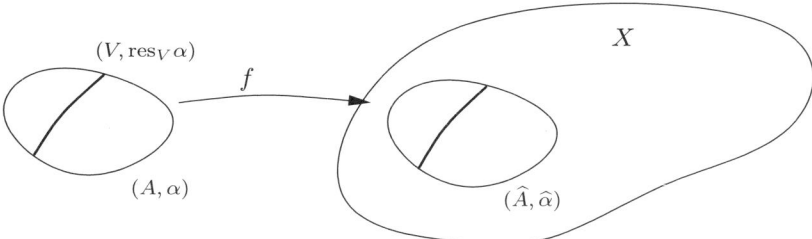

Fig. 4.4. A polar chain given by the triple (A, f, α) can be thought of as a pair $(\hat{A}, \hat{\alpha})$. Here $V = \mathrm{div}_\infty \alpha$.

We also note that the consideration of triples (A, f, α) instead of pairs $(\hat{A}, \hat{\alpha})$, which we used in the introduction, is similar to the definition of chains in singular homology theory. Indeed, in topology one considers the mappings of abstract simplices into the manifold, but one is interested only in the images of simplices. Here also comes a distinction: in contrast to topological homology, where in each dimension k one uses all continuous maps of one standard object (the standard k-simplex or the standard k-cell) to a given topological space, in polar homology we deal with complex analytic maps of a large class of k-dimensional varieties to a given one.

Definition 4.11 The *boundary operator* $\partial : \mathcal{C}_k(X) \to \mathcal{C}_{k-1}(X)$ is defined by

$$\partial(A, f, \alpha) = 2\pi i \sum_i (V_i, f_i, \mathrm{res}_{V_i}\, \alpha)\,,$$

where V_i are the components of the polar divisor of α, $\mathrm{div}_\infty\, \alpha = \cup_i V_i$, and the maps $f_i = f|_{V_i}$ are restrictions of the map f to each component of the divisor.

Theorem 4.12 ([195]) *The boundary operator ∂ is well defined, i.e., it is compatible with the relations* (R1), (R2), *and* (R3). *Furthermore,* $\partial^2 = 0$.

PROOF. We have to show that ∂ maps equivalent sums of triples to equivalent ones. It is trivial with (R1). For (R2) this follows from the commutativity of taking residue and pushforward.

To prove the compatibility of ∂ with (R3), consider first the case of a polar 1-chain, a complex curve with a meromorphic 1-form, that is mapped to a point. Then the image of the boundary of this 1-chain is zero. Indeed, this image must be the same point whose coefficient is equal to the sum of all residues of the meromorphic 1-form on the curve, i.e., zero. The general case is similar: the same phenomenon occurs along one of the coordinates. (We refer the interested reader to [195] for more details.)

As to the second part of the statement, we need to prove $\partial^2 = 0$ for triples $(A, f, \alpha) \in \mathcal{C}_k(X)$, i.e., for forms α with normal crossing divisors of poles. The repeated residue at pairwise intersections differs by the sign according to the order of taking the residues; see Remark 2.22. Thus the contributions to the repeated residue from different components cancel out (or, equivalently, the residue of a residue is zero). $\qquad\square$

Definition 4.13 For a smooth complex n-dimensional projective variety X, the chain complex

$$0 \to \mathcal{C}_n(X) \xrightarrow{\partial} \mathcal{C}_{n-1}(X) \xrightarrow{\partial} \cdots \xrightarrow{\partial} \mathcal{C}_0(X) \to 0$$

is called the *polar chain complex* of X. Its homology groups, $\mathrm{HP}_k(X), k = 0, \ldots, n$, are called the *polar homology groups* of X.

We found in Section 4.1 that the polar homology groups of a complex projective curve Z of genus $\varkappa$ are $\mathrm{HP}_0(Z) = \mathbb{C}$ and $\mathrm{HP}_1(Z) = \mathbb{C}^\varkappa$.

Exercise 4.14 Prove that for any n-dimensional X we have $\mathrm{HP}_n(X) = H^0(X, \Omega_X^n)$ and, if X is connected, also $\mathrm{HP}_0(X) = \mathbb{C}$.

Exercise 4.15 Prove the functoriality of polar homology, i.e., that a regular morphism of projective varieties $h: X \rightarrow Y$ defines a homomorphism $h_*: \mathrm{HP}_k(X) \rightarrow \mathrm{HP}_k(Y)$.

After the formal definitions have been given, we would like to show why the Dolbeault complex of $(0, k)$-forms should be related to the polar homology in the same way as the de Rham complex of smooth forms is related to the singular homology. First we show how the space of polar chains $\mathcal{C}_k(X)$ for a smooth projective variety X can be viewed as a subspace of currents, functionals on smooth differential forms on X.

Definition 4.16 A polar k-chain represented by a triple $a = (A, f, \alpha)$ defines the following linear functional on smooth $(0, k)$-forms. Its value on a smooth $(0, k)$-form u on X is given by the pairing

$$\langle a, u \rangle := \int_A \alpha \wedge f^* u . \tag{4.9}$$

The integral is well defined, since the meromorphic k-form α has only first-order poles on a normal crossing divisor.

Remark 4.17 It is straightforward to show that the pairing $\langle \, , \, \rangle$ descends to the space $\mathcal{C}_k(X)$ of equivalence classes of triples, i.e., that it is compatible with the relations (R1), (R2), (R3) of Definition 4.9. Indeed, (R1) is obvious, the compatibility with (R3) is a consequence of the equality $f^* u = 0$ if $\dim f(A) < k$, and the compatibility with (R2) follows from the relation $\int_A \alpha \wedge f^* u = \int_{f(A)} f_* \alpha \wedge u$ if $\dim f(A) = k$, where the last integral is taken over the smooth part of $f(A)$.

Proposition 4.18 *The pairing (4.9) defines the following homomorphism in (co)homology:*

$$\rho: \mathrm{HP}_k(X) \rightarrow H_{\bar{\partial}}^{n, n-k},$$

where $n = \dim X$.

PROOF. By Serre duality, ρ is a map $\mathrm{HP}_k(X) \rightarrow (H_{\bar{\partial}}^{0,k}(X))^*$, and it is sufficient to verify that the pairing vanishes if $\partial a = 0$ and $u = \bar{\partial} v$, or if $\bar{\partial} u = 0$ and $a = \partial b$. This follows immediately from the Cauchy–Stokes formula (Theorem 2.21):

$$\int_A \alpha \wedge f^*(\bar{\partial} u) = 2\pi i \int_{\mathrm{div}_\infty \alpha} (\mathrm{res}\, \alpha) \wedge f^*(u) ,$$

that is, $\langle a, \bar{\partial} u \rangle = \langle \partial a, u \rangle$. $\square$

One can prove that for smooth projective manifolds the homomorphism ρ is in fact an isomorphism:

Theorem 4.19 (Polar de Rham Theorem, [197]) *For a smooth projective manifold X the map $\rho \colon \mathrm{HP}_k(X) \to H_{\bar\partial}^{n,n-k}$ is an isomorphism of the polar homology and Dolbeault cohomology groups. Equivalently, $\mathrm{HP}^k(X) \cong H_{\bar\partial}^{0,k}(X)$ in terms of dual cohomology groups.*

Example 4.20 For a complex curve Z of genus $\varkappa$ one has $\mathrm{HP}_0(Z) \cong \mathbb{C} \cong H_{\bar\partial}^{1,1}(Z)$ and $\mathrm{HP}_1(Z) \cong \mathbb{C}^{\varkappa} \cong H_{\bar\partial}^{1,0}(Z)$.

Remark 4.21 As we mentioned in the introduction, one could consider more general polar (k,p)-chains (A, f, α), where α is a meromorphic p-form on A of not necessarily maximal degree, $p \leqslant k$, that can have only logarithmic singularities on a normal crossing divisor. The requirement of log-singularities is needed to have a convenient definition of the residue and hence the boundary operator ∂. The Cauchy–Stokes formula, the property $\partial^2 = 0$, and the definition of the polar homology groups $\mathrm{HP}_{k,p}(M)$ can be carried over to this, more general, situation.

As a consequence, the natural pairing between polar (k,p)-chains and smooth $(k-p, k)$-forms on X gives us the homomorphism

$$\rho \colon \mathrm{HP}_{k,p}(X) \to H_{\bar\partial}^{n-k+p,n-k}(X) \ .$$

However, unlike the case $p = k$, the map ρ is not, in general, an isomorphism for other values of p, $0 \leqslant p < k$. For instance, in the case of $p = 0$, the image consists of algebraic k-cycles (tensored with $\mathbb{C}$), while the full space $H_{\bar\partial}^{n-k,n-k}(X)$ can be much larger. It would be interesting to find an adjustment of the groups $\mathrm{HP}_{k,p}(M)$ to provide the isomorphism and hence to obtain a description of the chain complex for the Dolbeault cochains in all dimensions.

4.3 Polar Intersections and Linkings

We start by defining a polar analogue for the topological intersection number. Recall that in topology one considers a smooth oriented closed manifold M and two oriented closed submanifolds $A, B \subset M$ of complementary dimensions, i.e., $\dim A + \dim B = \dim M$. Suppose A and B intersect transversally at a finite number of points. Then to each intersection point P one assigns the local intersection index equal to ± 1 by comparing the mutual orientations of the tangent vector spaces $T_P A, T_P B$, and $T_P M$.

Now let X be a complex compact manifold of dimension n equipped with a nowhere-vanishing *holomorphic* n-form μ. Such a pair (X, μ) can be thought of as a *polar* analogue of an *oriented closed manifold*. (Recall that if μ were nonvanishing and meromorphic with first-order poles, this would be a polar analogue of an oriented manifold with boundary, while zeros of the n-form μ could be regarded as a complex analogue of singularities of a real manifold.)

Note that the existence of such a form requires the canonical bundle of the complex manifold X to be trivial, i.e., X can be a Calabi–Yau manifold, an abelian one, or a product of such.

Definition 4.22 Consider two polar cycles (A, α) and (B, β) of complementary dimensions that intersect transversally in the polar oriented complex manifold (X, μ). The *polar intersection number* of the cycles (A, α) and (B, β) is given by the following sum over the set of points in $A \cap B$:

$$(A, \alpha) \cdot (B, \beta) = \sum_{P \in A \cap B} \frac{\alpha(P) \wedge \beta(P)}{\mu(P)} \ .$$

Here $\alpha(P)$ and $\beta(P)$ are understood as exterior forms on $T_F M = T_P A \times T_P B$ obtained by the pullback from the corresponding factors.

Remark 4.23 At every intersection point P, the ratio in the right-hand side can be regarded as the comparison of the polar orientations of the cycles at that point with the orientation of the ambient manifold. Note that in the polar case, the intersection number does not have to be an integer. Rather, it should be viewed as a function of the "parameters" $(A, \alpha), (B, \beta)$, and (X, μ).

Remark 4.24 Essentially the same formula defines the intersection product of transversal polar cycles when they intersect over a manifold of positive dimension. Namely, for polar cycles (A, α) and (B, β) of dimensions p and q such that $p + q \geq n$ and transversal to each other, their intersection is a polar $(p+q-n)$-cycle $(C, \gamma) := (A, \alpha) \cdot (B, \beta)$, where $C = A \cap B$ and $\gamma := (\alpha \wedge \beta)/\mu$. In this way, one obtains the map

$$\mathrm{HP}_p(M) \otimes \mathrm{HP}_q(M) \to \mathrm{HP}_{p+q-n}(M) \ ,$$

upon finding smooth transverse representatives for every pair of homology classes. This map is well defined, as follows from the polar de Rham theorem; see [194] for more details.

Exercise 4.25 Check that the $(p+q-n)$-form $\gamma = (\alpha \wedge \beta)/\mu$ is indeed well defined on the complex manifold C. (Hint: this is a problem in linear algebra.)

A similar construction allows one to define a polar analogue of a linking number. Recall that the Gauss linking number of two *oriented closed* curves in $\mathbb{R}^3$ is an integer topological invariant equal to the algebraic number of crossings of one curve with a two-dimensional oriented surface bounded by the other curve; see Section 3.4. The linking number is a homology invariant in the following sense: it does not change if one of the curves is replaced by a homologically equivalent cycle in the complement to the other curve. More generally, the linking number can be defined for two oriented closed submanifolds of linking dimensions in any oriented (but not necessarily simply connected) manifold, provided that both submanifolds are homologous to zero.

To define the polar linking number we will "translate" the classical definition into the polar language. Let (A, α) and (B, β) be two polar smooth nonintersecting cycles of dimensions p and q in a polar oriented closed n-manifold (M, μ). Suppose that these cycles are polar boundaries (i.e., they are polar homologous to 0) and are of linking dimensions: $p + q = n - 1$.

Definition 4.26 The *polar linking number* of cycles (A, α) and (B, β) in (M, μ) is

$$\mathrm{lk}_{\mathrm{pol}}\left((A, \alpha), (B, \beta)\right) := \sum_{P \in A \cap S} \frac{\alpha(P) \wedge \sigma(P)}{\mu(P)},$$

where a chain (S, σ) has the polar boundary $(B, \beta) = \partial(S, \sigma)$. In other words, $\mathrm{lk}_{\mathrm{pol}}((A, \alpha), (B, \beta))$ is the intersection of the polar cycle (A, α) and the polar chain (S, σ), provided they intersect transversely.

Theorem 4.27 ([195, 196]) *The polar linking number* $\mathrm{lk}_{\mathrm{pol}}((A, \alpha), (B, \beta))$ *is*

   (i) *well defined, i.e., $\mathrm{lk}_{\mathrm{pol}}$ does not depend on the choice of the polar chain (S, σ), provided that $\partial(S, \sigma) = (B, \beta)$;*

   (ii) *(anti-)symmetric:*

$$\mathrm{lk}_{\mathrm{pol}}\left((A, \alpha), (B, \beta)\right) = (-1)^{(n-p)(n-q)}\, \mathrm{lk}_{\mathrm{pol}}\left((B, \beta), (A, \alpha)\right);$$

   (iii) *invariant when (A, α) is replaced by a cycle (A', α') polar homologous to (A, α) in the complement of $B \subset X$.*

Remark 4.28 In particular, for $p = q = 1$ we have defined a polar linking of complex curves in a complex threefold, all equipped with volume forms. The polar linking number is symmetric in this dimension. The simplest curves that can have nontrivial linking are elliptic curves: the linking number of a rational curve with any other curve is zero, since any holomorphic differential on a rational curve must vanish.

Exactly this polar linking number of complex curves appears from the holomorphic abelian Chern–Simons path integral, in the same way as the usual linking number comes out of the topological Chern–Simons path integral (see Section 3.4 and [134, 195, 366]).

The polar linking number is also closely related to the Weil pairing of functions on a complex curve and to the Parshin symbols, the higher-dimensional generalizations of the latter; see [309]. One intriguing open question in the area is to find a polar analogue of the *self-linking number* of a framed knot.

4.4 Polar Homology for Affine Curves

So far, we have studied polar homology of projective varieties, which play the role of compact manifolds in usual topology. It is natural to expect that the role of noncompact manifolds should be played by quasi-projective varieties, i.e., Zariski open subsets in projective varieties. In this section, following [195], we indicate how the polar homology theory can be extended to affine curves, one-dimensional quasi-projective varieties.

Let X be an affine curve and let $\bar{X}$ be its projective closure, i.e., X is Zariski open in $\bar{X}$. Denote by $D := \bar{X} \setminus X$ the compactification divisor. We shall define the polar chains for the quasi-projective variety X as a certain subset of polar chains for $\bar{X}$.

Definition 4.29 The space $\mathcal{C}_k(X)$ of polar k-chains in X is defined as the subspace in $\mathcal{C}_k(\bar{X})$ such that the corresponding k-forms vanish on D:

$$\mathcal{C}_k(X) = \{(A, f, \alpha) \in \mathcal{C}_k(\bar{X}) \mid \alpha(x) = 0 \text{ for all } x \in f^{-1}(D) \subset A\}.$$

In particular, for an affine curve X the space $\mathcal{C}_0(X)$ of polar 0-chains is the vector space formed by complex linear combinations of points in X, while for $\mathcal{C}_1(X)$ we consider smooth projective curves A and logarithmic 1-forms α that vanish at $f^{-1}(D) \subset A$.

Definition / Proposition 4.30 *For the affine curve X the spaces $\mathcal{C}_k(X), k = 0, 1$, form a subcomplex in the polar chain complex $(\mathcal{C}_\bullet(\bar{X}), \partial)$, which depends only on X and not on the choice of its compactification, the projective curve $\bar{X}$.*

The resulting homology groups of the chain complex $(\mathcal{C}_\bullet(X), \partial)$ are denoted, as before, by $\mathrm{HP}_k(X)$ and are called polar homology groups *of X.*

Example 4.31 Consider the case $X = \bar{X} \setminus \{P\}$ of a smooth projective curve of genus $\varkappa$ with one point removed. Then $\dim \mathrm{HP}_1(X) = \varkappa - 1$ if $\varkappa \geq 1$ and $\dim \mathrm{HP}_1(X) = 0$ if $\varkappa = 0$. Indeed, the space $\mathrm{HP}_1(X)$ is the space of holomorphic 1-forms on $\bar{X}$ that vanish at P.

To calculate $\mathrm{HP}_0(X)$ in the case $\varkappa \geq 1$ it is sufficient to notice that for any two points $Q_1, Q_2 \in \bar{X} \setminus \{P\}$, the 0-cycle $(Q_1, q_1) + (Q_2, q_2)$ is homologically equivalent to zero if and only if $q_1 + q_2 = 0$ (the same condition as in the case of a nonpunctured curve, which we discussed in the introduction). In the case of $\varkappa = 0$ an analogous statement requires three points to be involved (unlike the case of a nonpunctured projective line): the corresponding 1-form on $\mathbb{CP}^1$ has to have at least one zero, and hence at least three poles. Hence $\dim \mathrm{HP}_0(X) = 1$ if $\varkappa \geq 1$ and $\dim \mathrm{HP}_0(X) = 2$ if $\varkappa = 0$.

We collect the results about the curves in Figure 4.5, where we depict the complex curves by graphs such that polar homology groups of the curves coincide with singular homology groups of the corresponding graphs.

This way the divisor of zeros of a meromorphic form becomes a counterpart of punctures (or infinity) in an open real manifold. (Recall that in the introduction we also observed that the divisor of zeros can also stand for the points of nonsmoothness. These two points of view are consistent, since one can "puncture" a closed singular real variety at its singularities and make it smooth, but open.)

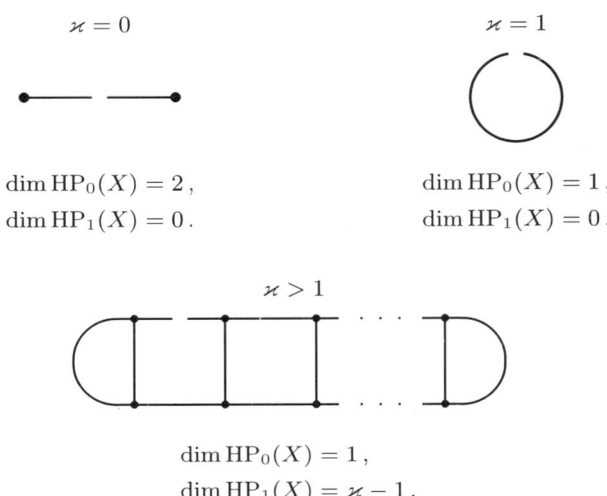

$\varkappa = 0$ $\varkappa = 1$

$\dim \mathrm{HP}_0(X) = 2$, $\dim \mathrm{HP}_0(X) = 1$,
$\dim \mathrm{HP}_1(X) = 0$. $\dim \mathrm{HP}_1(X) = 0$.

$\varkappa > 1$

$\dim \mathrm{HP}_0(X) = 1$,
$\dim \mathrm{HP}_1(X) = \varkappa - 1$.

Fig. 4.5. A smooth projective curve with one point removed, $X = \bar{X} \smallsetminus \{P\}$.

Example 4.32 In a similar way, one can deal with the case $X = \bar{X} \smallsetminus \{P, Q\}$ of a smooth projective curve with two points removed. We summarize the results in Figure 4.6. Here, one has to distinguish between the case of generic points P and Q and the case that $P + Q$ is a special divisor. In the latter case there exist more 1-differentials with zeros at P, Q than generically.

The consistency of the definitions of polar homology in the projective and quasi-projective cases is provided by the following Mayer–Vietoris sequence: the corresponding groups behave like ordinary homology groups with respect to taking unions of Zariski open subsets in the curve $\bar{X}$.

Theorem 4.33 *Let a complex curve* $X = U_1 \cup U_2$ *(either affine or projective) be the union of two Zariski open subsets* U_1 *and* U_2. *Then the following Mayer–Vietoris sequence of chains is exact:*

$$0 \to \mathcal{C}_k(U_1 \cap U_2) \xrightarrow{i} \mathcal{C}_k(U_1) \oplus \mathcal{C}_k(U_2) \xrightarrow{\sigma} \mathcal{C}_k(X) \to 0.$$

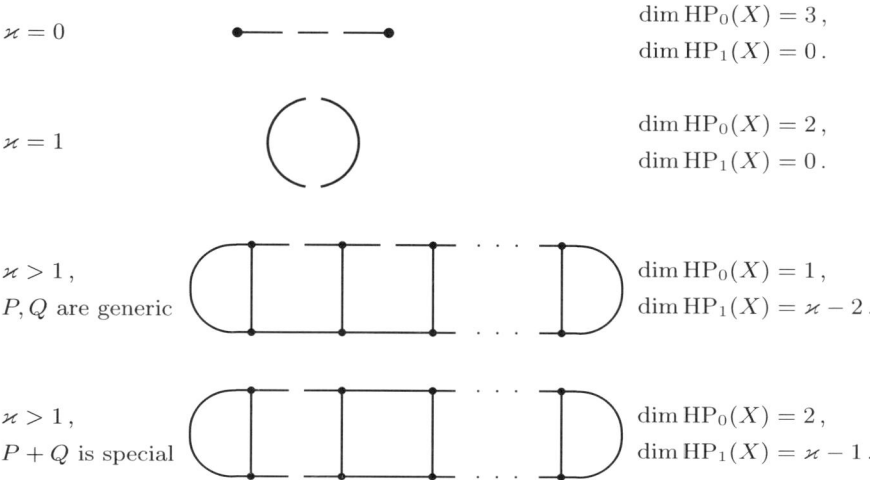

$\varkappa = 0$ $\dim \mathrm{HP}_0(X) = 3\,,$
$\dim \mathrm{HP}_1(X) = 0\,.$

$\varkappa = 1$ $\dim \mathrm{HP}_0(X) = 2\,,$
$\dim \mathrm{HP}_1(X) = 0\,.$

$\varkappa > 1\,,$
P, Q are generic $\dim \mathrm{HP}_0(X) = 1\,,$
$\dim \mathrm{HP}_1(X) = \varkappa - 2\,.$

$\varkappa > 1\,,$
$P + Q$ is special $\dim \mathrm{HP}_0(X) = 2\,,$
$\dim \mathrm{HP}_1(X) = \varkappa - 1\,.$

Fig. 4.6. A smooth projective curve with two points removed, $Z \smallsetminus \{P, Q\}$.

Here the map σ represents the sum of chains,

$$\sigma : a \oplus b \mapsto a + b,$$

and the map i is the embedding of the chain lying in the intersection $U_1 \cap U_2$ as a chain in each subset U_1 and U_2 :

$$i : c \mapsto (c) \oplus (-c).$$

This implies the following exact Mayer–Vietoris sequence in polar homology:

$$\cdots \to \mathrm{HP}_k(U_1 \cap U_2) \xrightarrow{\ i\ } \mathrm{HP}_k(U_1) \oplus \mathrm{HP}_k(U_2)$$
$$\xrightarrow{\ \sigma\ } \mathrm{HP}_k(X) \to \mathrm{HP}_{k-1}(U_1 \cap U_2) \to \cdots$$

Some of the properties of polar homology discussed above can be generalized from affine curves to higher-dimensional quasi-projective varieties, cf. [195], but most of the main features of the theory remain widely open problems in the general case.

4.5 Bibliographical Notes

In our exposition we follow the papers [195, 196]. The de Rham-type theorem for polar homology was proved in [197]. Related results on complex analogues of intersection and linking numbers can be found in [134] and [85], while the paper [26] was a motivation for considering the holomorphic linkings of this type.

The definition of polar chains and the boundary operator has similar features with Abel's theorem [149] and with Ishida's complexes for toric varieties [296].

For a thorough treatment of logarithmic forms see [78], while for multidimensional residues we refer the reader to the detailed monograph [369]; see also [150].

The term "polar homology" was coined in [195], and it originated from consideration of polar divisors of differential forms. One may think that the launch of the Antarctica Journal of Mathematics in 2004, soon after the appearance of the polar theory, was not a simple coincidence.

Appendices

A.1 Root Systems

1.1 Finite Root Systems

Let V be a finite-dimensional real vector space and let $(\, . \,)$ be a *positive definite* symmetric bilinear form on V. For an element $\alpha \in V$ we denote by r_α the reflection in the hyperplane orthogonal to α:

$$r_\alpha(\beta) = \beta - \frac{2(\beta, \alpha)}{(\alpha, \alpha)} \alpha$$

for any point $\beta \in V$.

Definition 1.1 A subset R of V is called a (finite) *root system* if the following axioms are satisfied:

(R1a) The set R is finite and it spans the vector space V.
(R1b) $0 \notin R$.
(R2) For any $\alpha \in R$ the reflection r_α maps the set R to itself.
(R3) If α, $\beta \in R$, then $2(\beta, \alpha)/(\alpha, \alpha) \in \mathbb{Z}$.
(R4) If $\alpha \in R$, then $2\alpha \notin R$.[14]

The elements of a root system R are called roots. Two root systems R and R' are called isomorphic if there exists an isomorphism φ of the corresponding vector spaces V and V' mapping R to R' and such that

$$(\varphi(\beta), \varphi(\alpha))/(\varphi(\alpha), \varphi(\alpha)) = (\beta, \alpha)/(\alpha, \alpha)$$

for all $\alpha, \beta \in R$.

---

[14] Sometimes this is called a *reduced root system*, as opposed to a *nonreduced* one, for which the axiom (R4) is dropped.

Definition 1.2 A root system R is called *irreducible* if it cannot be decomposed into two nonempty subsets R_1 and R_2 such that R_1 is orthogonal to R_2 and $R = R_1 \cup R_2$.

Axioms (R2) and (R3) restrict the relative lengths of and the angles between the roots. For example, in an irreducible root system, only roots of two different lengths can occur, and they are called the *short* and *long roots* respectively. Irreducible root systems can be classified. There are four infinite series (the root systems of types A_n, B_n, C_n, D_n) and five exceptional root systems (the root systems of types E_6, E_7, E_8, F_4, and G_2). Here the index denotes the dimension of the ambient vector space V. Figure 1.1 shows all two-dimensional root systems (note that the root system of type $A_1 \times A_1$ is not irreducible).

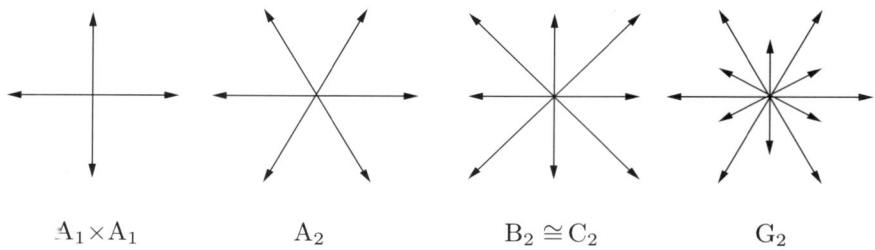

$A_1 \times A_1$ $\qquad\qquad$ A_2 $\qquad\qquad$ $B_2 \cong C_2$ $\qquad\qquad$ G_2

Fig. 1.1. The two-dimensional root systems.

Definition 1.3 Let $R \subset V$ be a root system. The subgroup $W \subset \mathrm{GL}(V)$ generated by the reflections r_α for all $\alpha \in R$ is called the *Weyl group* of the root system R.

Definition 1.4 A subset $\Pi \subset R$ of a root system $R \subset V$ is called a *basis* of R if Π is a basis of the vector space V and if every element $\beta \in R$ can be expressed as $\beta = \sum_{\alpha \in \Pi} m_\alpha \alpha$ such that either all m_α are nonnegative integers or all m_α are nonpositive integers.

Proposition 1.5 *Every root system has a basis. For a given root system R, the Weyl group of R permutes the set of all bases of R.*

After having fixed a basis Π of the root system R, we can decompose R into $R = R_+ \cup R_-$, where $R_+ = \{\beta \in R \mid \beta = \sum_{\alpha \in \Pi} m_\alpha \alpha \text{ with } m_\alpha \geq 0\}$ is the set of *positive roots* of R and $R_- = -R_+$ is the set of *negative roots* of R.

Example 1.6 Let $V = \{x = (x_1, \ldots, x_{n+1}) \in \mathbb{R}^{n+1} \mid x_1 + \cdots + x_{n+1} = 0\}$ with the inner product induced from the standard inner product on $\mathbb{R}^{n+1}$. Then the set

$$R = \{x \in \mathbb{Z}^{n+1} \mid x \in V \text{ and } (x, x) = 2\}$$

is a root system of type A_n. The Weyl group of this root system is given by the symmetric group S_{n+1}, which acts on the entries of the vectors in V by permutation. A basis of R is given by the set $\Pi = \{e_1 - e_2, e_2 - e_3, \dots, e_{n-1} - e_n\}$, where the vector $e_i = (0, \dots, 0, 1, 0, \dots, 0)$ has 1 at the ith entry and 0's everywhere else.

1.2 Semisimple Complex Lie Algebras

Here we shall briefly indicate how the finite root systems appear in the theory of semisimple Lie algebras.

Definition 1.7 A nonabelian Lie algebra $\mathfrak{g}$ is called *simple* if it does not have any nonzero ideals other than $\mathfrak{g}$ itself. A finite-dimensional *semisimple* Lie algebra $\mathfrak{g}$ is a direct sum of nonabelian simple Lie algebras; cf. Definition I.1.31.

Let $\mathfrak{g}$ be a complex finite-dimensional semisimple Lie algebra, and let $\mathfrak{h} \subset \mathfrak{g}$ be its *Cartan subalgebra*, that is, a maximal abelian subalgebra all of whose elements are diagonalizable by the adjoint action. Such subalgebras exist for dimensional reasons. Since the elements of $\mathfrak{h}$ commute, one can diagonalize their adjoint action on $\mathfrak{g}$ simultaneously. The latter allows one to write

$$\mathfrak{g} = \bigoplus_{\alpha \in \mathfrak{h}^*} \mathfrak{g}_\alpha \tag{1.10}$$

with

$$\mathfrak{g}_\alpha = \{X \in \mathfrak{g} \mid [H, X] = \alpha(H)X \text{ for all } H \in \mathfrak{h}\}.$$

One can show that $\mathfrak{g}_0 = \mathfrak{h}$ and that for $\alpha \neq 0$, each nonzero $\mathfrak{g}_\alpha$ is one-dimensional. The set R of all $\alpha \in \mathfrak{h}^*$ with $\alpha \neq 0$ and $\mathfrak{g}_\alpha \neq \{0\}$ is called the *root system of the Lie algebra* $\mathfrak{g}$. Hence, decomposition (1.10) is, in fact, as follows:

$$\mathfrak{g} = \mathfrak{h} \oplus \bigoplus_{\alpha \in R} \mathfrak{g}_\alpha. \tag{1.11}$$

It is called the *root space decomposition* of the semisimple Lie algebra $\mathfrak{g}$.

Example 1.8 For the Lie algebra $\mathfrak{sl}(2, \mathbb{C})$ the root space decomposition is as follows:

$$\mathfrak{sl}(2, \mathbb{C}) = \langle e \rangle \oplus \langle h \rangle \oplus \langle f \rangle.$$

Here the Cartan subalgebra $\mathfrak{h} = \langle h \rangle$ is one-dimensional, and the two roots corresponding to the subspaces $\langle e \rangle$ and $\langle f \rangle$ are opposite in sign: $[h, e] = 2h = -[h, f]$ and $[e, f] = h$.

It turns out that the root system R is a finite root system in the sense of Definition 1.1. Furthermore, the root system R of the Lie algebra $\mathfrak{g}$ determines the semisimple Lie algebra $\mathfrak{g}$ up to an isomorphism: given a root system R, one can construct a finite-dimensional complex semisimple Lie algebra $\mathfrak{g}$ with this root system. Finally, the Lie algebra $\mathfrak{g}$ is simple exactly if its root system is irreducible. So there is a one-to-one correspondence between complex finite-dimensional simple Lie algebras and irreducible root systems of types A_n, B_n, C_n, D_n, E_6, E_7, E_8, F_4, and G_2.

Example 1.9 Consider the Lie algebra $\mathfrak{g} = \mathfrak{sl}(n+1, \mathbb{C})$ with the Cartan subalgebra of diagonal matrices. For $i \neq j$ we define $\alpha_{ij} \in \mathfrak{h}^*$ by $\alpha_{ij}(\operatorname{diag}(q_1, \ldots, q_n)) = q_i - q_j$. Furthermore, let $\mathfrak{g}_{\alpha_{ij}}$ denote the subspace spanned by the matrix with 1 at the (i,j)th entry and 0's everywhere else. Then the root space decomposition of the Lie algebra $\mathfrak{sl}(n+1, \mathbb{C})$ is given by

$$\mathfrak{sl}(n+1, \mathbb{C}) = \mathfrak{h} \oplus \bigoplus_{\alpha_{ij} \in R} \mathfrak{g}_{\alpha_{ij}},$$

where the root system R is of type A_n.

The root system of type B_n corresponds to the Lie algebra $\mathfrak{so}(2n+1, \mathbb{C})$, the root system C_n is related to the Lie algebra $\mathfrak{sp}(2n, \mathbb{C})$, while D_n corresponds to the algebra $\mathfrak{so}(2n, \mathbb{C})$.

1.3 Affine and Elliptic Root Systems

Let V be a real finite-dimensional vector space with a *positive semidefinite* symmetric bilinear form $(\ ,\ )$ and let $\alpha \in V$ be a vector for which $(\alpha, \alpha) \neq 0$. Similarly to the positive definite case, the reflection r_α can be defined via

$$r_\alpha(\beta) := \beta - \frac{2(\alpha, \beta)}{(\alpha, \alpha)}\alpha.$$

Definition 1.10 A subset $R \subset V$ is called an *extended root system* if R satisfies the following axioms:

(ER1a) The additive subgroup $Q(R) = \sum_{\alpha \in R} \mathbb{Z}\alpha$ is a full lattice in V.
(ER1b) $(\alpha, \alpha) \neq 0$ for all $\alpha \in R$.
(ER2) For any $\alpha \in R$ the reflection r_α maps the set R to itself.
(ER3) If α, $\beta \in R$, then $2(\beta, \alpha)/(\alpha, \alpha) \in \mathbb{Z}$.
(ER4) If $\alpha \in R$, then $2\alpha \notin R$.

An extended root system R is called *irreducible* if it cannot be decomposed into two nonempty subsets R_1 and R_2 such that $(R_1, R_2) = 0$.

Axiom (ER1) means that the set $Q(R)$ is the $\mathbb{Z}$-span of an $\mathbb{R}$-basis of V, being an analogue of axiom (R1). For finite root systems, the fact that the set $Q(R)$ is a lattice in V is Proposition 1.5. Note, however, that unless the bilinear form on V is positive definite, the extended affine root systems are never finite.

Let R be an extended root system. The dimension of the radical

$$V_0 = \{v \in V \mid (v, w) = 0 \text{ for all } w \in V\}$$

is called the *nullity* of the root system R. It turns out that the extended root systems of nullity 0 are exactly the finite root systems. Other extended root systems playing a role in this book are those of nullity 1 and 2.

Definition 1.11 An irreducible extended root system of nullity 1 is called an *affine root system*. An extended root system of nullity 2 is called an *elliptic root system*.

Remark 1.12 We note that no isotropic roots are allowed in R. Our definition of extended root systems is taken from [334]. In the literature there exists another definition for extended root systems that includes isotropic roots (see, e.g., [8]), which is shown to be equivalent to the one above; see [30].

The Weyl groups for affine and elliptic root systems are defined in the same way as the Weyl groups for finite root systems.

Taking the quotient of an affine root system by the radical of the bilinear form gives a (possibly nonreduced) finite root system. Hence, one can attempt to classify affine root systems in terms of the underlying finite root systems. The full list of irreducible affine root systems up to isomorphism is $A_n^{(1)}$, $B_n^{(1)}$, $C_n^{(1)}$, $D_n^{(1)}$, $E_6^{(1)}$, $E_7^{(1)}$, $E_8^{(1)}$, $F_4^{(1)}$, $G_2^{(1)}$, $A_{2n+1}^{(2)}$, $A_{2n}^{(2)}$, $D_n^{(2)}$, $E_6^{(2)}$, $E_7^{(2)}$, $D_4^{(3)}$. (Here, the notation for the root systems is the same as in [178].) The root systems of type $X_n^{(1)}$ are called *untwisted affine* root systems; the systems of type $X_n^{(r)}$ with $r > 1$ are called *twisted affine* root systems.

Example 1.13 The untwisted affine root systems can be easily constructed from the corresponding finite root systems: Let $R \subset V$ be a finite root system of type X_n (for example, the root system of type A_n from Example 1.6). Extend the bilinear form (,) trivially from V to the space $\widetilde{V} = V \oplus \mathbb{R}$. Then the set

$$\widetilde{R} = \{(\alpha, n) \in \widetilde{V} \mid \alpha \in R \text{ and } n \in \mathbb{Z}\}$$

is an affine root system of type $X_n^{(1)}$. The construction of the twisted affine root systems is more complicated (see, e.g., [178]).

In the theory of loop groups and affine Lie algebras, the affine root systems play a role similar to that of the finite root systems in the theory of compact Lie groups and semisimple Lie algebras.

Example 1.14 Let $\mathfrak{g}$ be the Lie algebra $\mathfrak{sl}(n+1, \mathbb{C})$ and let $\mathfrak{h} \subset \mathfrak{g}$ denote the Cartan subalgebra of diagonal matrices. Consider the corresponding affine Lie algebra $\widetilde{L\mathfrak{g}}$ (or, rather, $(\widetilde{L\mathfrak{g}})_{\text{pol}}$) whose underlying vector space is given by

$$\widetilde{L\mathfrak{g}} = \mathfrak{g} \otimes \mathbb{C}[z, z^{-1}] \oplus \mathbb{C}\omega \oplus \mathbb{C}d \, ;$$

cf. Remark II.1.14. The Lie bracket on $\widetilde{L\mathfrak{g}}$ is given by

$$[\omega, X(z)] = [\omega, d] = 0, \qquad [d, X(z)] = z\frac{d}{dz}X(z),$$

and

$$[X(z), Y(z)] = [X, Y](z) + \frac{1}{2\pi i}\left(\int_{|z|=1}\left\langle X(z), \frac{d}{dz}Y(z)\right\rangle dz\right) \cdot \omega \, .$$

Here $[X, Y](z)$ denotes the pointwise commutator of X and Y, and $\langle \, , \, \rangle$ is the normalized Killing form on $\mathfrak{g}$. Now set $\widetilde{\mathfrak{h}} := \mathfrak{h} \oplus \mathbb{C}\omega \oplus \mathbb{C}d$ and choose an element $\delta \in (\mathfrak{h} \oplus \mathbb{C}\omega \oplus \mathbb{C}d)^*$ dual to the basis element d, i.e., such that $\delta(d) = 1$, $\delta(\omega) = 0$ and $\delta(X) = 0$ for all $X \in \mathfrak{h}$. Then the set $\widetilde{R} = \{\alpha + m\delta \mid \alpha \in R, \ m \in \mathbb{Z}\}$ is an affine root system of type $A_n^{(1)}$ for the Lie algebra $\widetilde{L\mathfrak{g}}$. In order to write out a root space decomposition for this Lie algebra $\widetilde{L\mathfrak{g}}$, it is convenient to use the notion of an extended root system that includes isotropic roots; see details in [178, 8].

Remark 1.15 The elliptic root systems can be classified in a way similar to affine root systems in terms of the underlying finite root systems (see [8], [334]) and they play a similar role in the theory of double loop algebras (elliptic or extended affine algebras) and double loop groups. Namely, consider the space $\widetilde{\widetilde{V}} := V \oplus \mathbb{R} \oplus \mathbb{R}$ with the bilinear form $(\, , \,)$ extended trivially and a finite root system $R \subset V$. Then the set

$$\widetilde{\widetilde{R}} = \{(\alpha, n, m) \in \widetilde{\widetilde{V}} \mid \alpha \in R \text{ and } n, m \in \mathbb{Z}\}$$

is an elliptic root system, usually denoted by $X_n^{(1,1)}$.

Exercise 1.16 Find a root space decomposition of the $\mathfrak{sl}(2, \mathbb{C})$-elliptic Lie algebra introduced in Definition II.5.1 of Chapter II.

1.4 Root Systems and Calogero–Moser Hamiltonians

In Section II.5.4 we considered a system of n interacting particles moving on the real line and governed by the Hamiltonian function

$$H(q_1, \ldots, q_n, p_1, \ldots, p_n) = \frac{1}{2}\sum_{i=1}^{n}p_i^2 + \sum_{i>j}U(q_i - q_j) \, , \qquad (1.12)$$

where the coordinates q_i denote the positions of the particles and p_i denote their momenta. We looked at the potentials of the form $U(\xi) = 1/\xi^2$ (the rational potential), $U(\xi) = 1/\sin^2 \xi$ or $U(\xi) = 1/\sinh^2 \xi$ (the trigonometric/hyperbolic potential), and, finally, the elliptic potentials $U(\xi) = \wp(\xi; \tau)$, where $\wp(\,.\,, \tau)$ denotes the Weierstrass $\wp$-function with the periods 1 and τ.

It has been observed in [298] that the Hamiltonian (1.12) can be generalized to the Hamiltonian

$$H(q,p) = \frac{1}{2}(p,p) + \frac{1}{2} \sum_{\alpha \in R} m_{|\alpha|} U(\alpha(q)), \qquad (1.13)$$

where $R \subset V$ is any finite root system in the vector space V with inner product (,), and $m_{|\alpha|}$ is a constant that depends only on the length of the root α. The momentum vector p is an element of the vector space V, and the position q is an element of the dual space V^*. The Hamiltonian systems (1.13) are integrable for any finite irreducible root system R [298, 50].

Here we would like to stress the fact that the three types of Hamiltonians (rational, trigonometric, and elliptic) are naturally related to the three types of root systems (finite, affine, and elliptic) considered in this section. Indeed, consider, for instance, the elliptic case. Let $R \subset V$ be an elliptic root system, and let V^* denote the dual space of the finite-dimensional vector space V. The radical V_0 of the bilinear form (,) has $\dim V_0 = 2$ in the elliptic case. Upon choosing a decomposition $V^* = \mathfrak{h} \oplus V_0^*$ one can denote elements of the space V^* by triples (q, a, b), where $q \in \mathfrak{h}$ and $(a, b) \in V_0^*$. Now we can write down the potential on $\mathfrak{h}$ corresponding to the elliptic root system R as follows:

$$U(q) = \frac{1}{2} \sum_{\alpha \in R} \left(\frac{1}{(\alpha(q, 1, \tau))^2} - \frac{1}{(\alpha(0, 1, \tau))^2} \right) \qquad (1.14)$$

for some fixed $\tau \in \mathbb{C}$ with $\operatorname{Im} \tau > 0$. Note that for the root system of type $A_{n-1}^{(1,1)}$ from Example 1.15, the potential U defined by the expression (1.14) can be rewritten via the Weierstrass $\wp$-function and coincides with the potential of the standard elliptic Calogero–Moser system. Using the classification of irreducible elliptic root systems, one shows that the potential (1.14) defines a holomorphic function on the complexified vector space $\mathfrak{h} \otimes \mathbb{C}$. It is known for many elliptic root systems (and, apparently, is the case for all of them) that the corresponding Hamiltonian systems are integrable [51].

Exercise 1.17 Write down the potential U for the affine root system $A_{n-1}^{(1)}$ and verify that it coincides with the potential of the trigonometric Calogero–Moser system.

More facts on root systems can be found in [56] and [168]. The latter is also a good reference for finite-dimensional semisimple complex Lie algebras. For the affine root systems and Lie algebras we refer the reader to [178], while the extended root systems of higher nullity (and, in particular, the elliptic ones), along with the corresponding Lie algebras, have been studied in [8] and [334].

A.2 Compact Lie Groups

In this appendix we recall some basic facts from the theory of compact Lie groups (see, e.g., [60] for more details and references). Throughout this section, G denotes a finite-dimensional compact connected Lie group with the Lie algebra $\mathfrak{g}$. The compact Lie groups that are abelian are the easiest ones to describe. Indeed, any n-dimensional compact connected abelian Lie group is isomorphic to the torus $T = \mathbb{R}^n/\mathbb{Z}^n$. The other extreme is formed by simple Lie groups.

2.1 The Structure of Compact Groups

Definition 2.1 A compact connected Lie group G is called *semisimple* if its center is finite. The group G (of dim > 1) is called *simple* if it does not have any nontrivial normal connected subgroups.

Recall that any Lie group G acts on its Lie algebra by the adjoint representation. If the group G is compact there exists a negative definite invariant symmetric bilinear form on the Lie algebra $\mathfrak{g}$. This form is, in general, not unique.

Definition / Proposition 2.2 *If the compact Lie group G is simple, there exists a unique (up to a scalar factor) symmetric and negative definite invariant bilinear form on its Lie algebra $\mathfrak{g}$, called the* Killing form.

Remark 2.3 For a finite-dimensional Lie algebra $\mathfrak{g}$, the *Killing form* is defined as $\langle X, Y \rangle := \mathrm{tr}(\mathrm{ad}_X \circ \mathrm{ad}_Y)$. The Cartan criterion states that the Killing form is nondegenerate exactly when the Lie algebra $\mathfrak{g}$ is semisimple. If $\mathfrak{g}$ is simple then any invariant symmetric bilinear form on $\mathfrak{g}$ is a scalar multiple of the Killing form. Furthermore, the Lie algebra $\mathfrak{g}$ corresponds to a compact Lie group exactly if the Killing form $\langle X, Y \rangle := \mathrm{tr}(\mathrm{ad}_X \circ \mathrm{ad}_Y)$ is negative definite. For matrix Lie algebras this form is a multiple of the trace form $\mathrm{tr}(XY)$. For most purposes in this book we use the trace version of the form. Often it is convenient to change the sign and think of the positive definite Killing form (see, for example, the discussion of roots in Appendix A.1, where the Killing form is used to measure length of the roots).

Example 2.4 The group $\mathrm{SU}(n)$ is simple. Its center is isomorphic to the cyclic group $\mathbb{Z}/n\mathbb{Z}$. The Killing form $\langle X, Y \rangle := \mathrm{tr}(\mathrm{ad}_X \circ \mathrm{ad}_Y)$ on the Lie algebra $\mathfrak{su}(n)$ is given by the trace: $\langle X, Y \rangle = -\mathrm{tr}(XY)$.

The importance of simple compact groups is clear from the following structure theorem.

Theorem 2.5 *Let G be a compact connected Lie group. Then G is isomorphic to an "almost direct product"*

$$G \cong (G_1 \times \cdots \times G_k \times T)/Z \,,$$

where the G_i are simply connected simple Lie groups, T is a compact connected abelian Lie group (i.e., a torus), and Z is a finite subgroup of the center of $G_1 \times \cdots \times G_k \times T$. In particular, every simply connected compact Lie group is a direct product of simply connected simple Lie groups.

An important role in the theory of compact Lie groups is played by certain abelian subgroups: a *maximal torus* of the compact connected Lie group G is a maximal connected abelian subgroup $T \subset G$. For dimensional reasons, maximal tori exist. Furthermore, one can show that every element of a compact connected Lie group G is contained in some maximal torus of G.

Theorem 2.6 *Any two maximal tori of a compact connected Lie group G are conjugate in G. In particular, every element of the group G is conjugate to an element inside a fixed maximal torus $T \subset G$.*

The exponential map for compact tori is surjective. Together with Theorem 2.6 this observation implies the following corollary.

Corollary 2.7 *For any compact connected Lie group G the exponential map $\exp : \mathfrak{g} \to G$ is surjective.*

Definition / Proposition 2.8 *Let $T \subset G$ be a maximal torus, and denote by $N(T)$ its normalizer in G, i.e., the elements of G that conjugate T to itself. Then the group $W(T) = N(T)/T$ is a finite group, called the* Weyl group *of G with respect to T.*

Theorem 2.9 *Let $T \subset G$ be a maximal torus of G. Then two elements t, $t' \in T$ are conjugate under G if and only if they are conjugate under $W(T)$. In particular, the set of conjugacy classes of a compact connected Lie group G can be identified with the set $T/W(T)$.*

The quotient $T/W(T)$ is called the *fundamental alcove* of the group G.

Example 2.10 A maximal torus T of the group $G = \mathrm{SU}(n)$ is given by the set of diagonal matrices. So in this case, Theorem 2.6 is merely a reformulation of the fact that every element of the group $\mathrm{SU}(n)$ is diagonalizable. The Weyl group $W(T)$ of $\mathrm{SU}(n)$ is isomorphic to the symmetric group S_n, which acts on the torus T by permuting the elements on the diagonal. Thus, defining $t := \mathrm{diag}(e^{2\pi i a_1}, \ldots, e^{2\pi i a_n})$, we can identify the set of conjugacy classes T/S_n with the set $\{(a_1, \ldots, a_n) \in \mathbb{R}^n \mid a_1 \geq \cdots \geq a_n, \ \sum a_i = 0 \text{ and } a_1 - a_n \leq 1\}$.

Our next goal is to describe the fundamental group of the compact Lie group G. In view of Theorem 2.5, it is enough to consider the case of a semisimple group G. First we sidestep slightly and introduce the root lattice of G. Let us fix a maximal torus $T \subset G$, and denote its Lie algebra by $\mathfrak{h}$; cf. Appendix A.1. (The latter is called a *Cartan subalgebra* of the Lie algebra $\mathfrak{g}$.) Recall that the group G acts on its Lie algebra $\mathfrak{g}$ and hence on the complexification $\mathfrak{g} \otimes \mathbb{C}$ by the adjoint action. Since the maximal torus $T \subset G$ is an abelian group, we can simultaneously diagonalize its action on the complexified Lie algebra $\mathfrak{g} \otimes \mathbb{C}$. This allows one to decompose the vector space $\mathfrak{g} \otimes \mathbb{C}$ into

$$\mathfrak{g} \otimes \mathbb{C} = \bigoplus_{\alpha \in \mathfrak{h}^*} \mathfrak{g}_\alpha \,,$$

where $\mathfrak{g}_\alpha$ denotes the space

$$\mathfrak{g}_\alpha = \{X \in \mathfrak{g} \otimes \mathbb{C} \,|\, [H, X] = \alpha(H)X \text{ for all } H \in \mathfrak{h}\} \,.$$

Here the nonzero elements $0 \neq \alpha \in \mathfrak{h}^*$ of the dual Cartan subalgebra $\mathfrak{h}$ for which $\mathfrak{g}_\alpha \neq \{0\}$ are called the *roots* of the Lie algebra $\mathfrak{g}$ (or of the corresponding Lie group G). We denote the set of roots of the group G by R. The set R spans a subspace of the dual space $\mathfrak{h}^*$.

Example 2.11 The compact group $G = \mathrm{SU}(2)$ has the Lie algebra of skew-Hermitian matrices $\mathfrak{g} = \mathfrak{su}(2)$. The complexification of the latter is the Lie algebra $\mathfrak{g} \otimes \mathbb{C} = \mathfrak{sl}(2, \mathbb{C})$ with the decomposition $\mathfrak{sl}(2, \mathbb{C}) = \langle e \rangle \oplus \langle h \rangle \oplus \langle f \rangle$. It has two opposite roots; see Example A.1.8.

Proposition 2.12 *If the group G is semisimple, the set R spans the whole of the Cartan dual $\mathfrak{h}^*$, and R is a finite root system in the sense of Definition A.1.1. Furthermore, in this case, the Weyl group $W(T)$ is isomorphic to the Weyl group of the root system R.*

Consider the roots of a semisimple group G, which span the space $\mathfrak{h}^*$. Denote by $Q^\vee \subset \mathfrak{h}$ the *co-root lattice*, i.e., the set

$$Q^\vee = \{\beta \in \mathfrak{h} \,|\, \langle \beta, \alpha \rangle \in \mathbb{Z} \text{ for all } \alpha \in R\} \,.$$

Furthermore, let $I \subset \mathfrak{h}$ denote the kernel of the exponential map $\exp : \mathfrak{h} \to T$. This is a lattice in the vector space $\mathfrak{h}$. The co-root lattice $Q^\vee$ is a subset of the lattice I, and we have the following theorem.

Theorem 2.13 *Suppose the compact Lie group G is semisimple. Then the fundamental group $\pi_1(G)$ is isomorphic to $I/Q^\vee$.*

Theorem 2.14 (Hopf) *The de Rham cohomology $H^*(G, \mathbb{R})$ of a compact semisimple Lie group G is isomorphic to the de Rham cohomology of a product of odd-dimensional spheres. In particular, if the group G is simple we have $H^2(G, \mathbb{R}) = 0$ and $H^3(G, \mathbb{R}) = \mathbb{R}$.*

Note also that for a simply connected Lie group G we always have $H^1(G, \mathbb{R}) = 0$.

Remark 2.15 The de Rham cohomology of the compact Lie group G is isomorphic to the Lie algebra cohomology of the Lie algebra $\mathfrak{g}$ with values in the trivial module. Actually, the Lie algebra cohomology can be defined as the cohomology of the complex of left-invariant differential forms on the corresponding group G. Note that cohomology classes of the compact group G can be represented by left-invariant forms by averaging over the group any closed form from a given cohomology class. We note also that any finite-dimensional Lie group can be contracted to (and consequently it has the same cohomology as) its maximal compact subgroup.

2.2 A Cohomology Generator for a Simple Compact Group

In Section II.1.3, the cohomology space $H^3(G, \mathbb{Z})$ for a simple compact Lie group G played an important role. We used an explicit form of the generator of $H^3(G, \mathbb{Z})$, which we recall below.

Proposition 2.16 *Suppose G is a simple compact simply connected Lie group. Then the integral cohomology $H^3(G, \mathbb{Z})$ is generated by the following left-invariant closed 3-form η:*

$$\eta = \frac{1}{24\pi^2} \operatorname{tr}(g^{-1}dg)^{\wedge 3}.$$

PROOF. We use the fact that $H_3(G, \mathbb{Z})$ is generated by embeddings of SU(2) into G corresponding to the roots of G (see, e.g., [52]). Furthermore, for the long roots, the Killing form of G restricts to the Killing form on SU(2), so it suffices to prove the proposition for SU(2). The generator of $H_3(\mathrm{SU}(2), \mathbb{Z})$ is given by the group SU(2) itself. Hence we just have to integrate the form η over SU(2) and verify that the result is equal to 1. Let us parametrize the group SU(2) by matrices

$$\begin{pmatrix} a & b \\ -\bar{b} & \bar{a} \end{pmatrix}$$

with

$$a = \cos\theta \cdot e^{i\phi} \quad \text{and} \quad b = \sin\theta \cdot e^{i\psi},$$

where the parameters ϕ, ψ assume values in $[0, 2\pi)$, and θ assumes values in $[0, \pi/2]$. Then a (lengthy) calculation shows that in these coordinates, the 3-form η becomes

$$\eta = \frac{1}{4\pi^2} \sin(2\theta) d\theta \wedge d\phi \wedge d\psi.$$

Integrating this over the range of our coordinates, we obtain $\int_{\mathrm{SU}(2)} \eta = 1$. $\square$

A.3 Krichever–Novikov Algebras

The Krichever–Novikov algebras are infinite-dimensional Lie algebras generalizing the Virasoro algebra and affine Lie algebras to higher-genus Riemann surfaces. The coadjoint orbits of the affine Krichever–Novikov algebras are related to holomorphic vector bundles on Riemann surfaces and have a description somewhat similar to the coadjoint orbits of the elliptic Lie groups from Section II.5.2. The place of the Krichever–Novikov algebras among other algebras discussed above is shown in the table at the end of this appendix. In a sense, they provide an (almost) graded version of affine and Virasoro algebras on graphs in the same way as the affine/Virasoro algebras on $\mathbb{C}^*$ are graded versions of the corresponding algebras on S^1.

3.1 Holomorphic Vector Fields on $\mathbb{C}^*$ and the Virasoro Algebra

To get started, recall the graded version of the definition of the Virasoro algebra, which is the universal central extension of the Lie algebra of vector fields on the circle. We restrict our attention to the algebra $\mathrm{Vect}_{\mathrm{pol}}(S^1)$ of polynomial vector fields on S^1, which form a dense subset in the space of all smooth vector fields on S^1. A basis of this space is given by the set $\{L_n = ie^{in\theta}\frac{d}{d\theta}\}_{n\in\mathbb{Z}}$ with the commutation relations

$$[L_n, L_m] = (m - n)L_{n+m} .$$

This shows that the Lie algebra of polynomial vector fields on the circle is a $\mathbb{Z}$-graded Lie algebra, i.e., it admits a decomposition

$$\mathrm{Vect}_{\mathrm{pol}} = \bigoplus_{n\in\mathbb{Z}} V_n \quad \text{with} \quad [V_n, V_m] \subset V_{n+m} .$$

Furthermore, its unique central extension is given by the Gelfand–Fuchs 2-cocycle

$$\omega\left(f\frac{d}{d\theta}, g\frac{d}{d\theta}\right) = \frac{1}{2\pi i}\int_{S^1} f'(\theta)g''(\theta)d\theta ;$$

see Section II.2.1. In the basis L_n the cocycle ω reads

$$\omega(L_n, L_m) = \delta_{n,-m}n^3 .$$

Identify the Lie algebra $\mathrm{Vect}_{\mathrm{pol}}(S^1)$ with the Lie algebra $\mathcal{L}$ of holomorphic vector fields on $\mathbb{C}^*$ that have at most finite-order poles at 0 and ∞. A standard basis for the latter Lie algebra is given by $\{e_n = z^{n+1}\frac{d}{dz}\}_{n\in\mathbb{Z}}$ with the Lie bracket $[e_n, e_m] = (m - n)e_{n+m}$, so that this algebra is indeed isomorphic to the algebra of polynomial vector fields on the circle. One can view $\mathbb{C}^*$ as the Riemann sphere $\mathbb{CP}^1$ with two points removed. Thus the Lie algebra $\mathrm{Vect}_{\mathrm{pol}}$ of polynomial vector fields on $\mathbb{C}^*$ can be viewed as the Lie algebra of meromorphic vector fields on $\mathbb{CP}^1$ that are holomorphic outside the points

0 and ∞. Krichever and Novikov [219, 220] studied the corresponding Lie algebra of meromorphic vector fields on a higher-genus Riemann surface Σ that are holomorphic outside two fixed points $P, Q \in \Sigma$. As we shall see, many basic properties of the Virasoro algebra carry over to these more general Lie algebras.

3.2 Definition of the Krichever–Novikov Algebras and Almost Grading

Now we consider a compact Riemann surface Σ of genus $\kappa \geq 2$. Let K be the canonical bundle of Σ and let $T := K^*$ denote the holomorphic tangent bundle of Σ. The sections of T are holomorphic vector fields on the surface Σ, while the sections of the line bundle K are holomorphic differentials, i.e., holomorphic 1-forms on Σ. It is known that the degree of the holomorphic tangent bundle is $\deg(T) = -2\kappa+2$, so that for $\kappa \geq 2$ there are no holomorphic vector fields on the surface Σ.

Definition 3.1 Let $P, Q \in \Sigma$ be two points in general position[15] on the surface Σ. The *Krichever–Novikov* algebra $\mathcal{L}$ is the Lie algebra of meromorphic vector fields on the surface Σ that are holomorphic outside the points $P, Q \in \Sigma$.

The Lie algebra $\mathcal{L}$ admits an almost grading, which generalizes the $\mathbb{Z}$-grading of the Lie algebra of polynomial vector fields on S^1. (This and many other results for the Krichever–Novikov algebras generalize to the case of more than two points, as well as to genus $\kappa = 1$; see, e.g., [337, 354] and references therein.)

Definition 3.2 A Lie algebra $\mathfrak{g}$ is called *almost graded* if it admits a direct sum decomposition $\mathfrak{g} = \bigoplus_{n \in \mathbb{Z}} \mathfrak{g}_n$ with finite-dimensional spaces $\mathfrak{g}_n$ and there exists a constant $N \in \mathbb{N}$ such that

$$[\mathfrak{g}_n, \mathfrak{g}_m] \subset \bigoplus_{l=n+m-N}^{n+m+N} \mathfrak{g}_l \quad \text{for all} \quad n, m \in \mathbb{Z}.$$

To introduce an almost graded structure on the Lie algebra $\mathcal{L}$, we set $\kappa_0 = 3\kappa/2$ and $J = \mathbb{Z} + \kappa/2$ (i.e., $J = \mathbb{Z}$ if the genus κ of the surface is even and $J = \mathbb{Z} + 1/2$ if κ is odd). For any $j \in J$ consider the line bundle

---

[15] This is a technical condition that is used in the proof of Proposition 3.3: By "general position," we mean that the points P, Q should not be 2-Weierstrass points. A point $P \in \Sigma$ is a 2-Weierstrass point if there exists a quadratic differential (i.e., a section of $K^{\otimes 2}$) with a zero of order $\geq 3\kappa - 3$ at the point P. It is known that there are only finitely many 2-Weierstrass points on a compact Riemann surface of genus $\kappa \geq 2$.

$$M_j = T \otimes L_P^{-j+\kappa_0-1} \otimes L_Q^{j+\kappa_0-1} \, ,$$

where L_P denotes the line bundle on Σ corresponding to the point $P \in \Sigma$. (Such a bundle is glued from the trivial bundles over a neighborhood U_0 of the point P and over $U_1 = \Sigma \setminus \{P\}$. Then the bundle L_P is defined by the transition function $g_{01} = z$, where z is a local coordinate around P in U_0.) Since the degree of the holomorphic tangent bundle T of the surface Σ is $\deg(T) = -2\kappa + 2$, we obtain $\deg(M_j) = \kappa$ for each $j \in J$.

Proposition 3.3 *For each $j \in J$ the space of holomorphic sections of the bundle M_j is one-dimensional, i.e., $\dim H^0(\Sigma, M_j) = 1$.*

The proof of this proposition is based on a repeated application of the Riemann–Roch theorem and it uses the fact that the points P and Q are in general position; see [219] or [336].

A holomorphic section of the line bundles M_j corresponds to a meromorphic vector field e_j on the surface Σ with a zero of order at least $j - \kappa_0 + 1$ at the point P and a pole of order at most $j + \kappa_0 - 1$ at the point Q. (As usual, a pole of negative order is a zero and vice versa.) In a local coordinate z around the point P, such a vector field can be written as

$$e_j = a_j z^{j-\kappa_0+1} \frac{d}{dz} + \text{terms of higher order in } z \,. \tag{3.15}$$

Similarly, in a local coordinate w around Q, the vector field e_j can be written as

$$e_j = b_j w^{-j-\kappa_0+1} \frac{d}{dw} + \text{terms of higher order in } w \,. \tag{3.16}$$

Fixing $a_j = 1$ determines the section e_j uniquely.

Theorem 3.4 ([219]) *The meromorphic vector fields e_j with $j \in J$ form a basis of the Lie algebra $\mathcal{L}$. In this basis, the Lie bracket is given by*

$$[e_i, e_j] = \sum_{k=i+j-\kappa_0}^{i+j+\kappa_0} c_{i,j}^k \, e_k \tag{3.17}$$

for some $c_{i,j}^k \in \mathbb{C}$. In particular, the Krichever–Novikov algebra $\mathcal{L}$ is almost graded.

PROOF. We first prove that the vector fields e_j form a basis of the Lie algebra $\mathcal{L}$. From formulas (3.15), (3.16) we note that the field e_j is holomorphic at the point P if $j \geq \kappa_0 - 1$ and it is holomorphic at the point Q if $j \leq -\kappa_0 + 1$. Now let v be an arbitrary meromorphic vector field with poles only at the points P and Q. By adding a linear combination of e_j with $j \geq \kappa_0 - 1$ we can achieve that the pole of v at the point Q has order $\leq 3\kappa - 3$ without changing the pole at P. Similarly, by adding multiples of the fields e_j with $j \leq \kappa_0 + 1$ we can achieve that the pole of v at the point P has order $\leq 3\kappa - 3$ without changing the pole at Q. Finally, by adding multiples of the fields e_j with $|j| \leq \kappa_0$, we can remove the pole of the field v at P, so that the order of the pole at Q is still of order $\leq 3\kappa - 3$. But then v has to vanish identically. Indeed, otherwise we set n to be the order of the pole of the field v at Q and note that $n \leq 3\kappa - 3$. We also have $n \neq 0$, since for $\kappa > 1$ any nonzero meromorphic vector field on Σ has at least one pole. So v cannot be a nonzero multiple of $e_{n-\kappa_0+1}$, which has a pole at P. Finally, multiplying the vector field v by holomorphic sections of $L_P^{-n+3\kappa-2}$ and L_Q^n, we obtain from it a holomorphic section $\tilde{v}$ of $M_{n-\kappa_0+1} = T \otimes L_P^{-n+3\kappa-2} \otimes L_Q^n$. But this is a contradiction to $\dim H^0(\Sigma, M_j) = 1$ for all $j \in J$. This shows that the vector field v can be written as a linear combination of the vector fields e_j.

To verify equation (3.17) for the commutator of two vector fields e_i and e_j, we calculate their commutator locally at the point P. Using equation (3.15), we get

$$[e_i, e_j] = (j - i)z^{(i+j-\kappa_0)-\kappa_0+1}\frac{d}{dz} + \text{terms of higher order in } z \,,$$

so that the commutator has to be a linear combination of e_k with $k \geq i+j-\kappa_0$. Similarly, at the point Q we get

$$[e_i, e_j] = b_i b_j (j - i)w^{(-i-j-\kappa_0)-\kappa_0+1}\frac{d}{dw} + \text{terms of higher order in } w \,.$$

Hence the fields e_k appearing in the commutator all have the indices $k \leq i + j + \kappa_0$, which proves the theorem. $\square$

3.3 Central Extensions

To define central extensions of the Krichever–Novikov algebras $\mathcal{L}$, we note that although there are several possible central extensions of them, essentially, just one of them behaves nicely with respect to the almost grading of $\mathcal{L}$. First we recall the notion of a projective connection.

Definition 3.5 Let (U_α, z_α) be a covering of the Riemann surface Σ by holomorphic coordinate charts with transition functions $z_\beta = \varphi_{\beta\alpha}z_\alpha$ on $U_\alpha \cap U_\beta$. A system of local holomorphic (meromorphic) functions $R_\alpha : U_\alpha \to \mathbb{C}$ is called a *holomorphic (meromorphic) projective connection* if it transforms according to the rule

$$R_\beta(z_\beta)(\varphi'_{\beta\alpha})^2 = R_\alpha(z_\alpha) + S(\varphi_{\beta\alpha}) \,, \tag{3.18}$$

where $S(\varphi)$ denotes the Schwarzian derivative

$$S(\varphi) = \frac{\varphi'''}{\varphi'} - \frac{3}{2}\left(\frac{\varphi''}{\varphi'}\right)^2 \,.$$

It is a classical result that on any compact Riemann surface there exists a holomorphic projective connection (see more detail in, e.g., [155]). Furthermore, equation (3.18) implies that the difference of two holomorphic projective connections is a quadratic differential.

Now define central extensions of the Krichever–Novikov algebra as follows. Fix a projective connection R on the Riemann surface Σ, and let C be any differentiable contour on Σ.

Exercise 3.6 Show that the map

$$\gamma_C(f, g) = \frac{1}{2\pi i} \int_C \left(\frac{1}{2}(g''' f - f''' g) - R \cdot (g' f - f' g) \right) dz \qquad (3.19)$$

defines a 2-cocycle on the Lie algebra $\mathcal{L}$. Prove that the cohomology class of the 2-cocycle γ_C does not depend on the projective connection R. (Hint: use the fact that the difference of two projective connections is a quadratic differential.)

It can be shown that every 2-cocycle of the Krichever–Novikov algebra $\mathcal{L}$ is cohomologous to a linear combination of the ones defined in Exercise 3.6 (see [379, 381] for the case of continuous cocycles and [359] for the general algebraic case). In general, these cocycles are not consistent with the almost grading of the Lie algebra $\mathcal{L}$. However, among them there are *local cocycles*, which respect the almost grading of $\mathcal{L}$ in the following sense.

Definition 3.7 Let $\mathfrak{g} = \bigoplus_{n \in \mathbb{Z}} \mathfrak{g}_n$ be an almost graded Lie algebra. A 2-cocycle γ on $\mathfrak{g}$ is called *local* if there exist numbers N_1, $N_2 \in \mathbb{Z}$ such that for all n, $m \in \mathbb{Z}$, $\gamma(\mathfrak{g}_n, \mathfrak{g}_m) \neq 0$ implies $N_1 \leq n + m \leq N_2$.

Exercise 3.8 Verify that given a local 2-cocycle on an almost graded Lie algebra $\mathfrak{g}$, the corresponding central extension $\widehat{\mathfrak{g}} = \mathfrak{g} \oplus \mathbb{C}$ defined by the local 2-cocycle has an almost graded structure for $\deg_{\widehat{\mathfrak{g}}}(x, a) := \deg_{\mathfrak{g}}(x)$.

Thus for local 2-cocycles the almost graded structure can be carried over to the central extension. Krichever and Novikov showed that there exists a *unique local* 2-*cocycle* on the Lie algebra $\mathcal{L}$. In order to construct it and the corresponding central extension $\widehat{\mathcal{L}}$, recall one more fact about differential forms on Riemann surfaces (see [336]):

Proposition 3.9 *On any compact Riemann surface Σ with two marked points P, $Q \in \Sigma$, there exists a meromorphic 1-form α that has first-order poles at the points P, Q, and whose residues at these points are $\mathrm{res}_P(\alpha) = -1$ and $\mathrm{res}_Q(\alpha) = +1$. Furthermore, the requirement of pure imaginary periods for the form α over all the homology cycles of Σ fixes this form uniquely.*

Now fix a point $z_0 \in \Sigma$ different from P and Q. Then we can define a function r on Σ by

$$r(z) := \operatorname{Re}\left(\int_{z_0}^{z} \alpha\right). \tag{3.20}$$

The function r is well defined (since the periods of α are purely imaginary), and its level sets

$$C_\tau := \{p \in \Sigma \mid r(p) = \tau\}$$

for $\tau \in \mathbb{R}$ define a (singular) foliation of the surface $\Sigma \setminus \{P, Q\}$.

Theorem 3.10 ([219]) *Fix a regular value τ of the function r. Then the corresponding 2-cocycle γ_{C_τ} on the Krichever–Novikov algebra $\mathcal{L}$ is local. Furthermore, every local 2-cocycle on $\mathcal{L}$ is cohomologous to a scalar multiple of the cocycle γ_{C_τ}.*

Remark 3.11 For $\tau \gg 0$ sufficiently large, the contour C_τ is a simple contour around the point P. One can consider the homomorphism of $\mathcal{L}$ to the algebra of smooth vector fields on $S^1 \approx C_\tau$. This map is injective and the image is dense [219]. One can see that the cocycle γ_{C_τ} descends to (a cocycle cohomologous to) the Gelfand–Fuchs cocycle on smooth vector fields on S^1. Since the latter cocycle is nontrivial, this implies the nontriviality of the central extension $\hat{\mathcal{L}}$.

Note that nontrivial *nonlocal* 2-cocycles may correspond to trivial 2-cocycles in the image under this homomorphism, and hence they might not be reducible to the ones for smooth fields (cf. the table at the end of this appendix).

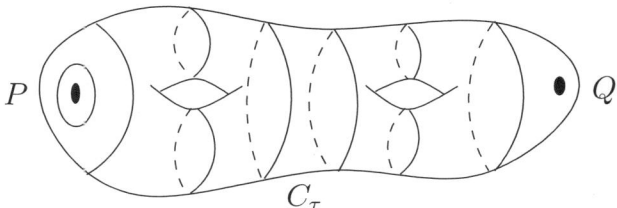

Fig. 3.1. Level sets C_τ of the function $r(z)$ define a foliation of the surface Σ. As $\tau \to \infty$ the curves C_τ become simple contours around P.

Remark 3.12 In a similar way one can define Krichever–Novikov-type algebras of differential operators and pseudodifferential symbols on a Riemann surface [86]. Instead of fixing a projective structure, it is convenient to consider a reference meromorphic vector field v on a surface Σ. Let D_v stand for the Lie derivative along the field v. In a neighborhood of the point P we expand meromorphic pseudodifferential symbols on Σ into a power series in D_v, similar to the case of symbols on the circle; cf. Section II.4. The coefficient at D_v^{-1} can be understood as a meromorphic 1-form on Σ, and let Tr_P be the residue of this 1-form at the point P. Then there is a nontrivial 2-cocycle on meromorphic pseudodifferential symbols defined by

$$c_P(A, B) = \mathrm{Tr}_P([\log D_v, A] \circ B),$$

similarly to the circle case. In particular, one recovers the 2-cocycle (3.19) of the Krichever–Novikov algebra $\mathcal{L}$ of meromorphic vector fields on Σ. These algebras of meromorphic differential operators and pseudodifferential symbols also have an analogue of the almost grading property of $\mathcal{L}$.

3.4 Affine Krichever–Novikov Algebras, Coadjoint Orbits, and Holomorphic Bundles

To describe the affine algebras of Krichever–Novikov-type we fix a compact Riemann surface Σ with two marked points $P, Q \in \Sigma$. Let $\mathcal{F}$ denote the associative algebra of meromorphic functions on Σ that can have poles only at the points P and Q. Furthermore, fix a finite-dimensional complex semisimple Lie algebra $\mathfrak{g}$. The *Krichever–Novikov current algebra* is defined as the Lie algebra of $\mathfrak{g}$-valued meromorphic functions on Σ:

$$\mathcal{KN} := \mathfrak{g} \otimes \mathcal{F}.$$

One can show that the Lie algebra $\mathcal{KN}$ admits an almost grading and there exists a unique local 2-cocycle on the Lie algebra $\mathcal{KN}$ given by

$$\gamma_{C_\tau}^{aff}(X, Y) = \frac{1}{2\pi i} \int_{C_\tau} \mathrm{tr}(X dY),$$

where the contour C_τ is a level set of the function r defined by formula (3.20). The *affine Krichever–Novikov algebra* $\widehat{\mathcal{KN}}$ is the central extension of the current algebra $\mathcal{KN}$ defined by means of the 2-cocycle $\gamma_{C_\tau}^{aff}$. As a vector space it is given by

$$\widehat{\mathcal{KN}} = \mathfrak{g} \otimes \mathcal{F} \oplus \mathbb{C}.$$

Denote by Ω the space of meromorphic 1-forms on the surface Σ that can have poles only at the points P and Q. The pairing $\mathcal{F} \times \Omega \to \mathbb{C}$ between the functions and the forms is given by

$$\langle f, \omega \rangle = \frac{1}{2\pi i} \int_{C_\tau} f\omega,$$

where $f \in \mathcal{F}$ and $\omega \in \Omega$. It turns out to be well defined (i.e., independent of τ) and nondegenerate. This allows one to view the space $\mathfrak{g} \otimes \Omega \oplus \mathbb{C}$ of $\mathfrak{g}$-valued

1-forms on Σ enlarged by a cocentral direction as a "regular part" $\widehat{\mathcal{KN}}^*_{\text{reg}}$ of the dual of the affine Krichever–Novikov algebra $\widehat{\mathcal{KN}} = \mathfrak{g} \otimes \mathcal{F} \oplus \mathbb{C}$. Here the pairing between $\widehat{\mathcal{KN}}$ and $\widehat{\mathcal{KN}}^*_{\text{reg}} = \mathfrak{g} \otimes \Omega \oplus \mathbb{C}$ is given by the natural pairing

$$\langle (X, c), (A, a) \rangle = \frac{1}{2\pi i} \int_{C_\tau} \text{tr}(XA) + c \cdot a \,,$$

for $X \in \mathfrak{g} \otimes \mathcal{F}$ and $A \in \mathfrak{g} \otimes \Omega$.

Let G denote the complex finite-dimensional simply connected Lie group corresponding to the Lie algebra $\mathfrak{g}$. We denote by $G^{\Sigma \backslash \{P, Q\}}$ the group of maps from the Riemann surface Σ to G that are holomorphic outside the points P and Q. The group $G^{\Sigma \backslash \{P, Q\}}$ transforms elements of the space $\widehat{\mathcal{KN}}^*_{\text{reg}} = \mathfrak{g} \otimes \Omega \oplus \mathbb{C}$ by means of the gauge transformations

$$g : (A, a) \mapsto (gAg^{-1} + a(dg)g^{-1}, a) \,.$$

However, the space $\mathfrak{g} \otimes \Omega \oplus \mathbb{C}$ is not invariant under this action. Define an orbit of the group $G^{\Sigma \backslash \{P, Q\}}$ in the space $\widehat{\mathcal{KN}}^*_{\text{reg}} = \mathfrak{g} \otimes \Omega \oplus \mathbb{C}$ to be the intersection of a "true" orbit of $G^{\Sigma \backslash \{P, Q\}}$ with the space $\widehat{\mathcal{KN}}^*_{\text{reg}}$; see [353]. Equivalently, one could consider the subgroup of $G^{\Sigma \backslash \{P, Q\}}$ that leaves the space $\widehat{\mathcal{KN}}^*_{\text{reg}}$ invariant. In any case, these orbits are the analogues of the coadjoint orbits of the centrally extended loop groups that we studied in Section II.1.2. Let $[C_\tau]$ denote the homotopy class of the level curve $C_\tau \subset \Sigma \backslash \{P, Q\}$ for $\tau \gg 0$.

Theorem 3.13 ([353, 354]) *Generic orbits of the group $G^{\Sigma \backslash \{P, Q\}}$ inside a fixed affine hyperplane $a = \text{const} \neq 0$ in $\widehat{\mathcal{KN}}^*_{\text{reg}}$ are in one-to-one correspondence with equivalence classes of representations $\psi : \pi_1(\Sigma \backslash \{P, Q\}) \to G$ of the fundamental group of the punctured surface $\Sigma \backslash \{P, Q\}$ such that $\psi([C_\tau])$ is a semisimple (i.e., diagonalizable) element of the group G.*

For the proof in one direction, one associates to an element $(A, a) \in \mathfrak{g} \otimes \Omega \oplus \mathbb{C}$ the monodromy group of the connection A meromorphic on the Riemann surface Σ and holomorphic on $\Sigma \backslash \{P, Q\}$. In the other direction, one has to check that any monodromy representation indeed comes from such a connection. This is where the assumption on the semisimplicity of $\psi([C_\tau])$ comes into play.

Theorem 3.13 links the theory of affine Krichever–Novikov algebras to the theory of holomorphic G-bundles (or moduli of flat connections) on the Riemann surface Σ, where a similar correspondence of symplectic leaves and monodromy groups arises; see [354] and the references therein for more detail.

Remark 3.14 To visualize the role of the Krichever–Novikov-type algebras among other algebras discussed in the book, we will allow nonlocal 2-cocycles. They can be defined as cocycles γ_C associated to arbitrary contours C on the Riemann surface Σ, not necessarily from the C_τ family. Independent cycles in $H_1(\Sigma \setminus \{P, Q\})$ lead to independent 2-cocycles γ_C. Note that all cycles C_τ define a one-dimensional subspace in the homology group $H_1(\Sigma \setminus \{P, Q\})$.

While the affine/Virasoro algebras on $\mathbb{C}^*$ are graded versions of the corresponding smooth affine/Virasoro algebras on S^1, the Krichever–Novikov type algebras provide an almost graded version of *affine and Virasoro algebras on arbitrary graphs* on Riemann surfaces. Different cycles in the graphs define non-cohomologous 2-cocycles of the corresponding current algebras on those graphs, similar to different 2-cocycles γ_C.

In this way, both the affine/Virasoro and the Krichever–Novikov-type algebras are Lie algebras of currents or vector fields on one-dimensional objects (either in the smooth or in the (almost) graded versions). These objects correspond to the $\mathbb{R}$-part in our complexification table (see Section III.4), since these currents depend on *one variable* (either real or complex). Meanwhile, the $\mathbb{C}$-part of the table corresponds to the Etingof–Frenkel-type algebras on an elliptic curve or on complex curves of higher genus. Here an elliptic curve is a complexification of a circle, while arbitrary complex curves of higher genus can be regarded as complexifications of graphs. In the Etingof–Frenkel algebras their elements, i.e., currents, depend on *two independent variables*. This discussion is summarized in the following table:

"Real" (almost) graded version	Real smooth version	Complex smooth version
Graded affine algebras on $\mathbb{C}^* = \mathbb{CP}^1 \setminus \{P, Q\}$	Smooth affine algebras on S^1	Etingof–Frenkel algebras on elliptic curves $\mathcal{E}$
Krichever–Novikov current algebras on Riemann surfaces $\Sigma \setminus \{P_1, \ldots, P_n\}$	Smooth affine algebras on graphs Γ	Etingof–Frenkel algebras on complex curves $\mathcal{C}$

The affine algebras on graphs and their relation to the Etingof–Frenkel algebras on curves of higher genus seem to be very interesting objects of study, where still most of the work is yet to be done.

For more facts on Krichever–Novikov algebras and the proofs of the theorems in this section we refer to the original papers by Krichever and Novikov [219, 220], the works [336, 337, 353, 354, 379], and the references therein.

A.4 Kähler Structures on the Virasoro and Loop Group Coadjoint Orbits

4.1 The Kähler Geometry of the Homogeneous Space $\mathrm{Diff}(S^1)/S^1$

Symplectic Structures

Consider the group $\mathrm{Diff}(S^1)$ of orientation-preserving diffeomorphisms of the circle S^1, and identify S^1 with the subgroup of rigid rotations $\mathrm{Rot}(S^1)$. In this section we study geometric structures on the quotient

$$\mathcal{S} = \mathrm{Diff}(S^1)/S^1$$

that are invariant under the natural (left) action of the group $\mathrm{Diff}(S^1)$ on $\mathcal{S}$.

The invariant structures are determined by their values at the point $e \in \mathcal{S}$. The tangent space at this point can be identified with the space of vector fields on S^1 with zero mean:

$$T_e\mathcal{S} \cong \mathrm{Vect}_0(S^1) = \left\{ \xi(\theta)\partial_\theta \in \mathrm{Vect}(S^1) \mid \int_{S^1} \xi(\theta)\, d\theta = 0 \right\}.$$

(As usual, we denote a coordinate on S^1 by θ and ∂_θ stands for $\frac{\partial}{\partial\theta}$.) Regard the vector field ξ as an element of the complexified tangent space $T_e\mathcal{S} \otimes \mathbb{C}$ and expand ξ into the Fourier series

$$\xi = \sum_{n \neq 0} \xi_n L_n \,,$$

where $\{L_n\}_{n\in\mathbb{Z}}$ denotes the standard basis of $\mathrm{Vect}_\mathbb{C}(S^1)$, i.e., $L_n = ie^{in\theta}\partial_\theta$. The real Lie algebra $\mathrm{Vect}(S^1) \subset \mathrm{Vect}_\mathbb{C}(S^1)$ is singled out by the relations $-\bar{\xi}_n = \xi_{-n}$.

Any invariant symplectic structure on the space $\mathcal{S}$ is also determined by its values at $e \in \mathcal{S}$. In particular, it is defined by a continuous 2-cocycle ω on the Lie algebra $\mathrm{Vect}_\mathbb{C}(S^1)$ that is invariant under rotations. More generally, invariant closed 2-forms on coadjoint orbits of a Lie group are related to 2-cocycles on the corresponding Lie algebra in the following way.

Exercise 4.1 Let G be a Lie group with Lie algebra $\mathfrak{g}$, and $\mathcal{O} \subset \mathfrak{g}^*$ is a group coadjoint orbit. Given a point $a \in \mathcal{O}$ of the orbit, show that G-invariant closed 2-forms ω on the orbit $\mathcal{O}$ are in one-to-one correspondence with 2-cocycles on the Lie algebra $\mathfrak{g}$ that vanish on the stabilizer of the point a. (Hint: Identify the tangent space $T_a\mathcal{O}$ for $a \in \mathcal{O}$ with a quotient of $\mathfrak{g}$. Use this to lift the G-invariant 2-form ω on $\mathcal{O}$ to a bilinear form on $\mathfrak{g}$. The closedness of ω on $\mathcal{O}$ is equivalent to the cocycle identity of this bilinear form on $\mathfrak{g}$.)

In the case of the algebra $\mathrm{Vect}(S^1)$, the proof of Proposition II.2.3 shows that any invariant 2-cocycle has the form

$$\omega_{\alpha,\beta}(L_m, L_n) := (\alpha m^3 + \beta m)\delta_{m,-n}$$

for some constants $\alpha, \beta \in \mathbb{R}$.

Exercise 4.2 Show that the 2-cocycle $\omega_{\alpha,\beta}$ is nondegenerate on $\mathrm{Vect}_0(S^1)$ provided that either $\alpha = 0$ and $\beta \neq 0$, or $-\beta/\alpha \neq k^2$ for some $k \in \mathbb{N}$.

As we have seen in the proof of Proposition II.2.3, the cocycle $\omega_{0,\beta}$ is exact, i.e., it is a coboundary. So to describe nontrivial 2-cocycles we can assume that $\alpha \neq 0$, and that $-\beta/\alpha$ is not an integer square.

The two-parameter family of symplectic structures on the space S defined by the 2-forms $\omega_{\alpha,\beta}$ has a natural interpretation in terms of coadjoint orbits of the Virasoro–Bott group. Indeed, the Virasoro–Bott coadjoint orbit containing the point $(p(d\theta)^2, a) \in \mathfrak{vir}^*$ is isomorphic to $\mathrm{Diff}(S^1)/S^1$ if $0 < 4p/a$ is not an integer square; see Section II.2.2. (If $4p/a = n^2$ for some $n \in \mathbb{Z}$, the stabilizer of the point $(p(d\theta)^2, a)$ is a three-dimensional subgroup of $\mathrm{Diff}(S^1)$.) One can see that the two-parameter family of symplectic structures on the space S comes from the identification of S with any coadjoint orbit from the two-parameter family of those in the Virasoro–Bott group. The natural symplectic structure of the coadjoint orbit through the point $(p(d\theta)^2, a) \in \mathfrak{vir}^*$ with $a \neq 0$ and $0 < 4p/a \neq n^2$ has two parameters: p and a.

The Complex Structure

A $\mathrm{Diff}(S^1)$-invariant almost complex structure on the space $S = \mathrm{Diff}(S^1)/S^1$ is given by an S^1-invariant automorphism I_e of the tangent space $T_e S \cong \mathrm{Vect}_0(S^1)$ such that $I_e^2 = -\mathrm{id}$. Such an automorphism can be described using the Fourier expansion of an element $\xi \in \mathrm{Vect}_0(S^1)$: for

$$\xi = \sum_{n \neq 0} \xi_n L_n$$

we set

$$I_e(\xi) = -i \sum_{n > 0} \xi_n L_n + i \sum_{n < 0} \xi_n L_n.$$

Exercise 4.3 Show that the almost complex structure I on the space S is compatible with the symplectic structure $\omega_{\alpha,\beta}$ in the sense that $\omega_{\alpha,\beta}$ is I-invariant, i.e., $\omega_{\alpha,\beta}(I_e\xi, I_e\eta) = \omega_{\alpha,\beta}(\xi, \eta)$ at $e \in S$, and the symmetric bilinear form $g_{\alpha,\beta}$ defined on the tangent space $T_e S$ by $g_{\alpha,\beta}(\xi, \eta) := \omega_{\alpha,\beta}(\xi, I_e\eta)$ is positive definite.

One can easily see that the almost complex structure I on the space $\mathcal{S}$ is "formally integrable": the bracket of two vector fields in the $+i$, respectively $-i$, eigenspace of the automorphism I is again in the $+i$, respectively $-i$, eigenspace [57]. In principle, this does not guarantee the existence of a complex structure in the infinite-dimensional context, since here the finite-dimensional Newlander–Nirenberg theorem does not work. The following theorem of Kirillov shows that the complex structure does exist in this infinite-dimensional case.

Theorem 4.4 ([204]) *There is a complex structure on the space* $\mathcal{S} = \mathrm{Diff}(S^1)/S^1$ *invariant with respect to the* $\mathrm{Diff}(S^1)$*-action.*

SKETCH OF PROOF.   To show that the almost complex structure on $\mathcal{S}$ is, in fact, integrable, one can give a different realization of the space $\mathcal{S}$ in terms of univalent functions on the unit disk. Namely, let $D \subset \mathbb{C}$ be the unit disk, and set $\mathcal{F}$ to be the space of holomorphic univalent functions on D that are continuous up to the boundary and normalized by the conditions at the origin:

$$\mathcal{F} = \{f : D \to \mathbb{C} \mid f \text{ holomorphic and continuous up to the boundary,}$$
$$f(0) = 0, \ f'(0) = 1, \text{ and } f(z_1) \neq f(z_2) \text{ whenever } z_1 \neq z_2\}.$$

This is an infinite-dimensional complex manifold. By expanding f in a series

$$f(z) = z + c_2 z^2 + c_3 z^3 + \cdots$$

one obtains a natural coordinate system on the space $\mathcal{F}$. De Brange's theorem (formerly the Bieberbach conjecture, [212]) implies that $|c_k| < k$, so that $\mathcal{F}$ can be regarded as an infinite-dimensional analogue of a bounded domain in $\mathbb{C}^n$, see [204].

The identification of $\mathcal{F}$ with the quotient $\mathcal{S} = \mathrm{Diff}(S^1)/S^1$ is implemented by means of the following geometric realization. Identify $\mathcal{F}$ with the space $\mathcal{K}$ of all simple contours around the origin that have conformal radius 1 with respect to the origin. Namely, a univalent function $f \in \mathcal{F}$ uniquely determines such a contour as the image $f(S^1)$ of the boundary $S^1 = \partial D$. Conversely, for any such contour $K \in \mathcal{K}$ there is a function f_K mapping its interior to the unit disk (the Riemann mapping theorem), which can be normalized so that $f_K \in \mathcal{F}$.

To see how one can associate a coset from $\mathrm{Diff}(S^1)/S^1$ to a function f_K, consider a function g_K mapping the exterior of the unit disk D to the exterior of the contour K and normalized by the condition $g_K(\infty) = \infty$. This function is defined modulo the rotation change of the variable $z \mapsto e^{i\alpha}z$. Then the map $\phi_K = f_K^{-1} \circ g_K$ defines a diffeomorphism of the unit circle $S^1 = \partial D$ modulo rotations, i.e., an element of $\mathcal{S} = \mathrm{Diff}(S^1)/S^1$.

This map is a bijection, and the definition of the inverse map from $\mathcal{S}$ to $\mathcal{F}$ is as follows. Given a circle diffeomorphism $\phi \in \mathrm{Diff}(S^1)$ construct a 2-sphere S^2_ϕ

by glueing together the boundaries of the two unit disks D_- and D_+ according to that diffeomorphism. There is a unique complex structure on S_ϕ^2 that coincides with the standard ones on D_- and D_+. Since all complex structures on S^2 are equivalent, there exists a holomorphic map $F : S_\phi^2 \to S^2 = \mathbb{CP}^1$ that can be normalized by the conditions $F(0) = 0, F'(0) = 1, F(\infty) = \infty$. In turn, the map F can be thought of as a pair of functions $F = (f, g)$ defined on D_- and D_+ (here $0 \in D_-$ and $\infty \in D_+$), and one can see that $f \in \mathcal{F}$. The corresponding contour $K \in \mathcal{K}$ is the image of the unit circle $f(S^1)$.

The group $\mathrm{Diff}(S^1)$ acts on the space of contours and hence on the space $\mathcal{F}$ by holomorphic transformations, and the stabilizer of the function $f(z) = z$ under this action is given by the group of rigid rotations $S^1 = \mathrm{Rot}(S^1) \subset \mathrm{Diff}(S^1)$. This identification $\mathcal{F} \cong \mathcal{S} \cong \mathcal{K}$ fixes the complex structure on the space $\mathcal{S}$ invariant with respect to the $\mathrm{Diff}(S^1)$-action. $\qquad\square$

The following theorem details and summarizes this section.

Theorem 4.5 ([204, 208]) *The space $\mathcal{S} = \mathrm{Diff}(S^1)/S^1$ is a Kähler Fréchet manifold with a two-parameter family of $\mathrm{Diff}(S^1)$-invariant Kähler metrics $g_{\alpha,\beta}$. These Kähler metrics originate from one invariant complex structure and a two-parameter family of invariant symplectic structures on $\mathcal{S}$.*

Note that the Virasoro group can be viewed as a holomorphic $\mathbb{C}^*$-bundle over $\mathcal{S}$ [236].

4.2 The Action of $\mathrm{Diff}(S^1)$ and Kähler Geometry on the Based Loop Spaces

Symplectic Structures

Let G be a simple compact simply connected Lie group with the Lie algebra $\mathfrak{g}$. The loop group of G is the group $LG = C^\infty(S^1, G)$ of smooth maps from the circle S^1 to G. The *based loop group* ΩG is the subgroup

$$\Omega G = \{g \in LG \mid g(0) = e\}$$

for a fixed point $0 \in S^1$. As a manifold, ΩG can be identified with the quotient LG/G, where G is thought of as the subgroup in LG consisting of constant loops in G. Note, however, that this identification is not canonical, since G is not a normal subgroup in LG.

An LG-invariant closed 2-form on the space ΩG is determined by its values on the tangent space $T_e \Omega G \cong L\mathfrak{g}/\mathfrak{g}$ or, equivalently, by a G-invariant 2-cocycle ω on the Lie algebra $L\mathfrak{g}$ that also satisfies $\mathfrak{g} \subset \ker \omega$. Proposition II.1.6 shows that any continuous G-invariant 2-cocycle on $L\mathfrak{g}$ has the form

$$\omega(X, Y) = \int_{S^1} \langle X(\theta), Y'(\theta) \rangle \, d\theta \,,$$

where $\langle \, , \, \rangle$ is an invariant inner product on the Lie algebra $\mathfrak{g}$. Also, ω satisfies the condition $\mathfrak{g} \subset \ker \omega$, since it vanishes on constant loops. The fact that the 2-form ω is nondegenerate, i.e., it defines a symplectic structure on ΩG, is equivalent to the nondegeneracy of the inner product $\langle \, , \, \rangle$.

For a simple Lie group G there is the Killing form, the unique (up to a scalar factor) invariant inner product on $\mathfrak{g}$. Hence, in this case there exists a one-parameter family of LG-invariant symplectic forms on the space ΩG. This family appears naturally from the identification of the quotient LG/G with the coadjoint orbits of the affine (i.e., centrally extended loop) group $\widehat{LG}$ that contain the points $(0, a) \in (\widehat{L\mathfrak{g}})^*$: the parameter in this family is the cocentral value $a \neq 0$; see Section II.1.2.

Exercise 4.6 The group $\mathrm{Diff}(S^1)$ acts on the space LG/G by symplectomorphisms. (Hint: the cocycle ω, and hence the corresponding symplectic structure, is invariant with respect to reparametrizations of S^1.)

Complex Structures

Any LG-invariant almost complex structure on the based loop group ΩG is given by fixing an automorphism J_e of the tangent space $T_e \Omega G$ such that $(J_e)^2 = -1$. In the complexification $T_e^{\mathbb{C}} \Omega G$ consider the set of Laurent polynomials (without the constant term) with values in the complexified Lie algebra $\mathfrak{g}^{\mathbb{C}}$:

$$X(z) = \sum_{n \neq 0} X_n z^n .$$

Now we define an LG-invariant almost complex structure J^0 on ΩG by assigning its value on such polynomials to be

$$J_e^0 X(z) = -i \sum_{n>0} X_n z^n + i \sum_{n<0} X_n z^n$$

and extending it to an automorphism of $T_e \Omega G$ by continuity.

One can see that this almost complex structure is integrable by identifying the based loop group ΩG with the quotient $LG^{\mathbb{C}}/L^+G^{\mathbb{C}}$. Here, $LG^{\mathbb{C}}$ denotes the group of smooth maps from the circle S^1 to the complexified Lie group $G^{\mathbb{C}}$, and $L^+G^{\mathbb{C}}$ denotes the subgroup of those maps that extend to holomorphic maps from the unit disk D, the interior of the circle, to the group $G^{\mathbb{C}}$. To obtain this identification one employs the fact that any loop $h \in LG^{\mathbb{C}}$ can be uniquely factorized into a product $h = h_b \cdot h_+$ with $h_b \in \Omega G$ and $h_+ \in L^+G^{\mathbb{C}}$; see [322]. Therefore the quotient $LG^{\mathbb{C}}/L^+G^{\mathbb{C}}$ carries an $LG^{\mathbb{C}}$-invariant complex structure that coincides with J_e^0 at the identity.

A diffeomorphism $\varphi \in \mathrm{Diff}(S^1)$ sends the complex structure J^0 to $J = \varphi_* \circ J^0 \circ \varphi_*^{-1}$, where φ_* denotes the action of the diffeomorphism φ on the corresponding tangent space.

Proposition 4.7 *The complex structure $J = \varphi_* \circ J^0 \circ \varphi_*^{-1}$ coincides with J^0 if and only if the diffeomorphism $\varphi \in \mathrm{Diff}(S^1)$ is a rotation.*

SKETCH OF PROOF.   Evidently, if φ is a rotation, then $J = J^0$. Conversely, assume that $J = J^0$. Recall that the complex structure J^0 on ΩG is defined by means of the identification of the latter with the quotient $\Omega G = LG^{\mathbb{C}}/L^+G^{\mathbb{C}}$. Then the action of the diffeomorphism φ preserves the representation of $\Omega G = LG^{\mathbb{C}}/L^+G^{\mathbb{C}}$ as a quotient, i.e., it sends the subgroup $L^+G^{\mathbb{C}}$ into itself.

Denote by $L_1^-G^{\mathbb{C}}$ the subset of $LG^{\mathbb{C}}$ consisting of loops in $G^{\mathbb{C}}$ that extend to holomorphic maps h from the *exterior* of the unit disk to $G^{\mathbb{C}}$ and that are normalized by the condition $h(\infty) = \mathrm{id} \in G^{\mathbb{C}}$. The multiplication map $L_1^-G^{\mathbb{C}} \times L^+G^{\mathbb{C}} \to LG^{\mathbb{C}}$ is a diffeomorphism to a dense subset of $LG^{\mathbb{C}}$; see [322, Section 8.6] One can show that in order to preserve the complex structure on ΩG, the diffeomorphism φ has to preserve both $L_1^-G^{\mathbb{C}}$ and $L^+G^{\mathbb{C}}$. This implies that the circle diffeomorphism φ extends to both a conformal automorphism of the unit disk D and to a conformal automorphism of the exterior of the disk D fixing ∞; cf. the proof of Theorem 4.4 above. Such a conformal automorphism of the Riemann sphere has to be a rotation, and so is φ itself. $\square$

The complex structures $J = \varphi_* \circ J^0 \circ \varphi_*^{-1}$ obtained from the complex structure J^0 by the action of the diffeomorphism group $\mathrm{Diff}(S^1)$ are called *admissible complex structures* on ΩG. By construction of J^0, all admissible complex structures are LG-invariant. From the proposition above, we directly obtain the following result:

Corollary 4.8 ([57, 312]) *Admissible complex structures on the based loop group ΩG are parametrized by the points of the infinite-dimensional manifold $\mathcal{S} = \mathrm{Diff}(S^1)/S^1$.*

Finally, one can verify that the complex structure J^0 is compatible with the symplectic structure ω in the sense that $\omega(J^0 X, J^0 Y) = \omega(X, Y)$, while $g^0(X, Y) := \omega(X, J^0 Y)$ is a positive definite quadratic form. Hence, g^0 defines an LG-invariant Kähler metric on ΩG. In view of the invariance of the symplectic form ω under the action of the group $\mathrm{Diff}(S^1)$, the complex structures $J = \varphi_* \circ J^0 \circ \varphi_*^{-1}$ obtained from J^0 are also compatible with the symplectic form ω and thus give rise to a family of LG-invariant Kähler metrics g on ΩG parametrized by the points of $\mathcal{S} = \mathrm{Diff}_+(S^1)/S^1$.

This description of Kähler metrics exhibits a kind of reciprocity between the homogeneous spaces LG/G and $\mathrm{Diff}_+(S^1)/S^1$. More details, the formulas for the corresponding curvature tensor, and further applications of the above can be found in [204, 208, 350].

A.5 Diffeomorphism Groups and Optimal Mass Transport

In this appendix we consider the Riemannian geometry of the group of *all* diffeomorphisms of a compact manifold. This large group, being equipped with a certain L^2-type metric, can be viewed as a unifying framework for problems of Euler hydrodynamics on the group of *volume-preserving* diffeomorphisms on the one hand and problems related to optimal mass transportation on *densities* on this manifold on the other hand.

5.1 The Inviscid Burgers Equation as a Geodesic Equation on the Diffeomorphism Group

Let M be a compact n-dimensional Riemannian manifold with metric $(\, , \,)$. Consider the group $\mathrm{Diff}(M)$ of smooth diffeomorphisms of M along with its subgroup $S\mathrm{Diff}(M)$ of diffeomorphisms preserving the (Riemannian) volume form μ (as before, we confine ourselves to the connected component of the identity diffeomorphism). For a curve $\{\eta(t) \mid t \in [0, a]\}$ in $\mathrm{Diff}(M)$ its L^2-energy is given by

$$E(\{\eta\}) = \frac{1}{2} \int_0^a \langle \dot{\eta}(t), \dot{\eta}(t) \rangle_{\mathrm{Diff}} \, dt \,,$$

where the (weak) Riemannian metric at each point $\eta \in \mathrm{Diff}(M)$ of the diffeomorphism group is defined in the following straightforward way: given $X, Y \in \mathrm{Vect}(M)$, the inner product of two vectors $X \circ \eta, Y \circ \eta \in T_\eta \mathrm{Diff}(M)$ is

$$\langle X \circ \eta, Y \circ \eta \rangle_{\mathrm{Diff}} = \int_M (X \circ \eta(x), Y \circ \eta(x)) \, \mu(x). \tag{5.21}$$

This metric is right-invariant when restricted to the subgroup $S\mathrm{Diff}(M)$ of volume-preserving diffeomorphisms, although it is not right-invariant on the whole group $\mathrm{Diff}(M)$. (Indeed, the change of variables in the integral (5.21) would give the Jacobian $\det[\partial \eta / \partial x]$ as an extra factor, which, however, is identically equal to 1 for a volume-preserving map $\eta \in S\mathrm{Diff}(M)$.) Note that for a *flat* manifold M this metric is a *flat* metric on $\mathrm{Diff}(M)$: a neighborhood of the identity $\mathrm{id} \in \mathrm{Diff}(M)$ with the metric $\langle \, , \, \rangle_{\mathrm{Diff}}$ is isometric to a neighborhood in the pre-Hilbert space of smooth "vector-functions" $\eta : M \to M$ with the L^2 inner product $\langle \, , \, \rangle_{L^2(M)}$.

Recall that the *Euler equation* $\partial_t u + (u, \nabla) u = -\nabla p$ on a divergence-free field u on M describes the motion of an ideal incompressible fluid filling M. It corresponds to the equation of the geodesic flow of the right-invariant L^2 metric on the group $S\mathrm{Diff}(M)$; see [12] and Section II.3.

Define the (inviscid) *Burgers equation* as the evolution equation

$$\partial_t u + (u, \nabla) u = 0 \tag{5.22}$$

for a vector field u on M, where $(u, \nabla)u$ stands for the covariant derivative $\nabla_u u$ on M. (This equation is often also called the *Hopf equation* or *compressible Euler equation*.) Consider the flow $(t, x) \mapsto \eta(t, x)$ corresponding to this velocity field:

$$\partial_t \eta(t, x) = u(t, \eta(t, x)), \qquad \eta(0, x) = x.$$

(Here we use the notation $\partial_t \eta(t, x)$ for the time derivative $\dot{\eta}$ to emphasize the dependence of η on both t and x.)

Proposition 5.1 ([96, 303]) *(1) Solutions of the Burgers equation are time-dependent vector fields on M that describe the following flows of fluid particles: each particle moves with constant velocity (defined by the initial condition) along a geodesic in M.*

(2) Geodesics in the group $\mathrm{Diff}(M)$ *with respect to the* L^2-*metric (5.21) correspond to solutions of the Burgers equation. Geodesics in* $\mathrm{Diff}(M)$ *normal to the submanifold* $S\mathrm{Diff}(M)$ *have potential initial conditions.*

PROOF. We first assume for simplicity that M is equipped with a flat metric. Then the geodesics in $\mathrm{Diff}(M)$ are given by the equation $\ddot{\eta}(t) = 0$, which describes flows of particles moving with constant velocity along their own geodesics, "straight lines," in M. This follows from the "flatness" of the L^2-metric (5.21). Indeed, consider a one-parameter variation $\eta_s(t)$ of a curve $\{\eta(t)\}$ with fixed ends for all s. Let $w(t) := \partial_s \eta_s(t)|_{s=0}$ be the variation vector field along the curve $\eta(t)$, and $w(0) = w(a) = 0$. Then the geodesic condition on $\eta(t)$, i.e., vanishing of the variation $\partial_s E(\eta_s) = 0$ at $s = 0$, gives

$$\begin{aligned}
0 &= \int_0^a \langle \partial_s \dot{\eta}_s(t), \dot{\eta}_s(t) \rangle_{L^2(M)} \, dt \Big|_{s=0} \\
&= \int_0^a \langle \dot{w}(t), \dot{\eta}(t) \rangle_{L^2(M)} \, dt = -\int_0^a \langle w(t), \ddot{\eta}(t) \rangle_{L^2(M)} \, dt.
\end{aligned}$$

This implies $\ddot{\eta}(t) = 0$ and proves (1) in the flat case.

In order to see that the corresponding velocity field u for the geodesic $\{\eta(t) \,|\, \partial_t^2 \eta = 0\}$ satisfies the Burgers equation, we apply the chain rule to the definition $\partial_t \eta(t, x) = u(t, \eta(t, x))$. This immediately gives

$$\partial_t^2 \eta = \partial_t u(t, \eta(t, x)) = (\partial_t u + \partial_x u \cdot \partial_t \eta)(t, \eta(t, x)) = \big(\partial_t u + (u \cdot \nabla)\, u\big)(t, \eta(t, x)),$$

and hence the geodesic condition $d^2 \eta / dt^2(t, x) = 0$ is equivalent to the Burgers equation on M:

$$\partial_t u + (u, \nabla)u = 0.$$

(One can compare the latter to the Euler equation on the subgroup $S\mathrm{Diff}(M) \subset \mathrm{Diff}(M)$, where the acceleration $\partial_t^2 \eta$ is L^2-orthogonal to the set $S\mathrm{Diff}(M)$ and hence is given by a gradient field, cf. Section II.3.2.)

In the general case, any Riemannian metric on M induces a unique Levi-Civita L^2-connection $\bar{\nabla}$ on $\mathrm{Diff}(M)$, which is determined pointwise by the

Riemannian connection ∇ on the manifold M itself (see for example [96] or [267]. Then the same chain rule leads to the Burgers equation (5.22) on velocity u with $(u, \nabla)u := \nabla_u u$. The latter is equivalent to the Riemannian version of freely flying non-interacting particles in M: $\bar{\nabla}_{\dot{\eta}}\dot{\eta} = 0$. The flow corresponding to the Burgers solution with the initial field $v(x)$ on M has the form $x \mapsto \exp_M(t \cdot v(x))$, where $\exp_M$ stands for the Riemannian exponential map on the manifold M. This is a family of diffeomorphisms (parametrized by time t) in which each point moves along its own geodesic on M with constant velocity.

Finally, observe that the tangent space to the subgroup $S\mathrm{Diff}(M)$ at the identity consists of divergence-free vector fields, and hence the space of normals is given by the gradients of all smooth functions on M. Thus horizontal geodesics are the ones whose initial velocities are gradient fields $u|_{t=0} = \nabla\phi$; see Figure 5.1. $\square$

Remark 5.2 For a smooth initial velocity field $u|_{t=0} = v$, the family $x \mapsto \exp(t \cdot v(x))$ consists of diffeomorphisms for small t, but at some moment particles can start colliding with each other. This moment corresponds to formation of a shock wave on M, and from this moment on, the map from initial to final positions of the particles ceases to be a diffeomorphism. In other words, the group $\mathrm{Diff}(M)$ is incomplete in the L^2-metric (5.21) and the corresponding geodesics "reach the boundary of the group" $\mathrm{Diff}(M)$ in finite time.

One can show that the shock wave formation for the Burgers equation with potential initial condition $u|_{t=0} = \nabla\phi$ corresponds to the first focal point in the direction $\nabla\phi$, which is normal to the set $S\mathrm{Diff}(M)$ regarded as a Riemannian submanifold of $\mathrm{Diff}(M)$; see [193] and Figure 5.1. Recall that a focal point of a submanifold S in a Riemannian manifold N in the direction ν normal to S is the point of intersection of normals to S that are infinitely close to ν.

Example 5.3 The Burgers equation on the line assumes the form $\partial_t u + uu' = 0$, where $u' := \partial_x u$. Faster particles start passing slower ones, and one can see that the shock wave solutions of the Burgers equation first arise from inflection points of the initial velocity profile of $u|_{t=0}$ (i.e., from the points corresponding to $u''|_{t=0} = 0$).

In higher dimensions the shock waves first arise from the special points of the initial potential $u|_{t=0} = \nabla\phi$, which are singularities of type A_3 modulo certain linear and quadratic terms in local orthogonal coordinate charts; see [46]. The list of initial singularities, possible bifurcations of the shock waves, and other related questions for the inviscid Burgers equations can be found in [21, 22, 137].

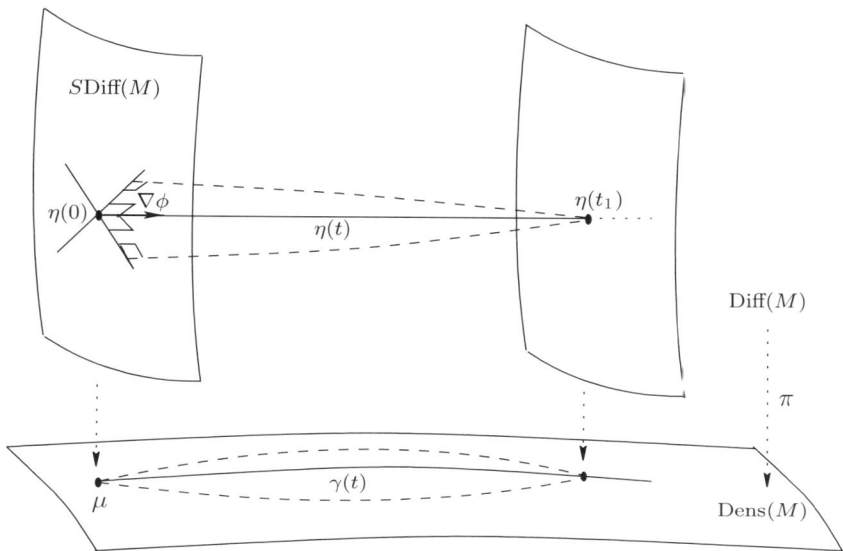

Fig. 5.1. Diffeomorphism group Diff(M) projects to the space of densities Dens(M) with the fiber SDiff(M); normals to SDiff(M) are horizontal geodesics $\eta(t)$, and focal points along them correspond to conjugate points along geodesics $\gamma(t)$ in Dens(M).

Remark 5.4 While first conjugate points along horizontal geodesics in Diff(M) naturally correspond to shock waves of the Burgers equation, the problem of description of conjugate points along vertical geodesics, i.e., geodesics in SDiff(M), was posed by Arnold back in the 1960s in the paper [12] on the geometry of the Euler equation. It is shown in [97] that for a compact surface M without boundary the exponential map of the L^2-metric on SDiff(M) is a nonlinear Fredholm map of index zero. In particular, this implies that conjugate points are isolated and of finite multiplicity along finite geodesic segments. In other words, in this case the Riemannian exponential map on SDiff(M) has the same structure of singularities as that on a finite-dimensional manifold. For a three-dimensional M the situation changes drastically: the set of conjugate points is not discrete. In particular, conjugate points cluster to the first one, while the exponential operator is not Fredholm, since its range is not closed [97, 323]. This is yet another manifestation of the difference between the geometries of the groups of volume-preserving diffeomorphisms in the two- and three-dimensional cases; cf. Section II.3.6.

5.2 Metric on the Space of Densities and the Otto Calculus

The differential geometry of diffeomorphism groups is closely related to the theory of optimal mass transport, and in particular, to the problem of moving one mass (or density) to another while minimizing a certain cost. In this section we discuss the relation of metric properties of the diffeomorphism group and the space of densities.

Let μ be a smooth reference volume form (or density) on M of unit total mass, and consider the projection $\pi : \mathrm{Diff}(M) \to \mathrm{Dens}(M)$ of diffeomorphisms onto the space $\mathrm{Dens}(M)$ of (normalized) smooth densities on M. The diffeomorphism group $\mathrm{Diff}(M)$ is fibered over $\mathrm{Dens}(M)$ by means of this projection π as follows: the fiber over a volume form ν consists of all diffeomorphisms η that push μ to ν, $\eta_* \mu = \nu$. In other words, two diffeomorphisms η_1 and η_2 belong to the same fiber if and only if $\eta_1 = \eta_2 \circ \varphi$ for some diffeomorphism φ preserving the volume form μ. Diffeomorphisms from $\mathrm{Diff}(M)$ act transitively on smooth densities, according to the Moser theorem [279].

Remark 5.5 Note that this projection π can be extended to more general (nonsmooth) maps and densities by tracing how they transport the density μ. More precisely, let μ and ν be two Borel measures of total volume 1 on a compact Riemannian manifold M (or, more generally, on a complete metric space) that are absolutely continuous with respect to the Lebesgue measure, and let $\mathrm{dist}(x, y)$ be the distance function on M. Consider the following *optimal mass transport problem*: Find a Borel map $\eta : M \to M$ that pushes the measure μ forward to ν and attains the minimum of the L^2-cost functional $\int_M \mathrm{dist}^2(x, \eta(x))\mu$ among all such maps. The minimal cost of transport defines a metric (often called the *Kantorovich* or *Wasserstein metric*) Dist on densities:

$$\mathrm{Dist}^2(\mu, \nu) := \inf_\eta \left\{ \int_M \mathrm{dist}^2(x, \eta(x))\mu \mid \eta_* \mu = \nu \right\}. \tag{5.23}$$

It turns out that this mass transport problem admits a unique solution (defined up to measure-zero sets), called the optimal map $\bar{\eta}$ (see [58] for $M = \mathbb{R}^n$ and [256] for any compact connected Riemannian manifold M without boundary). Furthermore, there exists a 1-parameter family of Borel maps $\eta(t)$ joining the identity map $\eta(0) = \mathrm{id}$ with the optimal map $\eta(1) = \bar{\eta}$ such that $\eta(t)$ pushes μ to $\eta(t)_* \mu$ in an optimal way for every t. Such a 1-parameter family of measures $\eta(t)_* \mu$ describes a geodesic between μ and ν in the space of densities with respect to the metric Dist; see [374] for details.

In what follows we consider a formal version of this problem, focusing on smooth densities. One can see that the Kantorovich metric Dist is formally generated by the (weak) Riemannian metric on the space Dens of smooth densities. Thus both Diff and Dens can be regarded as infinite-dimensional Riemannian manifolds.

Proposition 5.6 ([303]) *The bundle map* $\pi : \mathrm{Diff}(M) \to \mathrm{Dens}(M)$ *is a Riemannian submersion of the metric* $\langle \, , \, \rangle_{\mathrm{Diff}}$ *on the diffeomorphism group* $\mathrm{Diff}(M)$ *to the metric* Dist *on the density space* $\mathrm{Dens}(M)$. *The horizontal (i.e., normal to fibers) spaces in the bundle* $\mathrm{Diff}(M) \to \mathrm{Dens}(M)$ *are right-translated gradient fields.*

Recall that for two Riemannian manifolds P and B a *submersion* $\pi : P \to B$ is a mapping onto B that has maximal rank and preserves lengths of horizontal tangent vectors to P; see, e.g., [302]. For a bundle $P \to B$ this means that on P there is a distribution of horizontal spaces that are orthogonal to fibers and projected isometrically to the tangent spaces to B.

PROOF. Recall the Hodge decomposition $\mathrm{Vect} = S\mathrm{Vect} \oplus_{L^2} \mathrm{Grad}$ for vector fields on M: any vector field v decomposes uniquely into the sum $v = \xi + \nabla p$ of a divergence-free field ξ and a gradient field ∇p, which are L^2-orthogonal to each other: $\int_M (\xi, \nabla p) \, \mu = 0$; cf. Section II.3.[16]

For the fibration $\pi : \mathrm{Diff}(M) \to \mathrm{Dens}(M)$ the fiber passing through the identity diffeomorphism id $\in \mathrm{Diff}(M)$ is the subgroup $S\mathrm{Diff}(M)$, preserving the reference density μ itself. Thus the vertical tangent space at the identity coincides with $S\mathrm{Vect}(M)$, while the horizontal space is Grad. The L^2-metric on Grad projects isometrically to the tangent space to the base Dens at the point μ.

Other fibers of $\mathrm{Diff}(M) \to \mathrm{Dens}(M)$ are right cosets for the subgroup $S\mathrm{Diff}(M)$ in the group $\mathrm{Diff}(M)$. One can think of this fibration as an $S\mathrm{Diff}(M)$-principal bundle over $\mathrm{Dens}(M)$, where volume-preserving diffeomorphisms from $S\mathrm{Diff}(M)$ act on all diffeomorphisms by right translations $R_\varphi : \eta \mapsto \eta \circ \varphi$ for $\varphi \in S\mathrm{Diff}(M)$ and $\eta \in \mathrm{Diff}(M)$. This induces the $S\mathrm{Diff}$-action on the corresponding tangent spaces. The metric (5.21) is invariant with respect to this action of $S\mathrm{Diff}(M)$.

Now the proposition follows from the Hodge decomposition above, which is right-translated from id to any other point η of the diffeomorphism group $\mathrm{Diff}(M)$. The vertical space (tangent to a fiber) at a point $\eta \in \mathrm{Diff}(M)$ consists of divergence-free vector fields right-translated by the diffeomorphism η,

$$\mathrm{Vert}_\eta = \{X \circ \eta \mid X \in S\mathrm{Vect}(M)\},$$

while the horizontal space is given by the translated gradient fields,

$$\mathrm{Hor}_\eta = \{(\nabla p) \circ \eta \mid p \in C^\infty(M, \mathbb{R})\}.$$

The L^2-type metric $\langle \, , \, \rangle_{\mathrm{Diff}}$ on horizontal spaces for different points of the same fiber projects isometrically to one and the same metric on the base, due to the $S\mathrm{Diff}$-invariance of the metric. $\square$

---

[16] More generally, one can consider the volume form μ not related to the Riemannian metric $(\, , \,)$. The L^2-orthogonality still holds for μ-divergence-free fields and gradients with respect to this metric.

One of the main properties of a Riemannian submersion is the following feature of geodesics:

Corollary 5.7 *Any geodesic initially tangent to a horizontal space on the full diffeomorphism group* Diff(M) *remains horizontal, i.e., tangent to the gradient distribution* Hor$_\eta$ *on this group. There is a one-to-one correspondence between geodesics on the base* Dens(M) *starting at the density μ and horizontal geodesics in* Diff(M) *starting at the identity diffeomorphism* id.

Remark 5.8 In PDE terms, the horizontality of a geodesic means that any solution of the Burgers equation with a potential initial condition $u = \nabla\phi$ remains potential for all times. The corresponding potential ϕ satisfies the Hamilton–Jacobi equation $\partial_t\phi + (\nabla\phi, \nabla\phi)/2 = 0$ on M; see [303].

Thus horizontal geodesics in Diff(M), i.e., potential solutions of the Burgers equation, move the densities in Dens(M) in the fastest way. This statement is very natural geometrically: to obtain the fastest projection to the base, i.e., to move densities most effectively, one has to mod out the "volume-preserving" parts of the diffeomorphisms (since they are not moving the density, but preserve it), and use only their "gradient parts."

Remark 5.9 Consider the flow $\eta(t) : x \mapsto \exp_M(t \cdot \nabla\phi(x))$ corresponding to the potential Burgers solution with a given smooth potential ϕ. As we mentioned, the solution remains smooth for some time, and the appearance of a shock wave corresponds to the first moment at which the solution is nonsmooth. For $M = \mathbb{R}^n$ the map $x \mapsto \exp_M(t \cdot \nabla\phi(x))$ has the form $x \mapsto x + t \cdot \nabla\phi(x) = \nabla(x^2/2 + t \cdot \phi(x))$, and the loss of the map's smoothness corresponds to the loss of convexity of the potential $f(x) := x^2/2 + t \cdot \phi(x)$. On any M this loss of smoothness occurs when the potential $-t\,\phi$ ceases to be c-concave, which is an analogous notion for manifolds. (A function ψ on a manifold M is c-concave if $(\psi^c)^c = \psi$, where $\psi^c(y) := \inf\{c(x,y) - \psi(x) \mid x \in M\}$ for the square distance function $c(x,y) = \mathrm{dist}^2(x,y)/2$ on M.) As long as the potential remains c-concave, the curve $\{\eta(t)\}$ is the shortest curve in Diff(M) joining its endpoints, while its projection to Dens(M) gives the shortest curve joining the corresponding densities. This description is based on the following theorem on *polar decomposition*: every diffeomorphism $\eta \in$ Diff(M) has a unique decomposition $\eta = \mathrm{gr} \circ \varphi$ into a "gradient map" $\mathrm{gr}(x) := \exp_M(\nabla\phi(x))$ for a c-concave potential $-\phi$ and a volume-preserving map φ of M; see [61, 256].

We return to the problem of finding optimal maps for moving the density $\mu(x)$ to any other density $\nu(y) = h(y)\mu(y)$, where h is a function on M. The Jacobian of a map η that sends μ to $\nu = \eta_*\mu$ satisfies the relation

$$h(\eta(x)) \cdot \det[\partial\eta/\partial x] = 1\,.$$

The polar decomposition theorem implies that in $\mathbb{R}^n$ the optimal map is the gradient $\eta = \nabla f$ of a convex potential f, and the above relation assumes the form of the *Monge–Ampère equation* on the potential:

$$\det(\operatorname{Hess} f(x)) = \frac{1}{h(\nabla f(x))},$$

where $\operatorname{Hess} f := \partial(\nabla f)/\partial x$ is the Hessian matrix of the function f. For a manifold M the potential for the optimal map $\exp_M(\nabla \phi)$ also satisfies the Monge–Ampère-like equation

$$\det \left[\frac{\partial \exp_M(\nabla \phi)}{\partial x} \right] = \frac{1}{h(\exp_M(\nabla \phi(x)))},$$

where $\exp_M$ is the Riemannian exponential on the manifold M and $-\phi$ is a c-concave potential. We refer to the books [374, 375] and papers [58, 256, 303] for a comprehensive discussion of optimal transport.

5.3 The Hamiltonian Framework of the Riemannian Submersion

The Riemannian submersion property for the fibration $\operatorname{Diff}(M) \to \operatorname{Dens}(M)$ can be put in the framework of symplectic reduction, discussed in Section I.5; see [235]. Recall the following general construction in symplectic geometry. Let $\pi : P \to B$ be a principal bundle with the structure group G.

Lemma 5.10 (see, e.g., [23]) *The symplectic reduction of the cotangent bundle T^*P over the G-action gives the cotangent bundle $T^*B = T^*P/\!/G$.*

PROOF.   The moment map $\Phi : T^*P \to \mathfrak{g}^*$ associated with this action takes T^*P to the dual of the Lie algebra $\mathfrak{g}$ of the group G. For the G-action on T^*P the moment map Φ is the projection of any cotangent space $T^*_a P$ to cotangent space $T^*_a F \approx \mathfrak{g}^*$ for the fiber F through a point $a \in P$. The preimage $\Phi^{-1}(0)$ of the zero value is the subbundle of T^*P consisting of covectors vanishing on fibers. Such covectors are naturally identified with covectors on the base B. Thus factoring out the G-action, which moves the point a over the fiber F, we obtain the bundle T^*B.   □

Suppose now that P is equipped with a G-invariant Riemannian metric $\langle \, , \, \rangle_P$. Then it induces the metric $\langle \, , \, \rangle_B$ on the base B.

Lemma 5.11 *The Riemannian submersion of P $(P, \langle \, , \, \rangle_P)$ to the base B, equipped with the metrics $\langle \, , \, \rangle_P$ and $\langle \, , \, \rangle_B$ respectively, is the result of the symplectic reduction with respect to the G-action.*

PROOF. Indeed, the metric $\langle \, , \, \rangle_P$ gives a natural identification $T^*P \approx TP$ of the tangent and cotangent bundles for P, and the "projected metric" is equivalent to a similar identification for the base manifold B. In the presence of the metric in P, the preimage $\Phi^{-1}(0)$ is identified with the subbundle of horizontal spaces in TP: the zero value of the moment map stands for the orthogonality to the vertical (i.e., tangent to the fibers) spaces. Hence the symplectic quotient $\Phi^{-1}(0)/G$ can be identified with the tangent bundle TB. $\square$

Let $H^P : T^*P \to \mathbb{R}$ be a Hamiltonian function invariant under the G-action on the cotangent bundle of the total space P. The restriction of this function to the horizontal bundle $(\Phi^{-1}(0) \subset T^*P)$ is also G-invariant, and hence descends to a function $H^B : T^*B \to \mathbb{R}$ on the quotient, the cotangent bundle of the base B. One has the following reduction of Hamiltonian dynamics:

Proposition 5.12 ([23]) *The Hamiltonian flow of the function H^P preserves the preimage $\Phi^{-1}(0)$; i.e., trajectories with horizontal initial conditions stay horizontal. Furthermore, the Hamiltonian flow of the function H^P on the cotangent bundle T^*P of the total space descends to the Hamiltonian flow of the function H^B on the cotangent bundle T^*B of the base.*

Apply the above consideration to our setting, in which $P = \text{Diff}(M)$, $G = \text{SDiff}(M)$, and $B = \text{Dens}(M)$. The above lemmas give a Hamiltonian meaning to Proposition 5.6 on Riemannian submersion.

Now we would like to describe the geodesics on the spaces $\text{Diff}(M)$ and $\text{Dens}(M)$. Recall that one of possible definitions of geodesics in any Riemannian manifold M is that they are projections to M of trajectories of the Hamiltonian flow on T^*M, whose Hamiltonian function is the "kinetic energy" $K^M(p, q) := (p, p)/2$ given by this Riemannian metric. In the same way, geodesics on $\text{Diff}(M)$ and $\text{Dens}(M)$ are obtained by considering the corresponding Hamiltonians for the metric $\langle \, , \, \rangle_{\text{Diff}}$ and metric Dist, respectively.

It will be convenient for us to identify the tangent and cotangent spaces for the manifold M, as well as those for the group $\text{Diff}(M)$, using the corresponding metrics on them. Consider also a more general Hamiltonian function H^M on the (co)tangent bundle TM of the manifold M. The *averaged Hamiltonian function* is the function H^{Diff} on the (co)tangent bundle $T\text{Diff}(M)$ of the diffeomorphism group $\text{Diff}(M)$ obtained by averaging the corresponding Hamiltonian H^M over M in the following way: its value at a point $X \circ \eta \in T_\eta \text{Diff}(M)$ is

$$H^{\text{Diff}}(X \circ \eta) := \int_M H^M(X \circ \eta(x)) \, \mu(x)$$

for a vector field $X \in \text{Vect}(M)$ and a diffeomorphism $\eta \in \text{Diff}(M)$. For instance the energy Hamiltonian $K^{\text{Diff}}(X \circ \eta) := \frac{1}{2}\langle X \circ \eta, X \circ \eta \rangle_{\text{Diff}}$ on $T\text{Diff}(M)$ is the averaging of the "kinetic energy" Hamiltonian K^M defined on TM.

Consider the Hamiltonian flows for the Hamiltonian functions H^M and H^{Diff} on the (co)tangent bundles TM and $T\mathrm{Diff}(M)$, respectively, relative to the standard symplectic structures on the bundles.

Theorem 5.13 ([235]) *Each Hamiltonian trajectory for the averaged Hamiltonian functions H^{Diff} on $T\mathrm{Diff}(M)$ describes such a flow on the tangent bundle TM in which every tangent vector to M moves along its own H^M-Hamiltonian trajectory in TM.*

This theorem has the following simple geometric meaning for the energy Hamiltonians K^{Diff} and K^M. It implies that all geodesics on the diffeomorphism group $\mathrm{Diff}(M)$ (described by the Burgers equation; see Proposition 5.1) starting at the identity id with the initial velocity $X \in \mathrm{Vect}(M)$ are the flows that move each particle x on the manifold M along the geodesic with the initial direction $X(x)$. Such a geodesic is well defined on the diffeomorphism group $\mathrm{Diff}(M)$ as long as the particles do not collide.

Furthermore, Proposition 5.12 implies that a geodesic on $\mathrm{Diff}(M)$ with potential initial condition will stay potential (cf. Corollary 5.7). Finally, since the metric on $\mathrm{Diff}(M)$ is $S\mathrm{Diff}$-invariant, so is the energy Hamiltonian K^{Diff}, and hence it descends to the energy Hamiltonian on $T\mathrm{Dens}(M)$. The latter describes the geodesics on the density space $\mathrm{Dens}(M)$ with respect to the metric Dist. This way one recovers the geodesic properties of the group $\mathrm{Diff}(M)$ discussed above.

Remark 5.14 For a more general Hamiltonian H^M on the tangent bundle TM, each particle $x \in M$ with an initial velocity $X(x)$ will be moving along its *characteristic*, which is the projection to M of the corresponding Hamiltonian trajectory in the tangent bundle TM. This description for more general Hamiltonians allows one to extend the above description of geodesics to other situations, and in particular, to the case of nonholonomic distributions (i.e., to sub-Riemannian, or Carnot–Carathéodory, spaces); see [9, 188, 235].

Note that the above Hamiltonian framework is also valid for more general cost functions $c : M \times M \to \mathbb{R}$, which can replace dist^2 in the mass transport problem (5.23).

A.6 Metrics and Diameters of the Group of Hamiltonian Diffeomorphisms

6.1 The Hofer Metric and Bi-invariant Pseudometrics on the Group of Hamiltonian Diffeomorphisms

Let M be a symplectic manifold of dimension $2n$ with a symplectic form ω. Consider the space $\mathcal{F}(M)$ of Hamiltonian functions on M normalized in the following way. For a closed manifold M we define $\mathcal{F}(M)$ as the space of smooth functions with zero mean with respect to the canonical volume form ω^n. For an open M, the space $\mathcal{F}(M)$ consists of all smooth functions with compact support. In either case, $\mathcal{F}(M)$ is a Lie algebra with respect to the Poisson bracket related to ω or, equivalently, the Lie algebra of the corresponding Hamiltonian vector fields on M. (Note that this algebra can be endowed with an invariant inner product $\langle H, K \rangle := \int_M HK\,\omega^n$, where $H, K \in \mathcal{F}(M)$.)

Consider the corresponding group $\mathrm{Ham}_c(M)$ of all *compactly supported Hamiltonian diffeomorphisms* of M, which are time-one maps of time-dependent Hamiltonian fields with Hamiltonian functions from $\mathcal{F}(M)$. In Sections II.3.6 and II.6.2 we saw that the infinite-dimensional group $\mathrm{Ham}(M)$, as well as $\mathrm{Ham}_c(M)$, admits left-invariant l_p-metrics, which depend on the choice of a Riemannian metric on the manifold M. It turns out that the group $\mathrm{Ham}_c(M)$ does in fact admit a *bi-invariant* metric, which depends only on the symplectic structure on M.

Definition 6.1 Set the *energy* of a Hamiltonian diffeomorphism $\varphi \in \mathrm{Ham}_c(M)$ to be

$$E(\varphi) := \inf_H \left(\sup_{x,t} H(x,t) - \inf_{x,t} H(x,t) \right),$$

where $(x,t) \in M \times [0,1]$ and H runs over the set of all compactly supported time-dependent Hamiltonian functions $H : M \times [0,1] \to \mathbb{R}$ whose Hamiltonian vector field has the given diffeomorphism φ as the time-one map. The *Hofer metric* defines the distance on the group $\mathrm{Ham}_c(M)$ as

$$\rho_E(\psi, \varphi) = E(\psi\varphi^{-1})$$

for any two compactly supported Hamiltonian diffeomorphisms ψ and φ.

Exercise 6.2 Verify that

$$E(\varphi) = E(\varphi^{-1}) = E(\psi\varphi\psi^{-1}) \quad \text{and} \quad E(\psi\varphi) \leq E(\psi) + E(\varphi),$$

so that $\rho_E(\psi, \varphi)$ is a pseudometric, i.e., it is symmetric, bi-invariant, and satisfies the triangle inequality.

The following theorem is due to Hofer [166] in the case of $\mathbb{R}^{2n}$ with the standard symplectic form and to Lalonde–McDuff [227] in the case of general symplectic manifolds.

Theorem 6.3 *The pseudometric ρ_E is a genuine bi-invariant metric on the group* $\mathrm{Ham}_c(M)$. *That is, in addition to nonnegativity, symmetry, and the triangle inequality, the relation $\rho_E(\psi, \varphi) = 0$ implies $\psi = \varphi$.*

Remark 6.4 One can define the l_p-length of a curve in the group $\mathrm{Ham}_c(M)$ for any L^p-norm on the space $\mathcal{F}(M)$. Given the Hamiltonian function H_t of a path φ_t in $\mathrm{Ham}_c(M)$, we set

$$l_p(\varphi_t) = \int_0^1 \|H_t\|_{L_p} dt \,.$$

The length functional l_p generates a bi-invariant pseudometric ρ_p on the group $\mathrm{Ham}_c(M)$. It is shown in [98] that the pseudometric ρ_p is not a metric for any $p < \infty$. However, for $p = \infty$ one can show that ρ_∞ is indeed a metric, and it is equivalent to Hofer's metric ρ_E.

One of the key ingredients in the proofs of these statements is the notion of a displacement energy and its estimates. The *displacement energy* $e_\rho(S)$ of a subset $S \subset M$ is the (pseudo-) distance in $\mathrm{Ham}_c(M)$ from the identity diffeomorphism to the set of all Hamiltonian diffeomorphisms that push S away from itself:

$$e_\rho(S) := \inf\{\rho(\mathrm{id}, \varphi) \mid \varphi \in \mathrm{Ham}_c(M) \quad \text{such that} \quad \varphi(S) \cap S = \emptyset\} \,.$$

Theorem 6.5 ([98]) *If ρ is a bi-invariant nondegenerate metric on the group* $\mathrm{Ham}_c(M)$, *then the displacement energy $e_\rho(S)$ is positive for any open bounded set S.*

On the other hand, for the metric ρ_p with any $p < \infty$ one can show that the displacement energy of an embedded ball $B \subset M$ is zero. Indeed, suppose that the Hamiltonian flow g_t^H with the (compactly supported) Hamiltonian function $H : M \times [0,1] \to \mathbb{R}$ pushes B from itself: $g_{t=1}^H(B) \cap B = \emptyset$. Introduce the new Hamiltonian function $K(\cdot, t)$ by cutting off $H(\cdot, t)$ outside a neighborhood $U_t \subset M$ of the moving boundary $g_t^H(\partial B)$. The flows of K and H coincide on $(\partial B)_t$ for any t, and hence $g_{t=1}^K(B) \cap B = \emptyset$. Now we note that for any $p < \infty$, the L^p-norm of $K(\cdot, t)$ can be made arbitrarily small for all $t \in [0,1]$ by choosing a sufficiently narrow neighborhood U_t, and hence $\rho_p(\mathrm{id}, g_{t=1}^K)$ can likewise be made arbitrarily small. The latter implies that $e_{\rho_p}(B) = 0$ for $p < \infty$. Summarizing, in the L^p-norm one can push the ball from itself with arbitrarily small energy, but one must pay for this by fast rotation near the boundary $(\partial B)_t$. This fast rotation comes from the steep (skew-)gradients of the cutoff function $K(\cdot, t)$.

Of course, this consideration is not applicable to Hofer's ($p = \infty$) case, since the L^∞-norm of $K(\cdot, t)$ does not depend on the area of U_t. (See more details on the material of this section in [317].)

As to the diameter results in Hofer's metric, we just mention that for a symplectic manifold M with boundary, the group $\mathrm{Ham}_\partial(M)$ of Hamiltonian diffeomorphisms of M that are stationary on the boundary ∂M has infinite diameter. The case of the two-sphere was settled by Polterovich [316]: the group $\mathrm{Ham}(S^2)$ also has infinite diameter in Hofer's metric.

6.2 The Infinite L^2-Diameter of the Group of Hamiltonian Diffeomorphisms

In this section we prove a simplified version of Theorem II.3.46 of Eliashberg and Ratiu on the infinite diameter of the group of Hamiltonian diffeomorphisms in a right-invariant metric.

Let $B^{2n} \subset \mathbb{R}^{2n}$ be the $2n$-dimensional unit ball with the standard symplectic form $\omega = dx \wedge dy$. Denote by $\mathrm{Ham}_\partial(B^{2n})$ the group of its Hamiltonian diffeomorphisms that are the identity when restricted to the boundary ∂B^{2n}.

Theorem 6.6 ([99]) *The L^2-diameter of the group $\mathrm{Ham}_\partial(B^{2n})$ of Hamiltonian diffeomorphisms that are stationary on the boundary ∂B^{2n} is infinite.*

To prove this theorem we first need to define the Calabi invariant for such Hamiltonian diffeomorphisms. Fix some 1-form α such that $\omega = d\alpha$.

Definition / Proposition 6.7 *Given a 1-form α on B^{2n} and a Hamiltonian diffeomorphism $\varphi \in \mathrm{Ham}_\partial(B^{2n})$, there exists a unique function $h : B^{2n} \to \mathbb{R}$ vanishing with its gradient on ∂B^{2n} such that $dh = \varphi^*\alpha - \alpha$.*

The Calabi invariant of a Hamiltonian diffeomorphism $\varphi \in \mathrm{Ham}_\partial(B^{2n})$ is defined by

$$\mathrm{Cal}(\varphi) := \frac{1}{n+1} \int_{B^{2n}} h\,\omega^n\,.$$

It does not depend on the choice of α satisfying $\omega = d\alpha$.

PROOF. First, we prove the existence of the function h. To this end, observe that the 1-form $\varphi^*\alpha - \alpha$ is closed. Indeed, we have

$$d(\varphi^*\alpha - \alpha) = \varphi^*d\alpha - d\alpha = \varphi^*\omega - \omega = 0\,,$$

since φ preserves the symplectic form ω. The closed 1-form $\varphi^*\alpha - \alpha$ must be exact in the ball B^{2n}. So there exists a function h on B^{2n} such that $dh = \varphi^*\alpha - \alpha$. The fact that φ restricted to the boundary of B^{2n} is the identity gives the condition $\nabla h|_{\partial B^{2n}} = 0$. The condition $h|_{\partial B^{2n}} = 0$ fixes h uniquely.

To see that the Calabi invariant is indeed independent of the choice of the 1-form α, suppose that we have another 1-form $\widetilde{\alpha} = \alpha + df$. Then we obtain $\widetilde{h} = h + (\varphi^* f - f)$, and the required independence follows from

$$\int_{B^{2n}} (\varphi^* f)\omega^n = \int_{B^{2n}} f\omega^n \,,$$

which holds since φ preserves ω. $\qquad\qquad\qquad\qquad\qquad\qquad\qquad\square$

Remark 6.8 The Calabi invariant Cal : $\mathrm{Ham}_\partial(B^{2n}) \to \mathbb{R}$ is a homomorphism. Indeed, if $\varphi = \varphi_2 \circ \varphi_1$, then we have

$$dh = \varphi^* \alpha - \alpha = \varphi_2^*(\varphi_1^* \alpha) - \varphi_2^* \alpha + \varphi_2^* \alpha - \alpha = \varphi_2^* dh_1 + dh_2 \,,$$

so that we obtain

$$\int_{B^{2n}} h\omega^n = \int_{B^{2n}} h_1\omega^n + \int_{B^{2n}} h_2\omega^n \,.$$

There is an alternative definition of the Calabi invariant $\mathrm{Cal}(\varphi)$: Let $\{\varphi_t \mid 0 \le t \le T\}$ be a path in $\mathrm{Ham}_\partial(B^{2n})$ connecting $\varphi_0 = \mathrm{id}$, and $\varphi_T = \varphi$. This path may be regarded as the flow of a time-dependent Hamiltonian vector field on B^{2n}. We denote by $H_t : B^{2n} \times [0, T] \to \mathbb{R}$ the Hamiltonian function of this vector field at time t normalized by the condition that it vanish on ∂B^{2n} along with its differential.

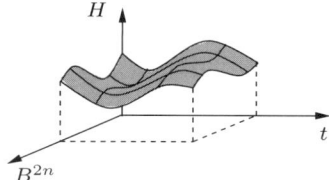

Fig. 6.1. Graph of the function $H_t : B^{2n} \times [0, T] \to \mathbb{R}$.

Theorem 6.9 *The Calabi invariant of a Hamiltonian diffeomorphism φ of B^{2n} is equal to the total integral of the Hamiltonian function H_t:*

$$\mathrm{Cal}(\varphi) = \int_0^T \left(\int_{B^{2n}} H_t\omega^n \right) dt \,.$$

In particular, this integral does not depend on the choice of the connecting path φ_t, i.e., on the choice of the Hamiltonian H_t, provided that the time-one map $\varphi_T = \varphi$ is fixed.

PROOF. Let φ_t be a path in $\mathrm{Ham}(B^{2n})$ such that $\varphi_0 = \mathrm{id}$ and $\varphi_T = \varphi$. For each $t \in [0, T]$, denote by h_t the corresponding function from Proposition 6.7. We have to show that

$$\int_{B^{2n}} h_T\, \omega^n = (n+1) \int_0^T \left(\int_{B^{2n}} H_t\, \omega^n \right) dt\,. \qquad (6.24)$$

Due to the homomorphism property of the Calabi invariant, it is enough to prove this equality for an infinitesimally short time T. That is, it suffices to show that the derivatives of both sides with respect to t at $t = 0$ coincide. We therefore need to prove the following identity:

$$\int_{B^{2n}} \left(\frac{d}{dt}\Big|_{t=0} h_t \right) \omega^n = (n+1) \int_{B^{2n}} H_0\, \omega^n\,.$$

Let ξ be the Hamiltonian vector field $\xi = \frac{d}{dt}\big|_{t=0}\varphi_t$. Then, by the definition of the function h_t in Proposition 6.7, we get

$$d\left(\frac{d}{dt}\Big|_{t=0} h_t \right) = L_\xi \alpha\,, \qquad (6.25)$$

where $L_\xi \alpha$ denotes the Lie derivative of the 1-form α with respect to the vector field ξ. Now apply Cartan's formula to obtain

$$L_\xi \alpha = d\iota_\xi \alpha + \iota_\xi d\alpha = d(\iota_\xi \alpha + H_0)\,, \qquad (6.26)$$

since H_0 is the Hamiltonian function corresponding to the vector field ξ, i.e., $dH_0 = \iota_\xi \omega = \iota_\xi d\alpha$. Equations (6.25) and (6.26) allow one to reconstruct $\frac{d}{dt}\big|_{t=0} h_t$:

$$\frac{d}{dt}\big|_{t=0} h_t = \iota_\xi \alpha + H_0\,.$$

(As a matter of fact, the above formulas allow one to find the derivative up to an additive constant. The vanishing boundary conditions for h_t, ξ, and H_t force this constant to be 0.) Then Theorem 6.9 would follow from the following lemma.

Lemma 6.10

$$\int_{B^{2n}} (\iota_\xi \alpha)\, \omega^n = n \int_{B^{2n}} H_0\, \omega^n\,.$$

PROOF OF LEMMA. We have

$$\int_{B^{2n}} (\iota_\xi \alpha)\, \omega^n = \int_{B^{2n}} \alpha \wedge \iota_\xi(\omega^n)$$

$$= n \int_{B^{2n}} \alpha \wedge \iota_\xi \omega \wedge \omega^{n-1} = n \int_{B^{2n}} \alpha \wedge dH_0 \wedge \omega^{n-1}$$

$$= n \int_{B^{2n}} H_0\, d\alpha \wedge \omega^{n-1} - n \int_{\partial B^{2n}} \alpha \wedge H_0 \wedge \omega^{n-1} = n \int_{B^{2n}} H_0\, \omega^n\,,$$

where in the second-to-last equality we used the Stokes formula. The boundary term vanishes due to the boundary conditions on the function H_0. This ends the proof of Lemma 6.10. $\square$

Now Theorem 6.9 follows from

$$\int_{B^{2n}} \left(\frac{d}{dt}\Big|_{t=0} h_t \right) \omega^n = \int_{B^{2n}} \left(\iota_\xi \alpha + H_0 \right) \omega^n = (n+1) \int_{B^{2n}} H_0\, \omega^n .$$

$\square$

Finally, we are ready to complete the proof of Theorem 6.6 for the group of Hamiltonian diffeomorphisms on $M = B^{2n}$ that are stationary on the boundary ∂B^{2n}.

PROOF OF THEOREM 6.6.   Let $\mu = \omega^n$ be the volume form corresponding to the standard symplectic structure on $\mathbb{R}^{2n}$. The L^2-length of a path φ_t joining the identity $\varphi_0 = \mathrm{id}$ and an arbitrary Hamiltonian diffeomorphism $\varphi_1 = \varphi$ is given by

$$\ell_2\{\varphi_t\} = \int_0^1 \|\partial_t\, \varphi_t\|_{L^2(B^{2n})}\, dt = \int_0^1 \|\nabla H_t(\varphi_t)\|_{L^2(B^{2n})}\, dt$$
$$= \int_0^1 \|\nabla H_t\|_{L^2(B^{2n})}\, dt ,$$

where the last equality holds since φ_t preserves μ. Now the estimate follows from the Poincaré and Schwarz inequalities:

$$\ell_2\{\varphi_t\} = \int_0^1 \|\nabla H_t\|_{L^2(B^{2n})}\, dt \geq c_1 \int_0^1 \|H_t\|_{L^2(B^{2n})}\, dt \geq c_2 \int_0^1 \|H_t\|_{L^1(B^{2n})}\, dt$$
$$\geq c_2 \left| \int_0^1 \left(\int_{B^{2n}} H_t\, \mu \right) dt \right| = c_2 \left| \operatorname{Cal}(\varphi) \right| .$$

Finally, since $\operatorname{Cal} : \operatorname{Ham}_\partial(B^{2n}) \to \mathbb{R}$ is a nontrivial homomorphism, one can find a Hamiltonian diffeomorphism of B^{2n} with an arbitrarily large value of the Calabi invariant. (For instance, one can take a multiple of the hill function shown in Figure II.3.6.) $\square$

A.7 Semidirect Extensions of the Diffeomorphism Group and Gas Dynamics

In this appendix we obtain the equations of motion of a barotropic fluid as the Hamiltonian equations on the dual Lie algebra to the semidirect product group $G = \mathrm{Diff}(M) \ltimes C^\infty(M)$. Although we keep the consideration below in the form of a sequence of exercises, it furnishes the necessary details for the outline in Section II.3.4; see also [253, 24, 90].

Let M be a compact Riemannian manifold. Recall that the group multiplication in the semidirect product $\mathrm{Diff}(M) \ltimes C^\infty(M)$ is given by

$$(\varphi, f) \cdot (\psi, g) = (\varphi \circ \psi, \varphi_* g + f),$$

where a pair (φ, f) consists of a diffeomorphism φ and a smooth function f on the manifold M, and $\varphi_* g = g \circ \varphi^{-1}$ denotes the pushforward of the function g by the diffeomorphism φ; see Figure II.3.2. The corresponding Lie algebra is the semidirect product $\mathrm{Vect}(M) \ltimes C^\infty(M)$, which is $\mathrm{Vect}(M) \oplus C^\infty(M)$ as a vector space.

Exercise 7.1 Show that the adjoint representation of the group $\mathrm{Diff}(M) \ltimes C^\infty(M)$ on its Lie algebra $\mathrm{Vect}(M) \ltimes C^\infty(M)$ is given by

$$\mathrm{Ad}_{(\varphi, f)}(\xi, a) = (\varphi_* \xi, L_\xi f + \varphi_* a).$$

Obtain from this formula that the infinitesimal adjoint action of the Lie algebra of $\mathrm{Diff}(M) \ltimes C^\infty(M)$ on $\mathrm{Vect}(M) \ltimes C^\infty(M)$ is given by

$$\mathrm{ad}_{(v,b)}(\xi, a) = (-L_v \xi, L_\xi b - L_v a), \tag{7.27}$$

where $L_v \xi$ denotes the usual bracket of two vector fields v and ξ on the manifold M.

Exercise 7.2 Show that the smooth part of the dual to the space $\mathrm{Vect}(M) \oplus C^\infty(M)$ can be identified with the space $(\Omega^1(M) \otimes_{C^\infty(M)} \Omega^n(M)) \oplus \Omega^n(M)$ via the pairing

$$\langle (\xi, a), (\alpha \otimes \mu, \nu) \rangle = \int_M (\iota_\xi \alpha) \cdot \mu + \int_M a\nu.$$

To shorten the notation we denote the latter space by $(\Omega^1 \otimes \Omega^n \oplus \Omega^n)(M)$.

Exercise 7.3 Verify that the coadjoint action of the group $\mathrm{Diff}(M) \ltimes C^\infty(M)$ on the space $(\Omega^1 \otimes \Omega^n \oplus \Omega^n)(M)$ is given by

$$\mathrm{Ad}^*_{(\varphi, f)^{-1}}(\alpha \otimes \eta, \nu) = (\varphi^* \alpha \otimes \varphi^* \eta + \varphi^* df \otimes \varphi^* \nu, \varphi^* \nu). \tag{7.28}$$

Hint: Recall that the coadjoint action of the group $\mathrm{Diff}(M) \ltimes C^\infty(M)$ on $(\Omega^1 \otimes \Omega^n \oplus \Omega^n)(M)$ is defined by

$$\langle(\xi, a), \mathrm{Ad}^*_{(\varphi, f)^{-1}}(\alpha \otimes \mu, \nu)\rangle = \langle\mathrm{Ad}_{(\varphi, f)}(\xi, a), (\alpha \otimes \mu, \nu)\rangle \,.$$

Use equation (7.28) to derive the following equation for the coadjoint action of the Lie algebra of $\mathrm{Diff}(M) \ltimes C^\infty(M)$ on $(\Omega^1 \otimes \Omega^n \oplus \Omega^n)(M)$:

$$\mathrm{ad}^*_{(v, b)}(\alpha \otimes \mu, \nu) = -\big((L_v\alpha) \otimes \mu + \alpha \otimes L_v\mu + db \otimes \nu, L_v\nu\big) \,.$$

Alternatively, one can use equation (7.27) and the relation

$$-\langle\mathrm{ad}^*_{(v, b)}(\alpha \otimes \mu, \nu), (\xi, a)\rangle = \langle(\alpha \otimes \mu, \nu), \mathrm{ad}_{(v, b)}(\xi, a)\rangle \,.$$

Remark 7.4 The above explicit expression for Ad^* helps in finding certain Casimir functions for this group. Note that the ratio $\alpha \otimes \mu / \nu$ has the "dimension" of a 1-form. One can see that the $\mathrm{Ad}^*_{(\varphi, f)^{-1}}$-action on the pair $(\alpha \otimes \mu, \nu)$ "changes the coordinates" by the diffeomorphism φ in this ratio 1-form modulo the differential of a function. This allows one to write out Casimir functions for the coadjoint action of the group $\mathrm{Diff}(M) \ltimes C^\infty(M)$ similar to the ones for $S\mathrm{Diff}(M)$; cf. Proposition II.3.9 and see [307, 187].

Now we are in a position to write down the Euler equation on the group $\mathrm{Diff}(M) \ltimes C^\infty(M)$. Recall that M is a Riemannian manifold with a Riemannian metric $(\ ,\ )$. Fix a volume form μ on M (not necessarily related to the metric). Finally, recall that in the equations of motion of a barotropic fluid on M we need to specify a function $h : C^\infty(M) \to C^\infty(M)$ relating density ρ to the pressure $p = h(\rho)$.

Define a Hamiltonian $\widetilde{H} : \mathrm{Vect}(M) \oplus C^\infty(M) \to \mathbb{R}$ by

$$\widetilde{H}(\xi, \rho) := \int_M \left(\frac{1}{2}(\xi, \xi)\rho + \rho\Phi(\rho)\right)\mu \,,$$

where $\Phi(\rho)$ is a function satisfying $\rho^2\Phi'(\rho) = h(\rho)$.

In order to write down the corresponding Euler equation we need to lift this Hamiltonian to the dual of the Lie algebra. Let us fix a (nonlinear!) inertia operator A from the Lie algebra $\mathrm{Vect}(M) \ltimes C^\infty(M)$ to its dual $(\Omega^1 \otimes \Omega^n \oplus \Omega^n)(M)$ via

$$A(\xi, \rho) = (\xi^\sharp \otimes \rho\mu, \rho\mu) \,,$$

where $\xi^\sharp$ is the 1-form on M that is obtained from the vector field ξ by "raising the indices" $\xi^\sharp = (\xi, .\,)$ with the help of the Riemannian metric.

Remark 7.5 From now on we restrict our attention to the (open) subset of the space $\mathrm{Vect}(M) \oplus C^\infty(M)$ consisting of pairs (ξ, ρ) such that $\rho > 0$ everywhere. The reason is that the inertia operator A restricted to this set is bijective: given any pair $(\alpha \otimes \nu, \theta)$ with $\theta = \rho\mu$, we can write $(\alpha \otimes \nu, \theta) = (\beta \otimes \rho\mu, \rho\mu)$ for some 1-form β provided that ρ is nonzero. This is consistent with the physical interpretation of ρ as the density of a fluid, which should be nowhere zero.

By pulling back the Hamiltonian $\widetilde{H}$ via the inertia operator A, we get a Hamiltonian function H on the image of A in $(\Omega^1 \otimes \Omega^n \oplus \Omega^n)(M)$. Explicitly it is given by

$$H(\alpha \otimes \theta, \theta) = \int_M \frac{1}{2}(\alpha, \alpha)\theta + \int_M \rho\Phi(\rho)\theta \,,$$

where $\rho \in C^\infty(M)$ is chosen such that $\theta = \rho\mu$.

Now let $m = (\alpha \otimes \theta, \theta)$ be a point in $(\Omega^1 \otimes \Omega^n \oplus \Omega^n)(M)$. Then the variational derivative $\delta H/\delta m$ of H at the point m is an element of the (smooth) dual of $(\Omega^1 \otimes \Omega^n \oplus \Omega^n)(M)$ and hence of the space $\mathrm{Vect}(M) \oplus C^\infty(M)$.

Exercise 7.6 Show that the variational derivative $\frac{\delta H}{\delta m}$ of H at the point $m = (\alpha \otimes \ell, \theta)$ is given by

$$\frac{\delta H}{\delta m} = \left(\alpha^\flat, \frac{1}{2}(\alpha, \alpha) + \rho\Phi'(\rho) + \Phi(\rho) \right) \,,$$

where $\alpha^\flat$ denotes the vector field on the manifold M obtained from the 1-form α by "lowering the indexes" with the help of the metric on M: $\alpha^\flat = (\alpha, \, . \,)$. As before, the function ρ is such that $\theta = \rho\mu$.

Hint: Use the definition of $\delta H/\delta m$ evaluated on a tangent vector v at the point m as the directional derivative:

$$\left\langle \frac{\delta H}{\delta m}, v \right\rangle = \frac{d}{dt}\bigg|_{t=0} H(m + tv) \,.$$

Show that the directional derivative of H at the point $m = (\alpha \otimes \theta, \theta)$ in the direction $(\beta \otimes \theta, 0)$ is given by

$$\frac{d}{dt}\bigg|_{t=0} H((\alpha + t\beta) \otimes \theta, \theta) = \langle (\alpha^\flat, 0), (\beta \otimes \theta, \theta) \rangle \,.$$

Similarly, the directional derivative of H in the θ-direction is given by

$$\frac{d}{dt}\bigg|_{t=0} H((\alpha \otimes (\theta + t\phi)), (\theta + t\phi)) = \left\langle \left(0, \frac{1}{2}(\alpha, \alpha) + \rho\Phi'(\rho) + \Phi(\rho) \right), (\alpha \otimes \phi, \phi) \right\rangle \,.$$

Finally, we are in a position to find explicitly the Euler equation

$$\dot{m} = -\,\mathrm{ad}^*_{\frac{\delta H}{\delta m}}\, m \,,$$

corresponding to the negative of the above Hamiltonian H on the dual to the Lie algebra $\mathrm{Vect}(M) \ltimes C^\infty(M)$.

Exercise 7.7 Verify that the Euler equation corresponding to the Hamiltonian $-H$ is the Euler equation of a barotropic (compressible) fluid

$$\begin{cases} \rho\,\partial_t\xi = -\rho\,(\xi, \nabla)\xi - \nabla h(\rho) \,, \\ \partial_t\rho + \mathrm{div}(\rho\,\xi) = 0 \,. \end{cases} \tag{7.29}$$

Find the equations of gas dynamics corresponding to $h(\rho) = \text{const} \cdot \rho^a$.

Hint: Use the fact that our choice of $\Phi : C^\infty(M) \to C^\infty(M)$ satisfies $\rho^2 \Phi'(\rho) = h(\rho)$. This implies

$$\rho \, d(\rho \Phi'(\rho) + \Phi(\rho)) = d \, h(\rho) \,.$$

A.8 The Drinfeld–Sokolov Reduction

Here we show that the quadratic (or second) Adler–Gelfand–Dickey Poisson structure on the space of smooth nth-order differential operators on the circle, which appeared in Section II.4.6 from the Poisson Lie group of pseudodifferential symbols, can also be obtained by a Hamiltonian reduction from the dual of the affine Lie algebra $\widehat{L\mathfrak{gl}_n}$. (To simplify notation we use $L\mathfrak{gl}_n$ for $L\mathfrak{gl}(n)$ in this appendix.)

8.1 The Drinfeld–Sokolov Construction

Recall that the (smooth) dual of the affine Lie algebra $\widehat{L\mathfrak{gl}_n}$ can be identified with the space of first-order differential operators $\{-a\partial + A \mid A \in L\mathfrak{gl}_n,\ a \in \mathbb{R}\}$; see Section II.1.2. (Here ∂ stands for $d/d\theta$ for a fixed parameter θ on the circle.) In this identification, the coadjoint action of the group LGL_n on the operator $-a\partial + A$ is simply the gauge action on differential operators:

$$g : -a\partial + A \mapsto -a\partial + gAg^{-1} + ag'g^{-1}.$$

For the rest of this section we consider the $(a = -1)$-hyperplane.

Let $\mathcal{N}_- \subset LGL_n$ be the subgroup of LGL_n consisting of loops with values in the *lower* triangular matrices with 1's on the diagonal. Denote by $\mathfrak{n}_- \subset L\mathfrak{gl}_n$ its Lie algebra of loops assuming values in *strictly lower* triangular matrices. Note that $\mathfrak{n}_-$ can also be regarded as a Lie subalgebra $\mathfrak{n}_- \subset \widehat{L\mathfrak{gl}_n}$ of the affine Lie algebra. (Indeed, the restriction of the 2-cocycle $\omega(X, Y) = \int_{S^1} \mathrm{tr}(X(\theta)Y'(\theta))\,d\theta$ defining the affine algebra vanishes if both X and Y are lower triangular, i.e., belong to $\mathfrak{n}_-$.)

The (smooth) dual space $\mathfrak{n}_-^*$ can be thought of as the space $\mathfrak{n}_+$ of loops in *strictly upper* triangular matrices with the nondegenerate pairing

$$\langle X, Y \rangle = \int_{S^1} tr(X(\theta)Y(\theta))\,d\theta$$

between $\mathfrak{n}_-$ and $\mathfrak{n}_+$.

Consider the affine hyperplane $\{\partial + A \mid A \in L\mathfrak{gl}_n\}$ in the dual space $\widehat{L\mathfrak{gl}_n}^*$. The group $\mathcal{N}_-$, as a subgroup of the extended loop group $\widehat{LGL_n}$, acts on this hyperplane by the coadjoint action.

Lemma 8.1 *The action of the group $\mathcal{N}_-$ on the affine hyperplane $\{\partial + A \mid A \in L\mathfrak{gl}_n\}$ is Hamiltonian with the moment map $\Phi : \{\partial + A \mid A \in L\mathfrak{gl}_n\} \to \mathfrak{n}_+$ given by the natural projection of a matrix A to its strictly upper triangular part (which is an element of $\mathfrak{n}_+$).*

PROOF. This is a manifestation of a general fact: For a Lie subalgebra $\mathfrak{n} \subset \mathfrak{g}$, the natural projection of the dual spaces $\mathfrak{g}^* \to \mathfrak{n}^*$ is the moment map for the coadjoint action of the corresponding group N on $\mathfrak{g}^*$. $\qquad\square$

Now we can perform the Hamiltonian reduction: Let $\Lambda \in \mathfrak{n}_-^* = \mathfrak{n}_+$ be the matrix with -1's on the superdiagonal and zeros everywhere else:

$$\Lambda := \begin{pmatrix} 0 & -1 & 0 & & 0 \\ & \ddots & \ddots & \ddots & \\ & & \ddots & \ddots & 0 \\ & & & \ddots & -1 \\ 0 & & & & 0 \end{pmatrix} .$$

The inverse image of Λ under the moment map Φ is the set of differential operators $\Phi^{-1}(\Lambda) = \{\partial + B + \Lambda\}$, where B runs over all loops in nonstrictly lower triangular matrices:

$$\Phi^{-1}(\Lambda) = \left\{ \partial + \begin{pmatrix} * & -1 & 0 & & 0 \\ & \ddots & \ddots & \ddots & \\ & & \ddots & \ddots & 0 \\ & & & \ddots & -1 \\ * & & & & * \end{pmatrix} \right\} .$$

Proposition 8.2 (*i*) *The matrix Λ is fixed under the (conjugation) action of the group $\mathcal{N}_-$ on the dual $\mathfrak{n}_-^* = \mathfrak{n}_+$. Hence the inverse image $\Phi^{-1}(\Lambda)$ is a union of $\mathcal{N}_-$-orbits.*

(*ii*) *In every $\mathcal{N}_-$-orbit in $\Phi^{-1}(\Lambda)$ there is a unique element $\partial + R$, where the matrix R is of the form*

$$R = \begin{pmatrix} 0 & -1 & 0 & & 0 \\ & \ddots & \ddots & \ddots & \\ \vdots & & \ddots & \ddots & 0 \\ 0 & \cdots & & 0 & -1 \\ u_0 & u_1 & \cdots & \cdots & u_{n-1} \end{pmatrix} \tag{8.30}$$

for some $u_0, \ldots, u_{n-1} \in C^\infty(S^1)$.

Proposition 8.2 shows that the quotient $\Phi^{-1}(\Lambda)/\mathcal{N}_-$ can be regarded as the set $\mathcal{L}_n$ of smooth monic nth-order differential operators on S^1 by identifying the differential operator $\partial + R$ with the nth-order differential operator

$$L = \partial^n + u_{n-1}\partial^{n-1} + \cdots + u_0 .$$

Namely, given an nth-order differential operator L as above, the ordinary differential equation

$$L\psi = 0$$

of order n is equivalent to the system of first-order differential equations

$$\partial\Psi + R\Psi = 0,$$

where R is the matrix of type (8.30) and Ψ is a vector-solution, $\Psi = (\psi_0, \ldots, \psi_{n-1})^t$. The equivalence of the two equations comes from setting $\psi_0 := \psi$. Then the first $n-1$ equations of the system imply $\psi_1 := \psi', \ldots, \psi_{n-1} := \psi^{(n-1)}$, while the last equation of the system reads $\psi^{(n)} + u_{n-1}\psi^{(n-1)} + \cdots + u_0\psi = 0$, i.e., $L\psi = 0$.

Return to the whole affine space $\{\partial + A \mid A \in L\mathfrak{gl}_n\}$ of all matrix differential operators, regarded as a hyperplane in the smooth dual of the affine Lie algebra $\widehat{L\mathfrak{gl}_n}$. It carries a natural Poisson structure, which is the linear Lie–Poisson structure on the dual $\widehat{L\mathfrak{gl}_n}^*$, restricted to this hyperplane. This linear Poisson structure produces a certain Poisson structure on the symplectic quotient $\Phi^{-1}(\Lambda)/\mathcal{N}_-$ as a result of the Hamiltonian reduction; see Section I.5. The theorem of Drinfeld and Sokolov states that this Poisson structure coincides with the quadratic Gelfand–Dickey Poisson structure on the space $\mathcal{L}_n = \{L\}$ of (monic) nth-order differential operators L:

Theorem 8.3 (Drinfeld–Sokolov [88]) *The Poisson structure on the quotient $\Phi^{-1}(\Lambda)/\mathcal{N}_-$ coincides with the quadratic (or second) Gelfand–Dickey structure on the space $\mathcal{L}_n$ of smooth monic nth-order differential operators on the circle.*

We refer to the original paper [88] or to the book [80] for a proof by direct calculation. Below we shall prove this fact using ideas from the Poisson Lie groups combined with the approach in [115].

Remark 8.4 The linear (or first) Gelfand–Dickey structure can also be obtained by the Hamiltonian reduction from the smooth dual of the affine Lie algebra $\widehat{L\mathfrak{gl}_n}$. Just as the *quadratic* Gelfand–Dickey structure is obtained from the reduction of the *linear* Lie–Poisson structure on $\widehat{L\mathfrak{gl}_n}^*$, the *linear* Gelfand–Dickey structure comes from a *constant* Poisson structure on $\widehat{L\mathfrak{gl}_n}^*$. Details on how to choose the freezing point and to perform the reduction can be found, for example, in [88, 31]. The Drinfeld–Sokolov reduction has also been generalized to pseudodifferential symbols of arbitrary *complex* degree $\alpha \in \mathbb{C}$ in [191]. In this case, the matrix algebra $\mathfrak{gl}_n$ is replaced by the "Lie algebra of matrices of complex size $\alpha \times \alpha$," which was introduced by Feigin in [116]. A q-analogue of the universal Drinfeld–Sokolov reduction for complex degrees is described in [314]. There is also a version of the Drinfeld–Sokolov reduction associated to any simple subalgebra of $\mathfrak{gl}_n$ [88].

8.2 The Kupershmidt–Wilson Theorem and the Proofs

We start with the proof for the explicit identification $\mathcal{L}_n \simeq \Phi^{-1}(\Lambda)/\mathcal{N}_-$ for the result of the Drinfeld–Sokolov Hamiltonian reduction.

PROOF OF PROPOSITION 8.2. (i) To show that the point $\Lambda \in \mathfrak{n}_+ = \mathfrak{n}_-^*$ is a one-point $\mathcal{N}_-$-coadjoint orbit, we note that the corresponding Lie algebra action on this point vanishes. Indeed, this action by an element $N_- \in \mathfrak{n}_-$ is the projection of $[N_-, \partial + \Lambda]$ to $\mathfrak{n}_+$, i.e., taking the upper triangular part of the commutator, while for a lower triangular current N_- this commutator is always zero above the diagonal.

(ii) Let $\partial + C$ be an element of the preimage $\Phi^{-1}(\Lambda)$ for Λ as above, i.e., the current C has the form $B + \Lambda$, where $B \in L\mathfrak{gl}_n$ is a loop in lower triangular matrices (including the diagonal). To the system of differential equations

$$\partial \Psi + C\Psi = 0$$

with $\Psi = (\psi_0, \ldots, \psi_{n-1})^t$ we associate the following nth-order differential equation $L_C \psi = 0$ for $\psi = \psi_0$. This system has the form

$$\psi_i' + \sum_{j=0}^{i} c_{i,j}\psi_j = \psi_{i+1} \quad \text{for} \quad i = 0, \ldots, n-2, \tag{8.31}$$

$$\psi_{n-1}' + \sum_{j=0}^{n-1} c_{n-1,j}\psi_j = 0. \tag{8.32}$$

By expressing in succession all the ψ_i in terms of ψ_0, the last relation leads to the equation $L_C \psi_0 = 0$. Note that different matrix operators $\partial + C$ can correspond to one and the same nth-order operator L_C.

First we show that two matrix differential operators $\partial + C$ and $\partial + \widetilde{C}$ that lie in the same $\mathcal{N}_-$-orbit in $\Phi^{-1}(\Lambda)$ give rise to the same nth-order differential operator: $L_C = L_{\widetilde{C}}$. Indeed, such matrix operators are gauge equivalent with the help of a lower triangular transformation $g \in \mathcal{N}_-$, and hence their vector-solutions are related by $\widetilde{\Psi} = g\Psi$. Since g has 1's on the diagonal, the first coordinate ψ_0 of the vector-solution Ψ, which defines the nth-order differential equation, does not change: $\widetilde{\psi}_0 = \psi_0$. Now it is enough to note that two monic nth-order differential operators with the same solution sets coincide, since ψ_0 was an arbitrary solution of $L_C \psi_0 = 0$.

In the other direction we have to show that if the nth-order differential operators L_C and $L_{\widetilde{C}}$ coincide, then there exists a gauge transformation $g \in \mathcal{N}_-$ that sends $\partial + C$ to $\partial + \widetilde{C}$, i.e., such that $\tilde{C} = gCg^{-1} - g'g^{-1}$. To this end, let Ψ be a vector-solution of $\partial \Psi + C\Psi = 0$ and let $\widetilde{\Psi}$ be a vector-solution of $\partial \widetilde{\Psi} + \widetilde{C}\widetilde{\Psi} = 0$. Since the corresponding nth-order equations $L_C \psi = L_{\widetilde{C}}\psi = 0$ coincide, we can assume that $\psi_0 = \widetilde{\psi}_0$. Now, comparing the vector-solutions Ψ and $\widetilde{\Psi}$ entry by entry and using the triangular form of the corresponding equations (8.31) they satisfy, we find that the transformation from Ψ to $\widetilde{\Psi}$ is also lower triangular with 1's on the diagonal: $\widetilde{\Psi} = g\Psi$ for $g \in \mathcal{N}_-$. This implies that $\partial + C$ and $\partial + \widetilde{C}$ lie in the same $\mathcal{N}_-$-orbit in $\Phi^{-1}(\Lambda)$. $\qquad \square$

Before proving the Drinfeld–Sokolov theorem we are going to derive the "multiplicative" property of the quadratic Gelfand–Dickey bracket known as the Kupershmidt–Wilson theorem.

For this purpose, consider the space of the first-order scalar differential operators $\mathcal{L}_1 := \{L = \partial + v \mid v \in C^\infty(S^1)\}$. We define the following constant Poisson bracket on the space of such operators (and discuss the origin for this Poisson structure a little later). Namely, if $F(v)$ and $G(v)$ are two functionals on the space $\mathcal{L}_1$, their Poisson bracket is

$$\{F, G\}(v) := \int_{S^1} \frac{\delta F}{\delta v} \left(\frac{\delta G}{\delta v} \right)' d\theta \,,$$

where $\delta F/\delta v$ is the variational derivative of F. This bracket is a constant Poisson bracket on $\mathcal{L}_1$ (cf. Definition I.4.20): the corresponding contraction of the variational derivatives does not depend on v itself. The Hamiltonian vector field on $\mathcal{L}_1$ corresponding to a linear functional $F_f(v) := \int f \cdot v \, d\theta$ assumes the value $V_f(L) := -f'$ at $L = \partial + v$.

Consider now the product $\mathcal{L}_1 \times \cdots \times \mathcal{L}_1 \to \mathcal{L}_n$ of several first-order differential operators to obtain one nth-order differential operator:

$$((\partial + v_{n-1}), \ldots, (\partial + v_0)) \mapsto L = (\partial + v_{n-1}) \circ \cdots \circ (\partial + v_0)$$
$$= \partial^n + u_{n-1}\partial^{n-1} + \cdots + u_0 \,.$$

Definition 8.5 The map $\{v_i\} \to \{u_j\}$ given by the product of differential operators $\mathcal{L}_1 \times \cdots \times \mathcal{L}_1 \to \mathcal{L}_n$ is called the *Miura transformation*.

Theorem 8.6 (Kupershmidt–Wilson [226]) *The Miura transformation sends the constant Poisson bracket on the product $\mathcal{L}_1 \times \cdots \times \mathcal{L}_1$ of first-order operators to the quadratic Gelfand–Dickey bracket on the space $\mathcal{L}_n$ of nth-order differential operators.*

PROOF. Recall that both the spaces $\mathcal{L}_1$ and $\mathcal{L}_n$ are Poisson submanifolds in the Poisson Lie group $\widetilde{G}_{\mathrm{INT}}$ of pseudodifferential symbols. In turn, the group multiplication on any Poisson Lie group maps the Poisson structure on the square of this group to the Poisson structure of this group itself. For the group $\widetilde{G}_{\mathrm{INT}}$ this implies that the Gelfand–Dickey bracket restricted to (several copies of) the Poisson submanifolds $\mathcal{L}_1$ is mapped by the group product, which is the multiplication of differential operators, to the Gelfand–Dickey bracket on $\mathcal{L}_n$.

Now the theorem follows from the observation that the quadratic Gelfand–Dickey structure restricted to the first-order operators $\mathcal{L}_1$ drastically simplifies

and becomes exactly the constant Poisson bracket discussed above. In order
to see the latter we consider a cotangent vector $X = \partial^{-1}a$ to the space of
first-order differential operators $L = \partial + v$. Then the corresponding Gelfand–
Dickey Hamiltonian field $V_X^2(L)$ is given by

$$V_X^2(L) = (L \circ X)_+ \circ L - L \circ (X \circ L)_+ = -a' \, ,$$

which is exactly the Hamiltonian field in the constant Poisson bracket
above.   □

Remark 8.7 The Hamiltonian property of the Miura map is important in
integrable systems. For instance, for $n = 2$ and $L = \partial^2 + u = (\partial - v)(\partial + v)$
the Miura map gives $u = v' - v^2$. If the function u satisfies the KdV equation
$u_t = -3uu' - (1/2)u'''$, then the function v satisfies the so called *modified
KdV* (or *m-KdV*) *equation* $v_t = 3v^2 v' - (1/2)v'''$.

Now we are ready to prove the Drinfeld–Sokolov reduction theorem.

PROOF OF THEOREM 8.3.   We would like to show that the linear Lie–
Poisson structure descends from $\widehat{L\mathfrak{gl}_n}^*$ to the quadratic Poisson structure on
the quotient of $\Phi^{-1}(\Lambda)$ over the $\mathcal{N}_-$-action:

$$\Phi^{-1}(\Lambda)/\mathcal{N}_- \simeq \{\partial + R\} = \mathcal{L}_n \, .$$

For this purpose we take an auxiliary step: we first restrict the linear Poisson
bracket from $\widehat{L\mathfrak{gl}_n}^*$ not to $\Phi^{-1}(\Lambda)$, but to a certain subspace, and after that
we take the quotient from this subspace to $\{\partial + R\}$.

This auxiliary (affine) subspace $\{\partial + P\} \subset \Phi^{-1}(\Lambda)$ consists of all currents
P of the form

$$P = \begin{pmatrix} v_0 & -1 & 0 & & & 0 \\ 0 & \ddots & \ddots & \ddots & & \\ & \ddots & \ddots & \ddots & 0 & \\ & & \ddots & \ddots & -1 \\ 0 & & & 0 & v_{n-1} \end{pmatrix} . \tag{8.33}$$

This subspace has the "same functional dimension" as $\mathcal{L}_n$. (Note, however,
that this subspace is not a section for the fibration $\Phi^{-1}(\Lambda) \to \mathcal{L}_n$ given by the
$\mathcal{N}_-$-action: different operators $\partial + P$ can belong to the same $\mathcal{N}_-$-orbit and
then correspond to the same differential operator $\partial + R$, as we discuss below.)

The restriction of the linear Lie–Poisson structure from $\widehat{L\mathfrak{gl}_n}^*$ to this subspace can be found explicitly.

Exercise 8.8 Verify that the linear Lie–Poisson structure from $\widehat{L\mathfrak{gl}_n}^*$ restricted to the subspace $\{\partial + P\}$ gives the constant Poisson structure on the

coefficients $\{v_j\}$ discussed above. (Hint: the Lie–Poisson bracket on this subspace is equivalent to the Lie structure on the centrally extended subalgebra of loops in diagonal matrices under the identification we considered. The only nontrivial term in the commutator between such "diagonal currents" comes from the 2-cocycle. Finally, the affine 2-cocycle restricted to the diagonal part can be thought of as the sum of several $\widehat{L\mathfrak{gl}_1}$-cocycles $c(a,b) = \int_{S^1}(a \cdot b')\,d\theta$. The latter Lie algebra is an infinite-dimensional Heisenberg algebra, and it corresponds to the constant Poisson bracket discussed above; see details in [115].)

Now we note that this subspace $\{\partial + P\}$ can be rewritten as the space of nth-order differential operators $\{L\}$ expressed as products of first-order operators

$$L = (\partial + v_{n-1}) \circ \cdots \circ (\partial + v_0)\,.$$

Indeed, the system of first-order differential equations $(\partial + P)\Psi = 0$ for a vector-solution $\Psi = (\psi_0, \ldots, \psi_{n-1})^t$ is equivalent to the scalar differential equation $L\psi = 0$ under an identification similar to the one above. Namely, we set $\psi_0 := \psi$, which forces $\psi_1 := (\partial + v_0)\psi_0$, $\psi_2 := (\partial + v_1)\psi_1$ etc., while $(\partial + v_{n-1})\psi_{n-1} = 0$.

Thus the passage to the quotient $\{\partial + P\} \to \{\partial + R\} = \mathcal{L}_n$ is the Miura transform, and the Kupershmidt–Wilson theorem completes the proof: the constant Poisson bracket on $\{\partial + P\}$ becomes the quadratic Gelfand–Dickey bracket on $\mathcal{L}_n = \Phi^{-1}(\Lambda)/\mathcal{N}_-$. $\square$

A.9 The Lie Algebra $\mathfrak{gl}_\infty$

In this section we introduce yet another Lie algebra, $\mathfrak{gl}_\infty$, along with its central extension $\widehat{\mathfrak{gl}}_\infty$, which connects some of the algebras we encountered throughout the book. In particular, the extended Lie algebra $\widehat{\mathfrak{gl}}_\infty$ contains as subalgebras the Virasoro algebra, the centrally extended Lie algebra $\widehat{\mathrm{DO}}$ of differential operators on the circle, and the affine Lie algebras $\widehat{L\mathfrak{gl}}_n$ for all $n \in \mathbb{N}$.

9.1 The Lie Algebra $\mathfrak{gl}_\infty$ and Its Subalgebras

Definition 9.1 The space $\mathfrak{gl}_\infty$ consists of doubly infinite matrices that can have nonzero entries only inside a strip of finite width around the main diagonal:

$$\mathfrak{gl}_\infty = \{A = (a_{ij})_{i,j \in \mathbb{Z}} \mid \exists\, N \text{ such that } a_{ij} = 0 \text{ for all } i, j \text{ with } |i - j| > N\}.$$

The multiplication of two such matrices is well defined, since for the product $C = AB$ of two matrices $A, B \in \mathfrak{gl}_\infty$, the sum $c_{ij} = \sum_k a_{ik} b_{kj}$ is finite. The *Lie algebra* $\mathfrak{gl}_\infty$ is the above space equipped with the Lie bracket $[A, B] = AB - BA$.

This large Lie algebra has many interesting subalgebras.

Example 9.2 (Differential operators on the circle) Consider the space $\mathbb{C}[z, z^{-1}]$ of Laurent polynomials with coefficients in $\mathbb{C}$. This space can be identified with the space of vectors $X = (x_i)_{i \in \mathbb{Z}}$ of infinite length with only finitely many nonzero entries. The Lie algebra $\mathfrak{gl}_\infty$ acts naturally on the latter space by left multiplication. In this identification, the derivative operator $z\frac{d}{dz}$ is identified with the diagonal matrix $D = (d_{ij})$, where $d_{ii} = i$ for all $i \in \mathbb{Z}$, and $d_{ij} = 0$ for all $i \neq j$. Furthermore, the operator Z of multiplication by the monomial z on $\mathbb{C}[z, z^{-1}]$ is given by the *shift matrix*, which has ones on the superdiagonal and zeros everywhere else: $Z = (a_{ij})$ with $a_{i\,i+1} = 1$ and $a_{ij} = 0$ for all i, j with $j \neq i + 1$. Similarly, the operator Z^{-1} is given by the matrix that has ones on the subdiagonal and zeros everywhere else.

The subspace in $\mathfrak{gl}_\infty$ generated by sums and compositions of the operators D and Z realizes the Lie algebra of polynomial differential operators on the circle as a subalgebra of $\mathfrak{gl}_\infty$. (We restrict our attention to differential operators whose coefficients are Laurent polynomials, instead of smooth functions on the circle, in order to avoid convergence issues.) As before, one can use the map $z \mapsto e^{i\theta}$ to identify the operator $iz\frac{d}{dz}$ with the usual derivative operator ∂_θ on the circle.

Note that the above embedding of differential operators to $\mathfrak{gl}_\infty$ does not extend to an embedding of (polynomial) pseudodifferential symbols. Indeed, in the identification described above, the inverse ∂_θ^{-1} of the derivative operator ∂_θ cannot be realized inside $\mathfrak{gl}_\infty$: the matrix $D = (d_{ij})$ with $d_{ii} = i$ has a zero row and is not invertible.

 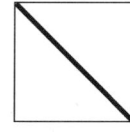 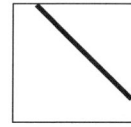

Fig. 9.1. Matrices in $\mathfrak{gl}_\infty$, the operator D realizing $z\frac{d}{dz}$, and the operator Z for multiplying by z.

Example 9.3 (Loop algebras inside $\mathfrak{gl}_n$) Fix some integer $n \in \mathbb{N}$. We call an element $A = (a_{ij})_{i,j \in \mathbb{Z}} \in \mathfrak{gl}_\infty$ periodic of order n if $a_{i+n,j+n} = a_{i,j}$ for all $i, j \in \mathbb{Z}$. The subalgebra of all periodic elements of order n can be identified with the algebra $\mathfrak{gl}_n \otimes \mathbb{C}[z, z^{-1}]$ of polynomial loops in $\mathfrak{gl}_n$. Indeed, a periodic matrix A of order n is completely determined by fixing the n rows $(a_{ij})_{0 \le i < n, j \in \mathbb{Z}}$. Associate to the periodic matrix A the sequence of $n \times n$ matrices A_k, $k \in \mathbb{Z}$, by $A_k := (a_{ij})$, where $0 \le i < n$ and $kn \le j < (k+1)n$. Note that only a finite number of the matrices A_k can be nonzero. We identify such a sequence with the Laurent series $\sum_k A_k z^k$. This gives a bijection between the set of periodic matrices of order n in $\mathfrak{gl}_\infty$ and polynomial loop algebra $(L\mathfrak{gl}_n)_{\text{pol}} = \mathfrak{gl}_n \otimes \mathbb{C}[z, z^{-1}]$.

Exercise 9.4 Show that the bijection between the set of periodic matrices of order n in $\mathfrak{gl}_\infty$ and the loop algebra $(L\mathfrak{gl}_n)_{\text{pol}}$ described above defines a Lie algebra isomorphism.

9.2 The Central Extension of $\mathfrak{gl}_\infty$

Let J be the diagonal matrix defined by $j_{ii} = -1$ if $i \ge 0$, and $j_{ii} = 1$ if $i < 0$.

Exercise 9.5 Given two elements $A, B \in \mathfrak{gl}_\infty$, show that the matrix $C := [J, A]E$ has only finitely many nonzero entries on the diagonal. (Hint: see Figure 9.2.)

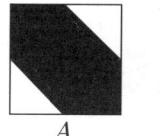 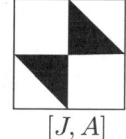

$$A \qquad\qquad [J, A]$$

Fig. 9.2. For every matrix $A \in \mathfrak{gl}_\infty$, the commutation with J "cuts out" the upper left and lower right quadrants of A and multiplies the off-diagonal elements by ± 2.

Definition 9.6 The *central extension* $\widehat{\mathfrak{gl}}_\infty$ of the Lie algebra $\mathfrak{gl}_\infty$ is defined by the 2-cocycle

$$\varpi(A, B) = \frac{1}{2} \operatorname{tr}([J, A], B). \qquad (9.34)$$

This cocycle has appeared in a series of papers [75] related to soliton equations and related algebraic constructions, and it is often referred to as the *Japanese cocycle*.

Proposition 9.7 *(i)* *When restricted to the Lie subalgebra* DO $\subset \mathfrak{gl}_\infty$ *of polynomial differential operators, the cocycle* ϖ *is cohomologous to (a multiple of) the Kac–Peterson 2-cocycle. In particular, restricted to the Lie algebra of vector fields on the circle,* ϖ *is cohomologous to (a multiple of) the Gelfand–Fuchs 2-cocycle.*
(ii) *Restricted to the subalgebra* $(L\mathfrak{gl}_n)_{\mathrm{pol}} \subset \mathfrak{gl}_\infty$, *the cocycle* ϖ *coincides with the cocycle defining the affine Lie algebra* $(\widehat{L\mathfrak{gl}_n})_{\mathrm{pol}}$.

In particular, the 2-cocycle ϖ is nontrivial on $\mathfrak{gl}_\infty$, and $\widehat{\mathfrak{gl}}_\infty$ is a nontrivial central extension of the latter algebra.

Exercise 9.8 Give a proof of Proposition 9.7. (Hint: For part (i) it is enough to prove the assertion for vector fields. Show that in the basis $L_n = z^{n+1} \frac{d}{dz}$, the cocycle ϖ has the explicit form

$$\varpi(L_n, L_m) = \delta_{n,-m} \sum_{i=1}^{n} i(i-1).$$

For part (ii) start with the case of $\mathfrak{gl}_1$, i.e., with ordinary Laurent polynomials. The general case is similar.)

Corollary 9.9 *The Virasoro algebra, the centrally extended Lie algebra* $\widehat{\mathrm{DO}}$, *as well as all the affine Lie algebras* $(\widehat{L\mathfrak{gl}_n})_{\mathrm{pol}}$, *are Lie subalgebras of* $\widehat{\mathfrak{gl}}_\infty$.

Remark 9.10 There is a generalization of $\mathfrak{gl}_\infty$ in which the strip can be unbounded downward: one requires only that $a_{ij} = 0$ for all i, j satisfying $j - i > N$, rather than $|i - j| > N$. The multiplication, the Lie bracket, and the 2-cocycle are still well defined on the space of such matrices.

9.3 q-Difference Operators and $\mathfrak{gl}_\infty$

Other interesting subalgebras of $\mathfrak{gl}_\infty$ are the algebras of q-versions of differential operators and pseudodifferential symbols. Consider the "shift operator" D_q on the space $\mathbb{C}[z, z^{-1}]$ of Laurent polynomials: $D_q f(z) := f(qz)$. Note that the finite difference operator

$$\partial_q := \frac{D_q - 1}{q - 1}$$

becomes the derivative operator $z\partial_z$ in the limit $q \to 1$, since $(f(qz)-f(z))/(q-1) \to zf'(z)$. Thus the operator $(D_q - 1)/(q - 1)$ (or the operator D_q itself) can be regarded as a *quantum version* of the derivative operator. The operator D_q also satisfies the q-analogue of the Leibniz rule $D_q \circ f = (D_q f)D_q$, similar to that for the derivative.

Now one can define the algebras of q-difference operators $\mathrm{DO}_q := \{\sum_0^n \iota_k(z)D_q^k\}$ and q-difference symbols $\psi\,\mathrm{DS}_q := \{\sum_{-\infty}^n u_k(z)D_q^k\}$. (One usually assumes that q is not a root of unity, i.e., $q^n \neq 1$ for any n, to avoid extra relations of $(D_q^n = 1)$-type.)

Furthermore, one can define the notion of $\log D_q$ for the algebra of q-difference symbols, and it turns out to have a very simple meaning. Since D_q is a shift operator, one can think of it as the exponent of the derivative operator $D_q = \exp(q \cdot z\partial_z)$; cf. Section II.4.2. Then $\log D_q$ is the derivative operator $q \cdot z\partial_z$ itself! Note that the derivative $z\partial_z$ does not belong to the algebra of q-difference symbols, and it defines an *outer* derivation of the algebra $\psi\,\mathrm{DS}_q$. One can proceed by defining the corresponding 2-cocycles, constructing the Lie group of fractional q-difference symbols, and the corresponding integrable systems; see [190, 313].

The trace on q-difference symbols is defined by $\mathrm{Tr}_q(\sum u_k(z)D_q^k) := \mathrm{tr}(u_0)$, where $\mathrm{tr}(u_0)$ is the constant term in the Laurent polynomial $u_0(z)$, i.e., its coefficient at z^0. It is an algebraic trace, since it satisfies $\mathrm{Tr}_q[A, B] = 0$, similar to the trace on pseudodifferential symbols. One should note, however, that the subalgebras of positive and negative powers of D_q are not isotropic with respect to the quadratic form related to this trace on $\psi\,\mathrm{DS}_q$, and thus they do not form a Manin triple. Although the corresponding Lie group is not a Poisson Lie group, it can still be equipped with a natural Poisson structure [313, 314].

Finally, we note that the algebras DO_q and $\psi\,\mathrm{DS}_q$ act on the space of Laurent polynomials $\mathbb{C}[z, z^{-1}]$, and hence their action can be defined by infinite matrices. One can easily see that the generators of $\psi\,\mathrm{DS}_q$ are described by matrices "of finite width," i.e., they belong to $\mathfrak{gl}_\infty$. This gives an embedding of $\psi\,\mathrm{DS}_q$ as a subalgebra of $\mathfrak{gl}_\infty$; see, e.g., [180].

Remark 9.11 The Lie algebra $\mathfrak{gl}_\infty$ and its central extension $\widehat{\mathfrak{gl}}_\infty$ by means of the cocycle ϖ have a long history and have been used to study the KdV equation, the KP hierarchy, and other integrable hierarchies (see, e.g., [372, 181] and references therein).

Among various relations between $\mathfrak{gl}_\infty$ and other algebras, including $\psi\,\mathrm{DS}_q$, it is worth mentioning that the algebra $\psi\,\mathrm{DS}_q$ is also isomorphic to the so-called *noncommutative torus*: the associative algebra generated by two unitary operators U_1 and U_2 satisfying the relations $U_2U_1 = qU_1U_2$ for $q = e^{ih}$; see [147, 130].

Another interesting version of this algebra is known as the *sine algebra* [111, 146, 147]. One can also think of it as an algebra with continuous root systems, as was defined in [335]. Finally, the limit of the structure constants of the sine algebra gives those of the Lie algebra $S\text{Vect}(T^2)$ of divergence-free vector fields on the two-dimensional torus [320, 49]. In this way, the algebra $\mathfrak{gl}_\infty$ provides a link to a number of the infinite-dimensional Lie algebras considered in this book. We also refer the interested reader to [118] for many other related constructions.

A.10 Torus Actions on the Moduli Space of Flat Connections

Let G be a compact or complex simple Lie group. In this appendix we discuss a torus action on the moduli space of flat connections on the trivial G-bundle over a Riemann surface Σ. We start with constructing a natural set of Poisson commuting functions on the moduli space of flat connections, and then turn to the case of $G = \mathrm{SU}(2)$, where this set of commuting functions gives rise to integrable systems on the symplectic leaves of the moduli space [175]. This action allows one to study the symplectic geometry of the leaves. Finally, following [120, 125], we consider the moduli space of flat $\mathrm{SL}(n, \mathbb{C})$-connections on the one-holed torus and show that the rational Ruijsenaars–Schneider system lives on a holomorphic symplectic leaf of this moduli space.

10.1 Commuting Functions on the Moduli Space

Let Σ be a compact Riemann surface of genus κ with d boundary components. Here we assume that G is a simply connected simple Lie group with the Lie algebra $\mathfrak{g}$. Denote by

$$\mathcal{A}^{\Sigma} = \{d + A \mid A \in \Omega^1(\Sigma, \mathfrak{g})\}$$

the affine space of connections in the trivial G-bundle over the surface Σ. We have seen in Section III.2.1 that the space $\mathcal{A}^{\Sigma}$ is an infinite-dimensional symplectic manifold with a Hamiltonian action of (the central extention $\widehat{G}^{\Sigma}$ of) the group $G^{\Sigma} = C^{\infty}(\Sigma, G)$ of gauge transformations. The moment map of this action is given by the curvature together with the restriction of the connection to the boundary $\partial\Sigma$ (see Proposition III.2.10 for the exact formulation). The corresponding symplectic quotient is the moduli space

$$\mathcal{M}_{\Sigma} = \mathcal{A}^{\Sigma}_{flat}/\widehat{G}^{\Sigma}$$

of flat connections on Σ. This is a finite-dimensional, possibly singular, Poisson manifold. The symplectic leaves of $\mathcal{M}_{\Sigma}$ are parametrized by fixing the conjugacy classes of the holonomies around the boundary circles of the surface Σ. (More precisely, in Section III.2.1 we considered the case of a compact Lie group G. But all we really need in the constructions is a nondegenerate invariant bilinear form on the Lie algebra $\mathfrak{g}$. In particular, we can take G to be any semisimple Lie group, in which case $\mathcal{M}_{\Sigma}$ denotes the space of representations $\mathrm{Rep}(\pi_-(\Sigma) \to G)$ modulo conjugation.)

A natural class of functions on the moduli space $\mathcal{M}_{\Sigma}$ can be constructed as follows. Let $f : G \to \mathbb{C}$ be a conjugation-invariant function on the group, and let Γ be a simple curve in Σ. Associate to them the function $f_{\Gamma} : \mathcal{M}_{\Sigma} \to \mathbb{C}$ by assigning to (the equivalence class of) a flat connection A on Σ the value of f on the holonomy of A around the curve Γ:

$$f_\Gamma([A]) = f(\mathrm{hol}_\Gamma(A)).$$

Goldman gave a formula for the Poisson bracket $\{f_{\Gamma_1}, g_{\Gamma_2}\}$ in terms of the intersection number of the curves Γ_1 and Γ_2; see [145]. Here we need only the following special case.

Proposition 10.1 *(i) Let Γ_1 and Γ_2 be two simple disjoint curves on the surface Σ. Then for any two conjugation-invariant functions f and g on the group G, we have*

$$\{f_{\Gamma_1}, g_{\Gamma_2}\} = 0.$$

(ii) For any simple closed curve Γ and any two conjugation-invariant functions f and g on the group G, we have

$$\{f_\Gamma, g_\Gamma\} = 0.$$

A *trinion* (or a *pair of pants*) is a Riemann surface of genus 0 with three boundary components. Recall that the genus of the Riemann surface Σ is denoted by κ, and the number of boundary components is d. From now on we assume that $d \geq 1$ if $\kappa = 1$, and $d \geq 3$ if $\kappa = 0$. In these cases the Riemann surface Σ has a *trinion decomposition* into $2\kappa - 2 + d$ trinions; see Figure 10.1.

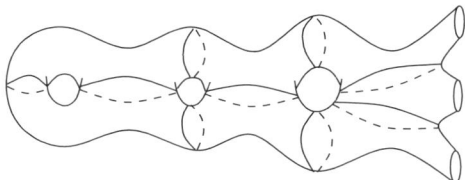

Fig. 10.1. A trinion decomposition of a Riemann surface with boundary.

This can be achieved by cutting Σ along $3\kappa - 3 + d$ disjoint curves. On the other hand, the number of independent conjugation-invariant functions on the group G is equal to its rank $\mathrm{rk}(G)$, i.e., to the dimension of a maximal torus $T \subset G$. Hence, the trinion decomposition of the surface Σ provides us with $(3\kappa - 3 + d) \cdot \mathrm{rk}(G)$ independent Poisson-commuting functions on the moduli space $\mathcal{M}_\Sigma$. For a closed surface Σ (i.e., for $d = 0$), the moduli space $\mathcal{M}_\Sigma$ is symplectic of dimension $(2\kappa - 2) \cdot \dim(G)$. The above functions f_Γ form an integrable system on the moduli space $\mathcal{M}_\Sigma$ whenever the number of independent functions is equal to half the dimension of $\mathcal{M}_\Sigma$, which for a closed surface happens exactly if $\dim(G)/\mathrm{rk}(G) = 3$. For simple groups this occurs only for $G = \mathrm{SU}(2)$.

10.2 The Case of SU(2)

Define a maximal torus T of the group SU(2) as the set of diagonal matrices

$$T = \left\{ \begin{pmatrix} e^{i\theta} & 0 \\ 0 & e^{-i\theta} \end{pmatrix} \right\} \cong S^1 \,.$$

Each element of SU(2) is conjugate to an element of the torus T. Furthermore, the matrices $\mathrm{diag}(e^{i\theta}, e^{-i\theta})$ and $\mathrm{diag}(e^{-i\theta}, e^{i\theta})$ are conjugate to each other, so that a conjugacy class of SU(2) is completely determined by the trace $\mathrm{tr}(\mathrm{diag}(e^{i\theta}, e^{-i\theta})) = 2\cos\theta$. The centralizers of the corresponding conjugacy classes are given by the torus T if $\theta \notin \pi\mathbb{Z}$. Otherwise, the centralizer is the whole group SU(2).

Recall that for a surface Σ with boundary a symplectic leaf in the moduli space $\mathcal{M}_\Sigma$ is singled out by fixing conjugacy classes of the holonomies around the boundary components of Σ.

Proposition 10.2 *Let $\theta_1, \ldots, \theta_d \in [0, \pi]$ be such that for all $\epsilon_j \in \{\pm 1\}$, we have $\sum \epsilon_j \theta_j \neq 0 \bmod 2\pi$. Then the symplectic leaf defined by the conjugacy classes of the diagonal matrices $\mathrm{diag}(e^{i\theta_j}, e^{-i\theta_j})$, $j = 1, \ldots, d$, is a smooth symplectic manifold of dimension $6\kappa - 6 + 2d$.*

Let $\Gamma_1, \ldots, \Gamma_{3\kappa-3+d}$ denote curves cutting the surface Σ into a trinion decomposition and denote by $\widetilde{\Gamma}_1 \ldots, \widetilde{\Gamma}_d$ the boundary circles of the surface Σ. Let ϕ_j be the functions on the moduli space M_Σ defined by

$$\phi_j([A]) := \cos^{-1}\left(\frac{1}{2} \mathrm{tr}(\mathrm{hol}_{\Gamma_j}(A)) \right) \,,$$

and similarly for $\widetilde{\phi}_k$ and $\widetilde{\Gamma}_k$. By Proposition 10.1, the functions ϕ_j and $\widetilde{\phi}_k$ all Poisson commute. Consider the map

$$\Phi = (\phi_1, \ldots, \phi_{3\kappa-3+d}, \widetilde{\phi}_1, \ldots, \widetilde{\phi}_d) : \mathcal{M}_\Sigma \to \mathbb{R}^{3\kappa-3+2d} \,.$$

Jeffrey and Weitsman showed that the restriction of the map Φ to the open and dense subset

$$\overset{\circ}{\mathcal{M}}_\Sigma := \bigcap_{j=1}^{3\kappa-3+d} \phi_j^{-1}(0, \pi) \cap \bigcap_{k=1}^{d} \widetilde{\phi}_k^{-1}(0, \pi) \subset \mathcal{M}_\Sigma$$

is the moment map of the Hamiltonian action of a $(3\kappa - 3 + d)$-dimensional torus [174]. This action is given by the *twist flows* defined as follows.

Denote by P_l, $l = 1, \ldots, 2\kappa - 2 + d$, the trinions in the above trinion decomposition of the surface Σ with the cutting curves Γ_j and boundary curves $\widetilde{\Gamma}_k$.

Definition 10.3 A connection $A \in \mathcal{A}^\Sigma$ is said to be *adapted to the trinion decomposition* if around each Γ_j, there exists a tubular neighborhood $\Gamma_j \times [-1,1]$ such that $A|_{\Gamma_j \times [-1,1]}$ can be written as

$$A|_{\Gamma_j \times [-1,1]} = X_j \, d\theta_j \,,$$

where X_j is an element of the Lie algebra $\mathfrak{h} = \mathrm{Lie}(T)$, and $d\theta_j$ is the coordinate 1-form on the circle Γ_j. Furthermore, we require the existence of a (semi-) neighborhood $\widetilde{\Gamma}_k \times [-1,0]$ of each of the boundary circles such that

$$A|_{\widetilde{\Gamma}_k \times [-1,0]} = \widetilde{X}_k \, d\widetilde{\theta}_k$$

with $\widetilde{X}_k$ and $d\widetilde{\theta}_k$ as above.

Exercise 10.4 Show that every flat connection $A \in \mathcal{A}^\Sigma_{flat}$ is gauge-equivalent to a connection adapted to the trinion decomposition.

To define an action of the torus $T^{3\kappa-3+2d} \cong \mathbb{R}^{3\kappa-3+2d}/\mathbb{Z}^{3\kappa-3+2d}$ on the moduli space $\mathcal{M}_\Sigma$ we fix an element

$$t := (t_1, \ldots, t_{3\kappa-3+d}, \widetilde{t}_1, \ldots, \widetilde{t}_d) \in T^{3\kappa-3+2d}$$

and take an element

$$(\psi_1, \ldots, \psi_{2\kappa-2+d}) \in \prod_{l=1}^{2\kappa-2+d} G^{P_l}$$

in the product of the corresponding current groups, such that the restrictions of ψ_l to different (semi-) neighborhoods of Γ_j and $\widetilde{\Gamma}_k$ are given by

$$\psi_l|_{\Gamma_j \times [0,1]} = t_j, \quad \psi_l|_{\Gamma_j \times [-1,0]} = \mathrm{id}, \quad \text{and} \quad \psi_l|_{\widetilde{\Gamma}_k \times [-1,0]} = \widetilde{t}_k$$

whenever any of the semi-neighborhoods $\Gamma_j \times [0,1]$, $\Gamma_j \times [-1,0]$, and $\widetilde{\Gamma}_k \times [-1,0]$ belongs to the pair of pants P_l. Such an element $(\psi_1, \ldots, \psi_{2\kappa-2+d})$ acts on the set of connections that are adapted to the trinion decomposition by the gauge action on each piece P_l:

$$A|_{P_l} \mapsto \bar{A}|_{P_l} := \psi_l \cdot A|_{P_l} \cdot \psi_l^{-1} - d\psi_l \cdot \psi_l^{-1} \,.$$

Indeed, the gauge transformation ψ_l does not change $A|_{P_l}$ in a neighborhood of the boundary of the trinion P_l, so that all $\bar{A}|_{P_l}$ fit together to form a connection $\bar{A}$ on the surface Σ. This action descends to an action of the element $t \in T^{3\kappa-3+2d}$ on the moduli space $\mathcal{M}_\Sigma$, since any two choices of $(\psi_1, \ldots, \psi_{2\kappa-2+d})$ for the same t differ by an element of the current group G^Σ. The action of the torus $T^{3\kappa-3+2d}$ preserves the holonomies of the connection A around the circles Γ_j and $\widetilde{\Gamma}_k$, and hence preserves the fibers of the map Φ.

Proposition 10.5 ([174]) *The action of the torus $T^{3\kappa-3+2d}$ on the open subset $\overset{\circ}{\mathcal{M}}_\Sigma$ is Hamiltonian with the moment map $\frac{1}{\pi}\Phi$.*

Example 10.6 For one trinion the image of the moment map Φ is as follows. Let Σ be a trinion, i.e., a sphere with three holes ($\kappa = 0$, $d = 3$), and let $\widetilde{\Gamma}_1$, $\widetilde{\Gamma}_2$, and $\widetilde{\Gamma}_3$ denote the boundary circles of Σ. For a connection A we denote by $M_k := \mathrm{hol}_{\widetilde{\Gamma}_k}(A)$, $k = 1, 2, 3$, the corresponding holonomies around the boundary components. The relation of the fundamental group of the trinion implies that these holonomies satisfy the condition $M_1 \cdot M_2 \cdot M_3 = \mathrm{id}$. As one can check, the latter implies that the functions $\theta_k := \cos^{-1}(\frac{1}{2}\,\mathrm{tr}(M_k))$ have to satisfy the inequalities

$$\theta_1 + \theta_2 - \theta_3 \geq 0\,,$$
$$\theta_1 - \theta_2 + \theta_3 \geq 0\,,$$
$$-\theta_1 + \theta_2 + \theta_3 \geq 0\,,$$
$$\theta_1 + \theta_2 + \theta_3 \leq 2\pi\,,$$

and these are the only restrictions on their possible values. Thus the image of the moment map Φ is the tetrahedron inscribed in the cube $[0, \pi]^3$; see Figure 10.2.

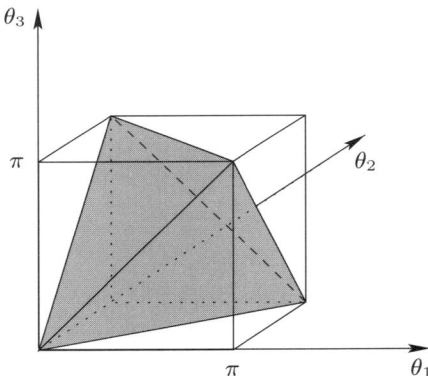

Fig. 10.2. The moment polytope of the sphere with three holes is a tetrahedron inscribed in the cube.

The general case follows from Example 10.6 by gluing the trinions:

Theorem 10.7 ([174]) *The closure of the image of the moment map*

$$\Phi : \overset{\circ}{\mathcal{M}}_\Sigma \to \mathbb{R}^{3\kappa-3+2d}$$

is given by the product of the tetrahedra corresponding to the trinion decomposition of the surface Σ, where one identifies the coordinates corresponding to the glued boundary components of the trinions.

Remark 10.8 In order to obtain the moment map image of a particular symplectic leaf of the moduli space $\mathcal{M}_\Sigma$, one has to intersect the moment polytope of $\mathcal{M}_\Sigma$ with the hyperplanes defined by fixing the holonomies around the boundary components of the surface Σ. This was used in [175] to compute the symplectic volume of the regular symplectic leaves of the moduli space $\mathcal{M}_\Sigma$.

Note also that the above torus action is defined on a dense open part of the moduli space and this construction provides us with an integrable system there.

10.3 SL$(n, \mathbb{C})$ and the Rational Ruijsenaars–Schneider System

Now we consider the case of a torus Σ with one hole (i.e., $\kappa = 1$, $d = 1$), and the group $G = \mathrm{SL}(n, \mathbb{C})$. There are $n - 1$ independent invariant functions on $\mathrm{SL}(n, \mathbb{C})$ given by $\mathrm{tr}(P^m)$ for $1 \le m \le n - 1$. So Goldman's result (Proposition 10.1) provides us with $n - 1$ Poisson-commuting functions on each symplectic leaf of the moduli space $\mathcal{M}_\Sigma$. This construction gives rise to an integrable system on a holomorphic symplectic leaf of complex dimension $2(n - 1)$, provided that the functions are independent on that leaf.

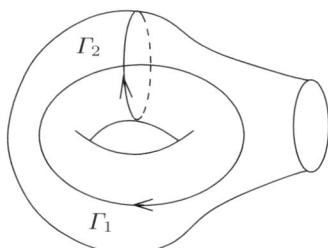

Fig. 10.3. A torus with one hole and generators of its fundamental group.

Theorem 10.9 ([120, 125]) *The corresponding Hamiltonian system is the rational Ruijsenaars–Schneider system defined below.*

To sketch the derivation of this system we fix two curves Γ_1 and Γ_2 on the surface Σ generating its fundamental group (see Figure 10.3). Let A be a flat connection on Σ, and denote by $P = \mathrm{hol}_{\Gamma_1}(A)$ the holonomy of A around the curve Γ_1, and by $Q = \mathrm{hol}_{\Gamma_2}(A)$ the holonomy of the connection A around the curve Γ_2. Then the symplectic leaf of the moduli space $\mathcal{M}_\Sigma$ determined by forcing the holonomy of the connection A around the boundary of Σ to lie in the conjugacy class $\mathcal{C}_X$ of an a priori fixed matrix X in $\mathrm{SL}(n, \mathbb{C})$ is given by

$$\mathcal{M}_\Sigma^X = \{P, Q \in \mathrm{SL}(n, \mathbb{C}) \mid QPQ^{-1}P^{-1} \in \mathcal{C}_X\}/\mathrm{SL}(n, \mathbb{C}),$$

where $\mathrm{SL}(n, \mathbb{C})$ acts on pairs (P, Q) by simultaneous conjugation.

A symplectic leaf of complex dimension $2(n-1)$ is given by fixing the matrix $X = \mathrm{diag}(a, \ldots, a, b)$, where a is any (nonzero) complex number, and $b = a^{-n+1}$. Indeed, a necessary condition for $QPQ^{-1}P^{-1}$ to be an element of the conjugacy class $\mathcal{C}_X$ is that

$$\mathrm{rk}(QPQ^{-1} - aP) \leq 1.$$

In particular, all 2×2 minors of $QPQ^{-1} - aP$ have to vanish. In a basis where Q is a diagonal matrix $Q = \mathrm{diag}(q_1, \ldots, q_n)$, we have

$$(QPQ^{-1} - aP)_{ij} = q_i q_j^{-1} p_{ij} - a p_{ij}.$$

Now by using the vanishing condition for the 2×2 minors of the matrix $QPQ^{-1} - aP$, we obtain

$$p_{ij} = \frac{\sqrt{p_{ii}p_{jj}}(1 - a)}{q_i q_j^{-1} - a}.$$

This shows that the symplectic leaf $\mathcal{M}_\Sigma^X$ is $2(n-1)$-dimensional: it is fixed by the parameters $(p_{11}, \ldots, p_{nn}, q_1, \ldots, q_n)$ subject to the conditions $\det P = 1$ and $\det Q = 1$.

In the new coordinates q_i and s_i, where

$$s_i := p_{ii} a^{(n-1)/2} \left(\prod_{j,\, j \neq i} \frac{(q_j - q_i)(q_i - q_j)}{(q_j - aq_i)(q_i - aq_j)} \right)^{\frac{1}{2}},$$

the Poisson brackets assume the form

$$\{q_i, q_j\} = 0,$$
$$\{s_i, s_j\} = 0,$$
$$\{q_i, s_j\} = \delta_{ij} p_i s_j;$$

see [120]. The functions s_i are independent and give rise to an integrable system on the symplectic leaf $\mathcal{M}_\Sigma^X$.

Finally, consider the Hamiltonian function $H(P, Q) = \mathrm{tr}(P + P^{-1})$ on the moduli space $\mathcal{M}_\Sigma^X$. In terms of the variables q_i and s_i it has the following explicit form:

$$H(s, q) = \sum_i (s_i + s_i^{-1}) a^{\frac{n-1}{2}} \left(\prod_{k,\, k \neq i} \frac{(q_k - q_i)(q_i - q_k)}{(q_k - aq_i)(q_i - aq_k)} \right)^{\frac{1}{2}},$$

where a is a parameter. This is the Hamiltonian of the *rational Ruijsenaars–Schneider system* [333, 332].

Note that the Hamiltonian $H(P, Q) = \text{tr}(P + P^{-1})$ can be regarded as a group version of the Hamiltonian that appeared in the construction of the rational Calogero–Moser system by symplectic reduction (see Section II.5.4). Indeed, by considering matrices $P = \text{id} + \widetilde{P}$ close to the identity in $\text{SL}(n, \mathbb{C})$ one can view the Hamiltonian $H(P, Q) = \text{tr}(P + P^{-1})$ as a deformation of the Calogero–Moser Hamiltonian $\text{tr}(\widetilde{P}^2) + \text{const}$.

Definition 10.10 The *elliptic Ruijsenaars–Schneider system* has the Hamiltonian

$$H(p, q) = \sum_i \cosh(\beta p_i) \left(\prod_{k,\ k \neq i} (1 - a^2 \wp(a(q_i - q_k); \tau)) \right)^{\frac{1}{2}},$$

which is a relativistic version of the elliptic Calogero–Moser Hamiltonian function.

Exercise 10.11 Derive the rational and trigonometric versions of the Ruijsenaars–Schneider Hamiltonian from the elliptic one by taking appropriate limits of the elliptic parameter τ. Compare the rational limit with the Hamiltonian above for $s_i = \exp(\beta p_i)$. (Hint: see [332].)

References

1. Adams, J.F., *Lectures on Lie Groups*, W. A. Benjamin, Inc., New York, Amsterdam, 1969.
2. Adams, M.R., Harnad, J., Hurtubise, J., *Liouville generating functions for isospectral flow in loop algebras*, in Integrable and Superintegrable Systems, 232–256, World Sci. Publ., Teaneck, NJ, 1990.
3. Adler, M., *On a trace functional for formal pseudo differential operators and the symplectic structure of the Korteweg-de Vries type equations*, Invent. Math. **50:3** (1978/79), 219–248.
4. Adler, M., van Moerbeke, P., *Completely integrable systems, Euclidean Lie algebras, and curves*, Adv. in Math. **38:3** (1980), 267–317.
5. Albeverio, S., Sengupta, A., *A mathematical construction of the non-abelian Chern-Simons functional integral*, Comm. Math. Phys. **186** (1997), 563–579.
6. Alekseev, A.Y., Malkin, A.Z., Meinrenken, E., *Lie group valued moment maps*, J. Differ. Geom. **48:3** (1998), 445–495.
7. Aleksovsky, V.A., Lukatsky, A.M., *Nonlinear dynamics of ferromagnet magnetization and the motion of a generalized rigid body with a current group*, Theoret. and Math. Phys. **85** (1990), 1090–1096.
8. Allison, B. S. Azam, S., Berman, S., Gao, Y., Pianzola, A., *Extended affine Lie algebras and their root systems*, Mem. Am. Math. Soc., **603** (1997), 1–122.
9. Ambrosio, L., Rigot, S., *Optimal mass transportation in the Heisenberg group*, J. Funct. Anal. **208** (2004), no. 2, 261–301.
10. Andersen, J.E., Mattes, J., Reshetikhin, N., *The Poisson structure on the moduli space of flat connections and chord diagrams*, Topology **35:4** (1996), 1069–1083.
11. Arnold, V.I., *A theorem of Liouville concerning integrable problems of dynamics*, Sibirsk. Mat. Zh. **4** (1963), 471–474.
12. Arnold, V.I., *Sur la géométrie différentielle des groupes de Lie de dimension infinie et ses applications à l'hydrodynamique des fluides parfaits*, Ann. Inst. Fourier (Grenoble) **16** (1966), 319–361;
13. Arnold, V.I., *The Hamiltonian nature of the Euler equation in the dynamics of a rigid body and of an ideal fluid*, Uspekhi Mat. Nauk **24:3** (1969), 225–226.
14. Arnold, V.I., *Arrangement of ovals of real plane algebraic curves, involutions of smooth four-dimensional manifolds, and on arithmetics of integral-valued quadratic forms*, Func. Anal. and Appl. **5:3** (1971), 169–176; *Remarks on eigenvalues and eigenvectors of Hermitian matrices, Berry phase, adiabatic*

connections and quantum Hall effect, Selecta Mathematica, New Series **1:1** (1995), 1–19.

15. Arnold, V.I., *Lectures on bifurcations and versal families*, Uspehi Mat. Nauk **27:5** (1972), 119–184.

16. Arnold, V.I., *The asymptotic Hopf invariant and its applications*, Proc. Summer School in Diff. Equations at Dilizhan, 1973, Erevan (in Russian); English transl.: Selecta Math. Soviet. **5:4** (1986), 327–345.

17. Arnold, V.I., *Geometrical methods in the theory of ordinary differential equations*, Grundlehren der Mathematischen Wissenschaften, **250**. Springer-Verlag, New York-Berlin, 1983; Second edition, 1988.

18. Arnold, V.I., *Mathematical methods of classical mechanics*, second edition, Springer-Verlag, New York, 1989.

19. Arnold, V.I., *Ordinary differential equations*, third edition, Springer-Verlag, Berlin, 1992.

20. Arnold, V.I., *On teaching mathematics*, talk at the Palais de Découverte in Paris, 1997, http://pauli.uni-muenster.de/~munsteg/arnold.html

21. Arnold, V.I., *Singularities of caustics and wave fronts*, Math. and Its Appl., **62**, Kluwer Acad. Publ., Dordrecht, 1990.

22. Arnold, V.I., Baryshnikov, Yu.M., Bogaevsky, I.A., *Singularities and bifurcations of potential flows*, Supplement 2 in the book: Gurbatov, S. N., Malakhov, A. N., Saichev, A. I., *Nonlinear random waves and turbulence in nondispersive media: waves, rays, particles*, Nonlinear Science: Theory and Applications. Manchester University Press, Manchester, 1991, 290–300.

23. Arnold, V.I., Givental, A. B., *Symplectic geometry*, Current problems in mathematics. Fundamental directions, **4** (1985), VINITI Moscow, 5–139; English transl.: Dynamical systems, IV, Encyclopaedia Math. Sci., **4** (2001), Springer, Berlin, 1–138.

24. Arnold, V.I., Khesin, B.A., *Topological Methods in Hydrodynamics*, Applied Math. Series **125**, Springer-Verlag, 1998.

25. Atiyah, M.F., *Vector bundles over an elliptic curve*, Proc. London Math. Soc. (3) **7** (1957), 414–452.

26. Atiyah, M.F., *Green's functions for self-dual four-manifolds*, Adv. in Math. Suppl. Stud. **7A**, Academic Press, New York-London (1981), 129–158.

27. Atiyah, M.F., *The Geometry and Physics of Knots*, Cambridge University Press, Cambridge, 1990.

28. Atiyah, M.F., Bott, R., *The Yang-Mills equations over Riemann surfaces*, Phibs. Trans. Roy. Soc. London **308** (1982), 523–615.

29. Audin, M., *Lectures on gauge theory and integrable systems. In: Gauge Theory and Symplectic Geometry (Montreal, PQ, 1995)*, 1–48, NATO Adv. Sci. Inst. Ser. C Math. Phys. Sci., 488, Kluwer Acad. Publ., Dordrecht, 1997.

30. Azam, S., *Extended affine root systems*, J. Lie Theory **12** (2002), 515–527.

31. Babelon, O., Bernard, D., Talon, M., *Introduction to Classical Integrable Systems*, Cambridge Monographs on Mathematical Physics. Cambridge University Press, Cambridge, 2003.

32. Balog, J., Fehér, L., Palla, L., *Coadjoint orbits of the Virasoro algebra and the global Liouville equation*, Internat. J. Modern Phys. A **13** (1998), no. 2, 315–362.

33. Banyaga, A., *The Structure of Classical Diffeomorphism Groups*, Mathematics and Its Applications, 400. Kluwer Academic Publishers Group, Dordrecht, 1997

34. Baranovski, V., Ginzburg, V., *Conjugacy classes in loop groups and G-bundles on elliptic curves*, Internat. Math. Res. Notices **15** (1996), 733–751.

35. Bar-Natan, D., *Perturbative Chern-Simons theory*, J. Knot Theory Ramifications **4** (1995), no. 4, 503–547.
36. Bar-Natan, D., *On the Vassiliev knot invariants*, Topology **34:2** (1995), 423–472.
37. Bazhanov, V.V., Lukyanov, S.L., Zamolodchikov, A.B., *Integrable structure of conformal field theory, quantum KdV theory and thermodynamic Bethe ansatz*, Comm. Math. Phys. **177** (1996), 381–398.
38. Beem, J.K., Low, R.J., Parker, Ph.E., *Spaces of geodesics: products, coverings, connectedness*, Geom. Dedicata **59** (1996), no. 1, 51–64.
39. Beem, J.K., Parker, Ph.E., *The space of geodesics*, Geom. Dedicata **38** (1991), no. 1, 87–99.
40. Berman, S., Billig, Yu., *Irreducible representations for toroidal Lie algebras*, J. Algebra **221:1** (1999), 188–231.
41. Bershadsky, M., Ooguri, H., *Hidden SL(n) symmetry in conformal field theories*, Comm. Math. Phys. **126** (1989), no. 1, 49–83.
42. Billig, Yu., *Magnetic hydrodynamics with asymmetric stress tensor*, J. Math. Phys. **46:4** (2005), 043101, 13 pp.
43. Billig, Yu., *A category of modules for the full toroidal Lie algebra*, Int. Math. Res. Not. (2006), Art. ID 68395, 46 pp.
44. Bloch, A.M., *Nonholonomic Mechanics and Control*, Interdisciplinary Applied Mathematics, **24**, Springer-Verlag, New York, 2003.
45. Bobenko, A.I., Suris, Yu.B., *Discrete Lagrangian reduction, discrete Euler-Poincaré equations, and semidirect products*, Lett. Math. Phys., **49:1** (1999), 79–93.
46. Bogaevsky, I.A., *Perestroikas of shocks and singularities of minimum functions*, Phys. D **173** (2002), no. 1-2, 1–28, math.AP/0204237
47. Bogoyavlensky, O.I., *The Euler equation on finite-dimensional Lie coalgebras arising in problems of mathematical physics*, Russian Math. Surveys, **47:1** (1992), 117–189.
48. Bolsinov, A. V., *Compatible Poisson brackets on Lie algebras and the completeness of families of functions in involution*, Math. USSR-Izv. **38:1** (1992), 69–90.
49. Bordemann, M., Hoppe, J., Schaller, P., Schlichenmaier, M., $\mathfrak{gl}_\infty$ *and geometric quantization*, Comm. Math. Phys. **138** (1991), no. 2, 209–244.
50. Bordner, A.J., Corrigan, E., Sasaki, R., *Calogero-Moser models I: A new formulation*, Progr. Theoret. Phys. **100** (1998), 1107–1129.
51. Bordner, A.J., Sasaki, R., *Calogero-Moser Models III: Elliptic potentials and twisting*, Progr. Theoret. Phys. **101** (1999), 799–829.
52. Bott, R., *An application of the Morse theory to the topology of Lie groups*, Bull. Soc. Math. France. **84** (1956), 251–281.
53. Bott, R., *On the characteristic classes of groups of diffeomorphisms*, Enseign. Math. (2), **23** (1977), 209–220.
54. Bott, R., Taubes, C., *On the self-linking of knots*, J. Math. Phys. **35:10** (1994), 5247–5287.
55. Bottacin, F., *Poisson structures on moduli spaces of sheaves over Poisson surfaces*, Invent. Math. **121:2** (1995), 421–436.
56. Bourbaki, N., *Groupes et algèbres de Lie, Ch. 4, 5 et 6*. Masson, Paris, 1981.
57. Bowick, M.J., Rajeev, S.G., *String theory as the Kähler geometry of loop space*, Phys. Rev. Lett. **58** (1987), 535–538; *The holomorphic geometry of closed bosonic string theory and* Diff S^1/S^1, Nuclear Phys. B **293** (1987), 348–384.
58. Brenier, Ya, *Polar factorization and monotone rearrangement of vector-valued functions*, Comm. Pure Appl. Math. **44:4** (1991), 323–351.

59. Brenier, Ya., *The least action principle and the related concept of generalized flows for incompressible perfect fluids*, J. Amer. Math. Soc. **2** (1989), no. 2, 225–255; *The dual least action problem for an ideal, incompressible fluid*, Arch. Rational Mech. Anal. **122** (1993), no. 4, 323–351. *Minimal geodesics on groups of volume-preserving maps and generalized solutions of the Euler equations*, Comm. Pure Appl. Math. **52** (1999), no. 4, 411–452.

60. Bröcker, T., tom Diek T., *Representations of Compact Lie Groups*, Grad. Texts in Math. **98**, Springer-Verlag, Heidelberg, New York, 1985.

61. Brüchert, G., *Trace class elements and cross sections in Kac-Moody groups*, Canad. J. Math. **50** (1998), 972-1006.

62. Brylinski, J.-L., *Loop spaces, characteristic classes and geometric quantization*, Progress in Math.: Birkhäuser **107** 1993, 300pp.

63. Brylinski, J.-L., *Coadjoint orbits of central extensions of gauge groups*, Comm. Math. Phys. **188** (1997), 351–365.

64. Calabi, E., *On the group of automorphisms of a symplectic manifold*, Problems in analysis, Lectures at the Sympos. in honor of Salomon Bochner, Princeton Univ., Princeton, NJ, (1970), 1–26.

65. Calini, A., *Recent developments in integrable curve dynamics*, in Geom. Approaches to Diff. Equations; Lect. Notes of the Australian Math. Soc., Cambridge University Press **15** (2000), 56–99.

66. Calogero, F., *Solution of the one-dimensional n-body problems with quadratic and/or inversely quadratic pair potentials*, J. Math. Phys., **12** (1971), 419–436.

67. Calogero, F., *Exactly solvable one-dimensional many-body problems*, Lett. Nuovo. Cim., **13** (1975), 411–416.

68. Calogero, F., Marchioro, C., Ragniosco, O., *Exact solutions of the classical and quantal one-dimensional many-body problems with the two-body potential $V_a(x) = g^2 a^2 / \sinh^2(x)$*, Lett. Nuovo. Cim., **13** (1975), 383–387.

69. Camassa, R. and Holm, D., *An integrable shallow water equation with peaked solutions*, Phys. Lett. Rev. **71** (1993), 1661–1664.

70. Casson, A., *Lecture notes*, MSRI Lectures, Berkeley, 1985.

71. Chas, M., Sullivan, D., *Closed string operators in topology leading to Lie bialgebras and higher string algebra*, in The legacy of Niels Henrik Abel, 771–784, Springer, Berlin, 2004.

72. Chern, S.S., Simons, J., *Characteristic forms and geometric invariants*, Ann. of Math. (2) **99** (1974), 48–69.

73. Constantin, A., Kappeler, T., Kolev, B., Topalov, P., *On geodesic exponential maps of the Virasoro group*, Ann. Global Anal. Geom. **31** (2007), no. 2, 155–180.

74. Constantin, A., Kolev, B., *Geodesic flow on the diffeomorphism group of the circle*, Comment. Math. Helv. **78** (2003), no. 4, 787–804.

75. Date, E., Kashiwara, M., Jimbo, M., Miwa, T., *Transformation groups for soliton equations*, Nonlinear integrable systems – classical theory and quantum theory (Kyoto, 1981), 39–119, World Sci. Publishing, Singapore, 1983; Proc. Japan Acad. Ser. A Math. Sci. **57** (1981), 342–347, 387–392; J. Phys. Soc. Japan **50** (1981), 3806–3812, 3813–3818; Physica D **4** (1981/82), no. 3, 343–365.

76. Deift, P., *Integrable Hamiltonian systems*, in Dynamical systems and probabilistic methods in partial differential equations (Berkeley, CA, 1994), 103–138, Lectures in Appl. Math., **31**, AMS, Providence, RI, 1996.

77. Deift, P., Li, L.C., Tomei, C., *Loop groups, discrete versions of some classical integrable systems, and rank 2 extensions*, Mem. Amer. Math. Soc. **100** (1992), no. 479.

78. Deligne, P., *Théorie de Hodge. II*, Inst. Hautes Études Sci. Publ. Math. **40** (1971), 5–57.

79. Deligne, P., Freed, D., *Classical Field Theory*, in Quantum fields and strings: a course for mathematicians, **1** (Princeton, NJ, 1996/1997), 137–225, Amer. Math. Soc., Providence, RI, 1999.

80. Dickey, L.A., *Soliton Equations and Hamiltonian Systems*, Advanced Series in Mathematical Physics, vol. 12. World Scientific, 1991.

81. Di Francesco, P., Itzykson, C., Zuber, J.-B., *Classical W-algebras*, Comm. Math. Phys. **140** (1991), no. 3, 543–567.

82. Dolzhansky, F.V., *On the mechanical prototypes of fundamental hydrodynamic invariants and slow manifolds*, Physics - Uspekhi, **48:12** (2005), 1205–1234.

83. Donaldson, S.K., *Boundary value problems for Yang-Mills fields*, J. Geom. Phys. **8:1-4** (1992), 89–122.

84. Donaldson, S.K., Kronheimer, P.B., *The Geometry of Four-Manifolds*. Oxford Mathematical Monographs. Oxford Science Publications. The Clarendon Press, Oxford University Press, New York, 1990.

85. Donaldson, S.K., Thomas, R.P., *Gauge theory in higher dimensions*, in The Geometric Universe (Oxford, 1996), 31–47, Oxford Univ. Press, Oxford, 1998.

86. Donin, D., Khesin, B., *Pseudodifferential symbols on Riemann surfaces and Krichever-Novikov algebras*, Comm. Math. Phys. **272:2** (2007), 507–527.

87. Drinfeld, V., *Quantum Groups*, Proc. Intern. Congress Math. (Berkeley, 1986), vol. 1, Amer. Math. Soc., Providence, RI, 1988, pp. 789–820.

88. Drinfeld, V.G., Sokolov, V.V., *Lie algebras and equations of Korteweg-de Vries type*, Current problems in mathematics (Moscow) Itogi Nauki i Tekhniki, vol 23, Akad. Nauk SSSR, Veseoyuz. Inst. Nauchn. i Tekhn. Inform., 1984, pp. 81–180 (Russian).

89. Dubrovin, B., *Geometry of 2D topological field theories*, in Integrable Systems and Quantum Groups (Montecatini Terme, 1993), 120–348, Lecture Notes in Math. **1620**, Springer, Berlin, 1996.

90. Dubrovin, B.A., Krichever, I.M., Novikov, S.P., *Integrable systems. I*, Itogi Nauki i Tekhniki, Akad. Nauk SSSR, **4**, (1985) VINITI, Moscow, 179–284. English transl.: Encyclopedia of Math.Sci **4** (1990), Springer-Verlag, 173–280.

91. Dubrovin, B.A., Novikov, S.P., *Hamiltonian formalism of one-dimensional systems of the hydrodynamic type and the Bogolyubov-Whitham averaging method*, Soviet Math. Dokl. **27:3** (1983), 665–669.

92. Dubrovin, B.A., Novikov, S.P., *Poisson brackets of hydrodynamic type*, Soviet Math. Dokl. **30:3** (1984), 651–654.

93. Duistermaat, J.J., Kolk, J.A.C., *Lie Groups*, Springer-Verlag, Berlin, 2000.

94. Dzhumadildaev, A.S., *Derivations and central extensions of the Lie algebra of formal pseudodifferential operators*, Algebra i Analiz **6:1** (1994), 140–158; translation in St. Petersburg Math. J. **6:1** (1995), 121–136.

95. Ebin, D.G., *On the space of Riemannian metrics*, Bull. Amer. Math. Soc. **74** (1968), 1001–1003; *The manifold of Riemannian metrics*, Global Analysis 1970, Proc. Sympos. Pure Math. **XV**, Berkeley, CA (1968), AMS, Providence, RI, 11–40.

96. Ebin, D.G., Marsden, J.E., *Groups of diffeomorphisms and the solution of the classical Euler equations for a perfect fluid*, Bull. Amer. Math. Soc. **75** (1969), 962–967; *Groups of diffeomorphisms and the notion of an incompressible fluid*, Ann. of Math. **92** (1970), 102–163.

97. Ebin, D.G., Misiołek, G., Preston, S.C., *Singularities of the exponential map on the volume-preserving diffeomorphism group*, Geom. Funct. Anal. **16** (2006), no. 4, 850–868.

98. Eliashberg, Y., Polterovich, L., *Bi-invariant metrics on the group of Hamiltonian diffeomorphisms*, Internat. J. Math. **4** (1993), 727–738.

99. Eliashberg, Y., Ratiu, T.S., *On the diameter of the symplectomorphism group of the ball*, Math. Sci. Res. Inst. Publ. **20** (1991), 169–172.

100. Eliashberg, Y., Ratiu, T.S., *The diameter of the symplectomorphism group is infinite*, Invent. Math. **103** (1991), 327–340.

101. Enriquez, B., Khoroshkin, S., Radul, A., Rosly, A., Rubtsov, V., *Poisson-Lie aspects of classical W-algebras*, Amer. Math. Soc. Transl. Ser. (2) **167**, AMS, Providence, RI, 1995, pp. 37–59.

102. Enriquez, B., Orlov, A.Yu., Rubtsov, V.N., *Dispersionful analogues of Benney's equations and N-wave systems*, Inverse Problems **12** (1996), no. 3, 241–250.

103. Enriquez, B.; Rubtsov, V., *Hitchin systems, higher Gaudin operators and R-matrices*, Math. Res. Lett. **3** (1996), no. 3, 343–357.

104. Enriquez, B., Rubtsov, V., *Quantizations of the Hitchin and Beauville-Mukai integrable systems*, Mosc. Math. J. **5** (2005), no. 2, 329–370.

105. Etingof, P., *Casimirs of the Goldman Lie algebra of a closed surface*, Int. Math. Res. Not. 2006, Art. ID 24894, 5 pp.

106. Etingof, P., Frenkel, I.B., *Central extensions of current groups in two dimensions*, Comm. Math. Phys. **165** (1994), 429–444.

107. Etingof, P., Khesin, B.A., *Affine Gelfand-Dickey brackets and holomorphic vector bundles*, Geom. and Funct. Anal. **4:4** (1994), 399–423.

108. Euler, L., *Principia motus fluidorum*, Novi Commentarii Acad. Sci. Petropolitanae **6** (1761), 271–311; *Commentationes mechanicae ad theoriam corporum fluidorum pertinentes*, Opera omnia. Series secunda. Opera mechanica et astronomica. Vol. XII, XIII. Soc. Sci. Natur. Helveticae, Edit. C.A. Truesdell, Lausanne 1954-1955.

109. Faddeev, L.D., Shatashvili, S., *Algebraic and Hamiltonian methods in the theory of nonabelian anomalies*, Teoret. Mat. Fiz. **60** (1984), 206–217.

110. Faddeev, L.D., Takhtajan, L.A., *Hamiltonian Methods in the Theory of Solitons*, Springer Series in Soviet Mathematics. Springer-Verlag, Berlin, 1987.

111. Fairlie, D.B., Zachos, C.K., *Infinite-dimensional algebras, sine brackets, and $SU(\infty)$*, Phys. Lett. B **224** (1989), no. 1-2, 101–107.

112. Fateev, V.A., Lukyanov, S.L., *The models of two-dimensional conformal quantum field theory with Z_n symmetry*, Internat. J. Modern Phys. A **3** (1988), no. 2 507–520.

113. Fedorov, Yu.N., Jovanovic, B., *Integrable nonholonomic geodesic flows on compact Lie groups*, in Topological methods in the theory of integrable systems, eds: Bolsinov A.V., Fomenko A.T., Oshemkov A.A., Cambridge Scientific Publ., 2005, pp. 115–152.

114. Fedorov, Yu.N., Kozlov, V.V., *Various aspects of n-dimensional rigid body dynamics*, AMS Transl., Ser. 2, **168**, 141–171, AMS Providence, RI, 1995.

115. Fehér, L., Harnad, J., Marshall, I., *Generalized Drinfeld-Sokolov reductions and KdV type hierarchies*, Commun. Math. Phys. **154:1** (1993), 181–214.

116. Feigin, B.L., *Lie algebras $\mathfrak{gl}(\lambda)$ and cohomology of the Lie algebra of differential operators*, Russian Math. Surveys **43:2** (1988), 169–170.

117. Feigin, B.L., Frenkel, E., *Integrals of motion and quantum groups*, in Integrable systems and quantum groups (Montecatini Terme, 1993), Lecture Notes in Math., **1620**, Springer, Berlin, 1996, 349–418.

118. Feigin, B.L., Fuchs, D.B., *Cohomologies of Lie groups and Lie algebras*, Lie groups and Lie algebras, II, Encyclopaedia Math. Sci., **21**, Springer, Berlin, 2000, 125–223.

119. Flaschka, H., Newell, A.C., Ratiu, T., *Kac-Moody Lie algebras and soliton equations*, Physica D **9:3** (1983), 300–332.

120. Fock, V.V., *Three remarks on group invariants related to flat connections*, in Geometry and Integrable Models (Dubna, 1994), 20–31, World Sci. Publishing, River Edge, NJ, 1996.

121. Fock, V.V., Goncharov, A., *Moduli spaces of local systems and higher Teichmüller theory*, Publ. Math. Inst. Hautes Études Sci. **103** (2006), 1–211.

122. Fock, V.V., Goncharov, A., *Moduli spaces of convex projective structures on surfaces*, Adv. Math. **208** (2007), no. 1, 249–273.

123. Fock, V.V., Gorsky, A., Nekrasov, N., Rubtsov, V., *Duality in integrable systems and gauge theories*, J. High Energy Phys. **7** (2000), Paper # 028.

124. Fock, V.V., Nekrasov, N.A., Rosly, A.A., Selivanov, K.G., *What we think about the higher-dimensional Chern-Simons theories*, Sakharov Memorial Lectures in Physics, **1,2** (Moscow, 1991), 465–471, Nova Sci. Publ., Commack, NY, 1992.

125. Fock, V.V., Rosly, A.A., *Flat connections and polyubles*, Theoret. and Math. Phys., **95**, (1993), 526–534; *Moduli space of flat connections as a Poisson manifold*, Advances in quantum field theory and statistical mechanics: 2nd Italian-Russian collaboration (Como, 1996). Internat. J. Modern Phys. B **11** (1997), 3195–3206.

126. Fock, V.V., Rosly, A.A., *Poisson structure on moduli of flat connections on Riemann surfaces and the r-matrix*, Moscow Seminar in Mathematical Physics, 67–86, Amer. Math. Soc. Transl. Ser. 2, **191**, AMS, Providence, RI, 1999.

127. Fokas, A. and Fuchssteiner, B., *Symplectic structures, their Bäklund transformations and hereditary symmetries*, Phys. D **4** (1981/82), 47–66.

128. Frahm, W., *Über gewisse Differentialgleichungen*, Math. Ann. **8** (1875), 35–44.

129. Frenkel, E., *Five lectures on soliton equations*, Surveys in differential geometry: integrable systems, 131–180, Surv. Differ. Geom., **IV**, Int. Press, Boston, MA, 1998.

130. Frenkel, E., Kac, V., Radul, A., Wang, W., $W_{1+\infty}$ and $W(\mathfrak{gl}_N)$ *with central charge N*, Comm. Math. Phys. **170** (1995), no. 2, 337–357.

131. Frenkel, E., Reshetikhin, N., Semenov-Tian-Shansky, M.A., *Drinfeld-Sokolov reduction for difference operators and deformations of W-algebras. I. The case of Virasoro algebra*, Comm. Math. Phys. **192** (1998), no. 3, 605–629.

132. Frenkel, I.B., *Orbital theory for affine Lie algebras*, Invent. Math. **77** (1984), 301–352.

133. Frenkel, I.B., Khesin, B.A., *Four dimensional realization of two dimensional current groups*, Comm. Math. Phys. **178** (1996), 541–562.

134. Frenkel, I.B., Todorov, A.N., *Complex counterpart of Chern-Simons-Witten theory and holomorphic linking*, Adv. Theor. Math. Phys. **11** (2007), no. 4, 531–590.

135. Friedman, R., Morgan, J.W., *Holomorphic principal bundles over elliptic curves. II. The parabolic construction*, J. Diff. Geom. **56** (2000), 301–379.

136. Friedman, R., Morgan, J.W., Witten, E., *Principal G-bundles over elliptic curves*, Math. Res. Lett. **5** (1998), no. 1-2, 97–118.

137. Frisch, U., Bec, J., *Burgulence*, in New trends in turbulence (Les Houches, 2000, eds: M. Lesieur, A. Yaglom, F. David,), Springer, Berlin 2001, 341–383.

138. Fuchs, D.B., *Cohomology of infinite-dimensional Lie algebras*, Contemp. Sov. Math., Consultants Bureau, New York, 1986. xii+339 pp.

139. Fukumoto, Y., Miyazaki, T., *Three-dimensional distortions of a vortex filament with axial velocity*, J. Fluid Mech. **222** (1991), 369–416.

140. Gardner, C.S., *Korteweg-de Vries equation and generalizations. IV. The Korteweg-de Vries equation as a Hamiltonian system*, J. Mathematical Phys. **12** (1971), 1548–1551.

141. Gelfand, I.M., Dickey, L.A., *Fractional powers of operators, and Hamiltonian systems*, Funkt. Anal. i Pril. **10:4** (1976), 13–29.

142. Gelfand, I.M., Dickey, L.A., *A family of Hamiltonian structures related to integrable nonlinear differential equations*, Akad. Nauk SSSR Inst. Prikl. Mat. Preprint **136** (1978), 1–41; English translation in: I.M. Gelfand, Collected papers, **1**, eds: S.G. Gindikin et al., Berlin–Heidelberg–New York: Springer (1987).

143. Gelfand, I.M., Fuchs, D.B., *Cohomology of the Lie algebra of vector fields on a circle*, Funkt. Anal. i Pril. **2:4** (1968), 92–93. English translation: Funct Anal. Appl. **2** (1969) 342–343.

144. Gelfand, I.M., Zakharevich, I., *Spectral theory of a pencil of third-order skew-symmetric differential operators on S^1*, Funct. Anal. Appl. **23:2** (1989) 85–93.

145. Goldman, W.M., *Invariant functions on Lie groups and Hamiltonian flows of surface group representations*, Invent. Math. **85** (1986), 263–302.

146. Golenishcheva-Kutuzova, M.I., Lebedev, D.R., *Z-graded trigonometric Lie subalgebras in $\hat{A}_\infty$, $\hat{B}_\infty$, $\hat{C}_\infty$, $\hat{D}_\infty$ and their representation by vertex operators*, Funct. Anal. Appl. **27** (1993), no. 1, 10–20

147. Golenishcheva-Kutuzova, M.I., Lebedev, D.R., Olshanetsky, M.A., *Between $\widehat{\mathfrak{gl}}_\infty$ and $\widehat{\mathfrak{sl}}_N$ affine algebras. I. Geometrical actions*, Theoret. and Math. Phys. **100** (1994), no. 1, 863–873.

148. Gorsky, A., Nekrasov, N., *Elliptic Calogero-Moser system from two dimensional current algebra*, e-print, arXiv: hep-th/9401021 (1994).

149. Griffiths, P.A., *Variations on a theorem of Abel*, Invent. Math., **35** (1976), 321–390.

150. Griffiths, P.A., Harris, J., *Principles of Algebraic Geometry*, Wiley, NY, 1978.

151. Griffiths, P.A., Schmid, W., *Recent developments in Hodge theory: a discussion of techniques and results*, in Discrete subgroups of Lie groups and applicatons to moduli (Internat. Colloq., Bombay, 1973), pp. 31–127. Oxford Univ. Press, Bombay, 1975.

152. Grothendieck. A., *Sur la classification de fibrés holomorphes sur la sphère de Riemann*, Am. J. Math., **79** (1957), 121–138.

153. Guieu, L., Roger, C., *L'Algèbre et le Groupe de Virasoro*, Les Publications CRM, Montreal, QC, 2007.

154. Guillemin, V., Sternberg, S., *Symplectic Techniques in Physics*, second edition, Cambridge University Press, Cambridge, 1990.

155. Gunning, R.C., *Lectures on Riemann Surfaces*, Princeton Mathematical Notes, Princeton University Press, Princeton, N.J., 1966.

156. Haller, S., Vizman, C., *Non-linear Grassmannians as coadjoint orbits*, Math. Ann. **329:4** (2004), 771–785

157. Hamilton, R.S., *The inverse function theorem of Nash and Moser*, Bull. Amer. Math. Soc. **7** (1982), 65–222.

158. Hasimoto, R., *A soliton on a vortex filament*, J. Fluid Mechanics **51** (1972), 477–485.
159. Helgason, S., *Differential Geometry and Symmetric Spaces*, Academic Press, New York, London, 1962.
160. Helmke, S., Slodowy, P., *Loop groups, principal bundles over elliptic curves and elliptic singularities* in Annual Meeting of the Math. Soc. of Japan, Hirshima, Sept. 1999, Abstracts, Section Infinite-dimensional analysis, 67–77. *Loop groups, elliptic singularities and principal bundles over elliptic curves*, in Geometry and topology of caustics, CAUSTICS '02, 87–99, Banach Center Publ., **62**, Warsaw, 2004.
161. Helmke, S., Slodowy, P., *On unstable principal bundles over elliptic curves*, Publ. Res. Inst. Math. Sci. **37** (2001), 349–395.
162. Hirsch, M.W., *Differential Topology*, Springer-Verlag, New York, 1976.
163. Hitchin, N.J., *Stable bundles and integrable systems*. Duke Math. J. **54** (1987), 91–114.
164. Hitchin, N.J., *Vector fields on the circle*, in Mechanics, Analysis and Geometry: 200 years after Lagrange, 359–378, North-Holland Delta Ser., North-Holland, Amsterdam, 1991
165. Hitchin, N.J. *Riemann surfaces and integrable systems*, in Hitchin, N.J., Segal, G.B., Ward, R., *Integrable Systems*, Oxf. Grad. Texts Math., **4**, Oxford Univ. Press, 1999.
166. Hofer, H., *On the topological properties of symplectic maps*, Proc. Roy. Soc. Edinburgh Sect. A **115** (1990), 25–38.
167. Holm, D.D., Kupershmidt, B.A., *Poisson brackets and Clebsch representations for magnetohydrodynamics, multifluid plasmas, and elasticity*, Phys. D **6** (1983), 347–363.
168. Humphreys, J.E., *Introduction to Lie Algebras and Representation Theory*, Springer-Verlag, Berlin, Heidelberg, New York, 1972.
169. Hurtubise, J.C., Markman, E., *Calogero-Moser systems and Hitchin systems*, Comm. Math. Phys. **223** (2001), 533–552.
170. Huybrechts, D., Lehn, M., *The Geometry of Moduli Spaces of Sheaves*, Aspects of Mathematics, **E31**, Friedr. Vieweg & Sohn, Braunschweig, 1997.
171. Huybrechts, D., *Complex Geometry. An Introduction*, Universitext, Springer-Verlag, Berlin, 2005.
172. Ismagilov, R.S., *Representations of Infinite-Dimensional Groups*, Transl. of Math. Monographs, **152**, AMS, Providence, RI, 1996, x+197 pp.
173. Jeffrey, L.C., *Extended moduli spaces of flat connections on Riemann surfaces* Math. Ann. **298** (1994), 667–692.
174. Jeffrey, L.C., Weitsman, J., *Bohr-Sommerfeld orbits in the moduli space of flat connections and the Verlinde dimension formula*, Comm. Math. Phys. **150** (1992), 593–630.
175. Jeffrey, L. C., Weitsman, J., *Toric structures on the moduli space of flat connections on a Riemann surface: volumes and the moment map*, Adv. Math. **106** (1994), 151–168.
176. Jimbo, M., Miwa, T., *Solitons and infinite-dimensional Lie algebras*, Publ. Res. Inst. Math. Sci. **19:3** (1983), 943–1001.
177. Kac, V.G., *Simple irreducible Lie algebras of finite growth*, Izv. Akad. Nauk SSSR Ser. Mat. **32** (1968), 1323–1367.
178. Kac, V.G., *Infinite-Dimensional Lie Algebras*, third edition, Cambridge University Press, Cambridge, 1990.

179. Kac, V.G., Peterson, D.H., *Spin and wedge representations of infinite-dimensional Lie algebras*, Proc. Nat. Acad. Sci. USA **78** (1981), 3308–3312.

180. Kac, V., Radul, A., *Quasifinite highest weight modules over the Lie algebra of differential operators on the circle*, Comm. Math. Phys. **157** (1993), no. 3, 429–457; *Representation theory of the vertex algebra $W_{1+\infty}$*, Transform. Groups **1** (1996), no. 1-2, 41–70.

181. Kac, V.G., Raina, A.K. *Bombay Lectures on Highest Weight Representations of Infinite-Dimensional Lie Algebras*, Advanced Series in Mathematical Physics, **2**, World Scientific Publishing Co., Inc., Teaneck, NJ, 1987.

182. Kappeler, T., Pöschel, J., *KdV & KAM*, Ergebnisse der Mathematik und ihrer Grenzgebiete (3), **45**, Springer-Verlag, Berlin, 2003.

183. Karshon, Y., *An algebraic proof for the symplectic structure of moduli space*, Proc. Amer. Math. Soc. **116** (1992), 591–605.

184. Kashaev, R.M., *Coordinates for the moduli space of flat $PSL(2,\mathbb{R})$-connections*, Math. Res. Lett. **12** (2005), no. 1, 23–36.

185. Kazhdan, D., Kostant, B., Sternberg, S., *Hamiltonian group actions and dynamical systems of Calogero type*, Comm. Pure Appl. Math. **31** (1978), 481–507.

186. Khesin, B.A., *A Poisson-Lie framework for rational reductions of the KP hierarchy*, Letters in Math. Physics **58** (2001), 101–107.

187. Khesin, B.A., Chekanov, Yu.V., *Invariants of the Euler equations for ideal or barotropic hydrodynamics and superconductivity in D dimensions*, Physica D, **40** (1989), 119–131.

188. Khesin, B.A., Lee, P., *A nonholonomic Moser theorem and optimal mass transport*, preprint arXiv:0802.1551 (2008), 31pp.

189. Khesin, B.A., Lenells, J., Misiołek, G., *Generalized Hunter-Saxton equation and the geometry of the group of circle diffeomorphisms*, Mathem. Annalen (2008), DOI 10.1007/s00208-008-0250-3, 40pp.

190. Khesin, B.A., Lyubashenko, V., Roger, C., *Extensions and contractions of the Lie algebra of q-pseudodifferential symbols on the circle*, J. Funct. Anal. **143** (1997), no. 1, 55–97.

191. Khesin, B.A., Malikov, F., *Universal Drinfeld-Sokolov reduction and matrices of complex size*, Comm. Math. Phys. **175** (1996), 113–134.

192. Khesin, B.A., Misiołek, G., *Euler equations on homogeneous spaces and Virasoro orbits*, Advances in Math. **176** (2003), 116–144.

193. Khesin, B.A., Misiołek, G., *Shock waves for the Burgers equation and curvatures of diffeomorphism groups*, Proc. Steklov Math. Inst. **259** (2007), 73–81.

194. Khesin, B.A., Rosly, A.A., *Symplectic geometry on moduli spaces of holomorphic bundles over complex surfaces*, in The Arnoldfest (Toronto, ON, 1997), 311–323, Fields Inst. Commun., **24**, AMS, Providence, RI, 1999.

195. Khesin, B.A., Rosly, A.A., *Polar homology and holomorphic bundles*, Phil. Trans. Roy. Soc. London Ser. A, **359** (2001), 1413-1427; *Polar homology*, Canad. J. Math. **55** (2003), 1100–1120.

196. Khesin, B.A., Rosly, A.A., *Polar linkings, intersections, and Weil pairing*, Proc. of Royal Soc. London A, **461** (2005), 3505–3524.

197. Khesin, B.A., Rosly, A.A., Thomas, R.P., *Polar de Rham theorem*, Topology **43** (2004), 1231–1246.

198. Khesin, B.A., Zakharevich, I., *The Lie-Poisson group of pseudodifferential symbols and fractional KP-KdV hierarchies*, C.R. Acad. Sci. **316** (1993), Serie I, 621–626; *Poisson-Lie group of pseudodifferential symbols*, Comm. Math. Phys. **171** (1995), 475–530.

199. Kirillov, A.A., *Local Lie algebras*, Uspehi Mat. Nauk **31:4** (1976), 57–76; **32:1** (1977), 268.

200. Kirillov A.A., *Elements of the Theory of Representations*, Springer-Verlag, Berlin, Heidelberg, New York, 1976.

201. Kirillov, A.A., *Unitary representations of the group of diffeomorphisms and of some of its subgroups*, Selecta Math. Soviet. **1:4** (1981), 351–372.

202. Kirillov A.A., *The orbits of the group of diffeomorphisms of the circle, and local Lie superalgebras*, Funct. Anal. Appl. **15:2** (1981), 135–136.

203. Kirillov, A.A., *Infinite-dimensional Lie groups: their orbits, invariants and representations. The geometry of moments*, in Twistor geometry and nonlinear systems (Primorsko, 1980), Lecture Notes in Math., **970**, Springer, Berlin, 1982, 101–123.

204. Kirillov A.A., *Kähler structure on the K-orbits of a group of diffeomorphisms of the circle*, Funct. Anal. Appl. **21:2** (1987), 122–125.

205. Kirillov A.A., *The orbit method, II: Infinite-dimensional Lie groups and Lie algebras*, Contemp. Math., **145** (1993), 33–63.

206. Kirillov A.A., *Merits and demerits of the orbit method*, Bull. Amer. Math. Soc. **36** (1999), 433–488.

207. Kirillov A.A., *Lectures on Infinite-Dimensional Groups*, CRM Montreal, 2001.

208. Kirillov, A.A., Yuriev, D.V., *Kähler geometry of the infinite-dimensional homogeneous space $M = \mathrm{Diff}_+(S^1)/\mathrm{Rot}(S^1)$*, Funct. Anal. Appl. **21:4** (1987), 284–294.

209. Kobayashi, S., *Differential Geometry of Complex Vector Bundles*, Publ. Math. Soc. Japan (Iwanami Shoten and Princeton Univ. Press), 1987.

210. Kohno, T., *Conformal Field Theory and Topology*, Transl. of Math. Monographs, **210**, AMS, Providence, RI, 2002.

211. Kontsevich, M., Vishik, S., *Geometry of determinants of elliptic operators*, in Functional analysis on the eve of the 21st century, Vol. 1, Progr. Math., **131**, Birkhauser Boston, MA, 1995, 173–197; *Determinants of elliptic pseudo-differential operators*, preprint MPI 94-30, 1994.

212. Korevaar, J., *Ludwig Bieberbach's conjecture and its proof by Louis de Branges*, Amer. Math. Monthly **93** (1986), no. 7, 505–514.

213. Kostant, B., *The solution to a generalized Toda lattice and representation theory*, Adv. in Math. **34:3** (1979), 195–338.

214. Kozlov, V.V. *General Theory of Vortices*, "R & C Dynamics" IV., Izhevsk, 1998. Dynamical Systems **X**, Encyclop. of Math. Sciences **67**, Springer-Verlag, 2003.

215. Kravchenko, O.S., Khesin, B.A., *A central extension of the algebra of pseudo-differential symbols*, Funct. Anal. Appl. **25:2** (1991), 152–154.

216. Krichever, I.M., *Methods of algebraic geometry in the theory of nonlinear equations*, Uspehi Mat. Nauk **32:6** (1977), 183–208.

217. Krichever, I.M., *Linear operators with self-consistent coefficients and rational reductions of KP hierarchy*, Physica D **87:1-4** (1995), 14–19; *General rational reductions of the Kadomtsev-Petviashvili hierarchy and their symmetries*, Funct. Anal. Appl. **29:2** (1995), 75–80.

218. Krichever, I.M., *Isomonodromy equations on algebraic curves, canonical transformations and Whitham equations*, Mosc. Math. J. **2:4** (2002), 717–752.

219. Krichever, I.M., Novikov, S.P., *Algebras of Virasoro type, Riemann surfaces and structures of the theory of solitons*, Funct. Anal. Appl., **21:2** (1987), 46–63.

220. Krichever, I.M., Novikov, S.P., *Virasoro-type algebras, Riemann surfaces and strings in Minkowski space*, Funct. Anal. Appl., **21**:4 (1987), 47–61.

221. Krichever, I.M., Phong, D.H., *Symplectic forms in the theory of solitons*, Surv. Differ. Geom., IV, Int. Press, Boston, MA, 1998, 239–313.

222. Kriegl, A., Michor, P.W., *The convenient setting of global analysis*, Mathematical Surveys and Monographs, **53**, AMS, Providence, RI, 1997.

223. Kuiper, N.H., *Locally projective spaces of dimension one*, Michigan Math. J. **2** (1954) 95–97.

224. Kuksin, S.B., *KAM-persistence of finite-gap solutions*, Dynamical systems and small divisors (Cetraro, 1998), 61–123, Lecture Notes in Math., **1784**, Springer, Berlin, 2002.

225. Kumar, S., *Kac-Moody Groups*, Progress in Mathematics, Birkhäuser, Boston 2003.

226. Kupershmidt, B.A., Wilson, G., *Modifying Lax equations and the second Hamiltonian structure*, Invent. Math. **62** (1981), 403–436.

227. Lalonde, F., McDuff, D., *The geometry of symplectic energy*, Ann. of Math. **141** (1995), 349–371

228. Lamb, G.L., *Elements of Soliton Theory*, J.Wiley & Sons, New York, 1980.

229. Landau, L.D., Lifschitz, E.M., *Fluid Mechanics*, Course of Theoretical Physics, Vol. 6, Pergamon Press, Addison-Wesley Publishing Co., Inc., Reading, MA, 1959.

230. Lang, S., *Elliptic Functions*, second edition, Graduate Texts in Mathematics **112**, Springer-Verlag, New York, 1987.

231. Lang, S., *Introduction to Differentiable Manifolds*, second edition, Springer-Verlag, New York, 2002.

232. Langer, J., Perline, R., *Poisson geometry of the filament equation*, J. Nonlinear Science, **1** (1991), 71–93.

233. Lazutkin, V.F., Pankratova, T.F., *Normal forms and versal deformations for Hill's equation*, Funct. Anal. Appl., **9**:4 (1975), 41–48.

234. Lebedev, D.R., Manin, Yu.I., *The Gelfand-Dikii Hamiltonian operator and the coadjoint representation of the Volterra group*, Funct. Anal. Appl. **13**:4 (1979), 40–46, 96.

235. Lee, P., *Nonholonomic mass transport*, Ph.D. thesis, Univ. of Toronto, in preparation.

236. Lempert, L., *The Virasoro group as a complex manifold*, Math. Res. Lett., **2** (1995), no. 4, 479–495.

237. Lenells, J., *The Hunter-Saxton equation describes the geodesic flow on a sphere*, J Geom. Phys., **57** (2007), no. 10, 2049–2064.

238. Lenells, J., *Weak geodesic flow and global solutions of the Hunter-Saxton equation*, Discrete Contin. Dyn. Syst. **18** (2007), no. 4, 643–656.

239. Le Potier, J., *Lectures on Vector Bundles*, Cambridge Studies in Advanced Mathematics, **54**, Cambridge University Press, Cambridge, 1997.

240. Looijenga, E., *Root systems and elliptic curves*, Invent. Math. **38** (1976), 17–32.

241. Losev, A., Moore, G., Nekrasov, N., Shatashvili, S., *Central extensions of gauge groups revisited*, Selecta Math. (N.S.) **4**:1 (1998), 117–123.

242. Lu, J.H., Weinstein, A., *Poisson Lie groups, dressing transformations, and Bruhat decompositions*, J. Diff. Geom. **31** (1990), 501–526.

243. Lukatsky, A.M., *On the curvature of the diffeomorphisms group*, Ann. Global Anal. Geom. **11** (1993), no. 2, 135–140.

244. Lukatsky, A.M., *On the geometry of current groups and a model of the Landau-Lifschitz equation*, Lie groups and Lie algebras, 425–433, Math. Appl., **433**, Kluwer Acad. Publ., Dordrecht, 1998.

245. Lukatsky, A.M., *On an application of a class of infinite-dimensional Lie groups to the dynamics of an incompressible fluid*, J. Appl. Math. Mech. **67** (2003), no. 5, 693–702; *A group approach to the dynamics of incompressible fluid*, Nauchny Vestnik MGTU GA, Mathematics and Physics, **114** (2007), 42–49.

246. Magri, F., *A simple model of the integrable Hamiltonian equation*, J. Math. Phys. **19** (1978), 1156–1162.

247. Maier, P., *Central extensions of topological current algebras*, Geometry and analysis on finite- and infinite-dimensional Lie groups (Bledlewo, 2000), 61–76, Banach Center Publ., **55**, Polish Acad. Sci., Warsaw, 2002.

248. Maier, P., Neeb, K.-H., *Central extensions of current groups*, Math. Ann. **326** (2003), 367–415.

249. Manakov, S., *A note on the integration of Euler's equation of the dynamics of an n-dimensional rigid body*, Funct. Anal. Appl. **10:4** (1976), 93–94.

250. Manin, Yu.I., *Algebraic aspects of nonlinear differential equations*, Current problems in mathematics, Vol. 11, pp. 5–152. Akad. Nauk SSSR, VINITI, Moscow, 1978.

251. Manin, Yu.I., *Frobenius Manifolds, Quantum Cohomology, and Moduli Spaces*, AMS Colloquium Publications, **47**, AMS, Providence, RI, 1999.

252. Marsden, J.E., Ratiu, T.S., *Introduction to Mechanics and Symmetry*, second edition, Springer-Verlag, New York, 1999.

253. Marsden, J.E., Ratiu, T.S., Weinstein, A., *Semidirect products and reduction in mechanics*, Trans. Amer. Math. Soc. **281** (1984), 147–177.

254. Marsden, J.E., Weinstein, A., *Reduction of symplectic manifolds with symmetry*, Rep. Math. Phys., **5** (1974), 121–130.

255. Marsden, J.E., Weinstein, A., *Coadjoint orbits, vortices, and Clebsch variables for incompressible fluids*, Phys. D **7** (1983), 305–323.

256. McCann, R., *Polar factorization of maps in Riemannian manifolds*, Geom. Funct. Anal., **11:3** (2001), 589–608.

257. McDuff, D., *Lectures on groups of symplectomorphisms*, Rend. Circ. Mat. Palermo **2** Suppl. No. 72 (2004), 43–78

258. McDuff, D., Salamon, D., *Introduction to Symplectic Topology*, second edition. Oxford Mathematical Monographs. The Clarendon Press, Oxford University Press, New York, 1998.

259. Meinrenken, E., *Symplectic Geometry*, Lecture notes, University of Toronto, 2000.

260. Meinrenken, E., Woodward, C., *Hamiltonian loop group actions and Verlinde factorization*, J. Differential Geom. **50** (1998), 417–469.

261. Meyer, K.R., *Symmetries and integrals in mechanics*, in Dynamical Systems, ed. M. Peixoto, Academic Press, New York (1973), 259–273.

262. Mickelsson J., *Two-cocycle of a Kac-Moody group*, Phys. Rev. Lett. **55** (1985), 2099–2101.

263. Mickelsson J., *Kac-Moody groups, topology of the Dirac determinant bundle and fermionization*, Comm. Math. Phys. **110** (1987), 173–183.

264. Mickelsson J., *Current Algebras and Groups*, Plenum Monographs in Nonlinear Physics. Plenum Press, New York, 1989.

265. Milnor, J., *Remarks on infinite-dimensional Lie groups*, in *Proc. Summer School on Quantum Gravity*, 1008–1057, B. De Witt ed., 1983.

266. Mishchenko, A.S. *Integrals of geodesic flows on Lie groups*, Funct. Anal. Appl. **4 3** (1970), 232–235.

267. Misiołek, G., *Stability of flows of ideal fluids and the geometry of the group of diffeomorphisms*, Indiana Univ. Math. J. **42** (1993), no. 1, 215–235.

268. Misiołek, G., *A shallow water equation as a geodesic flow on the Bott-Virasoro group*, J. Geom. Phys. **24** (1998), 203–208; *Classical solutions of the periodic Camassa-Holm equation*, Geom. Funct. Anal. **12:5** (2002), 1080–1104.

269. Moffatt, H.K. *The degree of knottedness of tangled vortex lines*, J. Fluid. Mech. **106** 1969, 117–129.

270. Mohrdieck, S., Wendt, R., *Integral conjugacy classes of compact Lie groups*, Manuscripta Math. **113:4** (2004), 531–547.

271. Mohrdieck, S., Wendt, R., *Conjugacy classes in Kac-Moody groups and principal G-bundles over elliptic curves*, Preprint.

272. Mokhov, O.I., *Symplectic and Poisson Geometry on Loop Spaces of Smooth Manifolds and Integrable Equations*, Reviews in Mathematics and Mathematical Physics **11, 2**, Harwood Academic Publishers, Amsterdam, 2001.

273. Montgomery, R., *A Tour of Subriemannian Geometries, Their Geodesics and Applications*, Mathematical Surveys and Monographs, **91**, AMS, Providence, RI, 2002.

274. Moody, R.V., *A new class of Lie algebras*, J. Alg. **10** (1968), 211–230.

275. Moody, R.V., *Euclidian Lie algebras*, Canad. J. Math. **21** (1969), 1432–1454.

276. Moody, R.V., Pianzola, A., *Lie Algebras with Triangular Decomposition*, Wiley Interscience, New York, 1995.

277. Moody, R.V., Rao, S.E., Yokonuma, T., *Toroidal Lie algebras and vertex representations*, Geometricae Dedicata **35** (1990), 238–307.

278. Moreau, J.-J., *Constantes d'un îlot tourbillonnaire en fluide parfait barotrope*, C R. Acad. Sci. Paris **252** (1961), 2810–2812.

279. Moser, J., *On the volume elements on a manifold*, Trans. Amer. Math. Soc. **120** (1965), 286–294.

280. Moser, J., *Three integrable Hamiltonian systems connected with isospectral deformations*, Adv. Math., **16**, (1975), 197.

281. Moser, J. *Geometry of quadrics and spectral theory*, in The Chern Symposium (Proc. Internat. Sympos., Berkeley, Calif., 1979), pp. 147–188, Springer, New York-Berlin, 1980.

282. Moser, J., Veselov, A.P., *Discrete versions of some classical integrable systems and factorization of matrix polynomials*, Comm. Math. Phys., **139:2** (1991), 217–243.

283. Mukai, S., *Symplectic structure of the moduli space of stable sheaves on an Abelian or K3 surface*, Invent. Math., **77** (1984), 101–116.

284. Nag, S., Sullivan, D., *Teichmüller theory and the universal period mapping via quantum calculus and the $H^{1/2}$ space on the circle*, Osaka J. Math. **32:1** (1995), 1–34.

285. Nag, S., Verjovsky, A., *Diff(S^1) and the Teichmüller spaces*, Comm. Math. Phys. **130** (1990), no. 1, 123–138.

286. Narasimhan, M.S. Seshadri, C.S., *Holomorphic vector bundles on a compact Riemann surface*, Math. Ann. **155** (1964) 69–80.

287. Neeb, K.-H., *Nancy lectures on infinite-dimensional Lie groups*, Preprint Nr. 2203, Technische Universität Darmstadt, 2002.

288. Neeb, K.-H., *Current groups for non-compact manifolds and their central extensions*, Infinite dimensional groups and manifolds, 109–183, IRMA Lect. Math. Theor. Phys. **5**, de Gruyter, Berlin, 2004.

289. Nekrasov, N., *Holomorphic bundles and many-body systems*, Comm. Math. Phys. **180** (1996), 587–603.

290. Nekrasov, N., *Infinite-dimensional algebras, many-body systems and gauge theories*, in Moscow Seminar in Mathematical Physics, 263–299, Amer. Math. Soc. Transl. Ser. 2, **191**, AMS, Providence, RI, 1999.

291. Neretin, Yu.A., *On a complex semigroup containing the group of diffeomorphisms of the circle*, Funct. Anal. Appl. **21:2** (1987), 82–83.

292. Neretin, Yu.A., *Representations of Virasoro and affine Lie algebras*, Representation theory and noncommutative harmonic analysis I, Encyclopaedia Math. Sci., **22**, 157–234, Springer, Berlin, 1994.

293. Neretin, Yu.A., *Categories of Symmetries and Infinite-Dimensional Groups*, London Mathematical Society Monographs. New Series, **16**, The Clarendon Press, Oxford University Press, New York, 1996.

294. Newell, A.C., *Solitons in Mathematics and Physics*, CBMS-NSF Regional Conference Series in Applied Mathematics, **48**, Society for Industrial and Applied Mathematics, Philadelphia, PA, 1985.

295. Novikov, S.P., *The Hamiltonian formalism and a multivalued analogue of Morse theory*, Russian Math. Surveys **37:5** (1982), 1–56.

296. Oda, T., *Convex Bodies and Algebraic Geometry. An introduction to the theory of toric varieties*, Ergebnisse der Mathematik und ihrer Grenzgebiete (3) **15**, Springer-Verlag, Berlin, 1988.

297. Okonek, C., Schneider, M., Spindler, H., *Vector Bundles on Complex Projective Spaces*, Progress in Mathematics, **3**, Birkhauser, Boston, Mass., 1980.

298. Olshanetsky, M.A., Perelomov, A.M., *Completely integrable Hamiltonian systems connected with semisimple Lie algebras*, Invent. Math., **37** (1976), 93–108.

299. Olshanetsky, M.A., Perelomov, A.M., *Integrable systems and Lie algebras*, Mathematical physics reviews **3**, 151–220, Soviet Sci. Rev. Sect. C: Math. Phys. Rev., 3, Harwood Academic, Chur, 1982.

300. Olshanetsky, M.A., Perelomov, A.M., Reyman, A.G., Semenov-Tian-Shansky, M.A., *Integrable Systems. II*, Current problems in mathematics. Fundamental directions, Vol. 16, 86–226, Itogi Nauki i Tekhniki, Akad. Nauk SSSR, VINITI, Moscow, 1987; English transl.: *Integrable Systems. Nonholonomic Dynamical Systems*, Encyclopaedia of Mathematical Sciences **16** (Dynamical Systems VII) Springer-Verlag 1994.

301. Omori, H., *Infinite-Dimensional Lie Groups*, Transl. of Math. Monographs **158**, AMS, Providence, RI, 1997.

302. O'Neill, B., *Submersions and geodesics*, Duke Math. J. **34** (1967), 363–373.

303. Otto, F., *The geometry of dissipative evolution equations: the porous medium equation*, Comm. in PDE **26** (2001), no. 1-2, 101–174.

304. Ovsienko, V.Yu. *On the Denogardus great number and Hooke's law*, Kvant (1989), **8**, 8–16. Kvant selecta: algebra and analysis, II, 153–159, Math. World, **15**, AMS, Providence, RI, 1999.

305. Ovsienko, V.Yu., Khesin, B.A., *The (super) KdV equation as an Euler equation*, Funct. Anal. Appl. **21** (1987), 81–82.

306. Ovsienko, V.Yu., Khesin, B.A., *Symplectic leaves of the Gelfand-Dikii brackets and homotopy classes of non-degenerate curves*, Funct. Anal. Appl. **24:1** (1990), 33–40.

307. Ovsienko, V.Yu., Khesin, B.A., Chekanov, Yu.V., *Integrals of the Euler equat-ons in multidimensional hydrodynamics and superconductivity*, Diff. Geom., Lie Groups, and Mechanics, Zap. Sem. LOMI **172** (1988), 105–113; Engl. transl: J. Soviet Math. **59** (1992), 1096–1101.

308. Palais, R.S., *The symmetries of solitons*, Bull. Amer. Math. Soc. (N.S.) **34:4** (1997), 339–403.

309. Parshin, A.N., *Residues and duality on algebraic surfaces*, Uspehi Mat. Nauk **32:2** (1977), 225–226.

310. Parshin, A.N., *On a ring of formal pseudo-differential operators*, Proc. Steklov Inst. Math. 1999, **1** (224), 266–280.

311. Perelomov, A.M., *Integrable Systems of Classical Mechanics and Lie Algebras*, Birkhäuser Verlag, Basel, 1990.

312. Pilch, K., Warner, N.P., *Holomorphic structure of superstring vacua*, Classical Quantum Gravity **4** (1987), 1183–1192.

313. Pirozerski, A.L., Semenov-Tian-Shansky, M.A., *Generalized q-deformed Gelfand-Dickey structures on the group of q-pseudodifference operators*, L.D. Faddeev's Seminar on Mathematical Physics, 211–238, AMS Transl. Ser. 2, **201**, AMS, Providence, RI, 2000.

314. Pirozerski, A.L., Semenov-Tian-Shansky, M.A., *Q-pseudodifference universal Drinfeld-Sokolov reduction*, Proceedings of the St. Petersburg Mathematical Society, Vol. VII, 169–199, AMS Transl. Ser. 2, **203**, AMS, Providence, RI, 2001

315. Pollmann, U., *Realisation der biaffinen Wurzelsysteme von Saito in Lie-Algebren*, Hamburger Beiträge zur Mathematik, 1995.

316. Polterovich, L., *Hofer's diameter and Lagrangian intersections*, Internat. Math. Res. Notices **4** (1998), 217–223.

317. Polterovich, L., *The Geometry of the Group of Symplectic Diffeomorphisms*, Lectures in Mathematics ETH Zürich. Birkhäuser Verlag, Basel, 2001.

318. Polyakov, A.M., *Fermi-Bose transmutations induced by gauge fields*, Modern Phys. Lett. A **3:3** (1988), 325–328.

319. Polyakov, A.M., Wiegmann, B.P., *Goldstone fields in two dimensions with mul-tivalued actions*, Phys. Lett. B, **141** (1984), 223–228.

320. Pope, C.N., Stelle, K.S., $SU(\infty)$, $SU_+(\infty)$ *and area-preserving algebras*, Phys. Lett. B **226** (1989), no. 3-4, 257–263.

321. Praught, J., Smirnov, R.G., *Andrew Lenard: a mystery unraveled*, SIGMA (Symmetry Integrability Geom. Methods Appl.) **1** (2005), Paper 005.

322. Pressley, A., Segal, G., *Loop Groups*, Oxford University Press, Oxford, 1986.

323. Preston, S.C., *On the volumorphism group, the first conjugate point is always the hardest*, Comm. Math. Phys. **267** (2006), no. 2, 493–513.

324. Radul, A.O., *Lie algebras of differential operators, their central extensions and W-algebras*, Funct. Anal. Appl. **25:1** (1991), 25–39.

325. Ramanathan, A., *Stable principal bundles on a compact Riemann surface*, Math. Ann. **213** (1975), 129–152.

326. Ratiu, T., Todorov, A., *An infinite-dimensional point of view on the Weil-Petersson metric*, in Several complex variables and complex geometry (Santa Cruz, CA, 1989), 467–476, Proc. Sympos. Pure Math., **52**, Part 2, AMS Providence, RI, 1991.

327. Reshetikhin, N., Turaev, V.G., *Invariants of 3-manifolds via link polynomials and quantum groups*, Invent. Math. **103:3** (1991), 547–597.

328. Reyman, A.G., Semenov-Tian-Shansky, M.A., *Reduction of Hamiltonian systems, affine Lie algebras and Lax equations*, Invent. Math. **54:1** (1979), 81–100; Invent. Math. **63:3** (1981), 423–432.

329. Reyman, A.G., Semenov-Tian-Shansky, M.A., *Current algebras and nonlinear partial differential equations*, Sov. Math. Doklady **251:6** (1980), 1310-1314.

330. Reyman, A.G., Semenov-Tian-Shansky, M.A., *Integrable Systems: Group-Theoretical Approach*, Moscow-Izhevsk, 2003.

331. Ricca, R.L., *The contributions of Da Rios and Levi-Civita to asymptotic potential theory and vortex filament dynamics*, Fluid Dynam. Res. **18:5** (1996), 245–268.

332. Ruijsenaars, S.N.M., *Complete integrability of relativistic Calogero-Moser systems and elliptic function identities*, Comm. Math. Phys. **110** (1987), 191–213.

333. Ruijsenaars, S.N.M., Schneider, H., *A new class of integrable systems and its relation to solitons*, Ann. Physics **170** (1986), 370–405.

334. Saito, K., Yoshii, D., *Extended affine root system. IV. Simply-laced elliptic Lie algebras*, Publ. Res. Inst. Math. Sci. **36** (2000), 385–421.

335. Saveliev, M.V., Vershik, A.M., *Continuum analogues of contragredient Lie algebras (Lie algebras with a Cartan operator and nonlinear dynamical systems)*, Comm. Math. Phys. **126** (1989), no. 2, 367–378.

336. Schlichenmaier, M., *Introduction to Riemann Surfaces, Algebraic Curves and Moduli Spaces*, Lecture notes in Physics, **322**, Springer-Verlag, Berlin, Heidelberg, New York, 1990.

337. Schlichenmaier, M., *Local cocycles and central extensions for multi-point algebras of Krichever-Novikov type*, J. Reine Angew. Math. **559** (2003), 53–94.

338. Schonbek, M.E., Todorov, A.N., Zubelli, J.P., *Geodesic flows on diffeomorphisms of the circle, Grassmannians, and the geometry of the periodic KdV equation*, Adv. Theor. Math. Phys. **3:4** (1999), 1027–1092.

339. Schottky, F., *Über das analytische Problem der Rotation eines starren Körpers in Raume von vier Dimensionen*, Sitzungsber. Königl. Preuss. Akad. Wiss. Berlin, **13** (1891), 227–232.

340. Schwarz, A., *The partition function of degenerate quadratic functional and Ray-Singer invariants*, Lett. Math. Phys. **2:3** (1977/78), 247–252.

341. Schwarz, A., *Symplectic formalism in conformal field theory*, Symétries quantiques (Les Houches, 1995), 957–977, North-Holland, Amsterdam, 1998.

342. Segal, G., *Unitary representations of some infinite-dimensional groups*, Comm. Math. Phys. **80** (1981), 301–342.

343. Segal, G., *The definition of conformal field theory*, in Diff. Geom. Methods in Theor. Physics (Como, 1987), 165–171, NATO Adv. Sci. Inst. Ser. C Math. Phys. Sci., **250**, Kluwer Acad. Publ., Dordrecht, 1988.

344. Segal, G., *The geometry of the KdV equation*, in Topological methods in quantum field theory (Trieste, 1990) Internat. J. Modern Phys. A **6:16** (1991), 2859–2869.

345. Segal, G., *Lie groups*, in Carter, R., Segal, G., Macdonald, I., Lectures on Lie Groups and Lie Algebras. London Math. Soc. Student Texts, **32**, Cambridge Univ. Press, Cambridge, 1995.

346. Segal, G., *Integrable systems and inverse scattering*, Integrable systems (Oxford, 1997), 53–119, Oxf. Grad. Texts Math., **4**, Oxford Univ. Press, New York, 1999.

347. Segal, G., Wilson, G., *Loop groups and equations of KdV type*, Inst. Hautes Etudes Sci. Publ. Math. **61** (1985), 5–65.

348. Semenov-Tian-Shansky, M.A., *What a classical r-matrix is*, Functional Anal. Appl. **17:4** (1983), 259–272.
349. Semenov-Tian-Shansky, M.A., *Dressing transformations and Poisson group actions*, Publ. RIMS **21** (1985), 1237–1260.
350. Sergeev, A.G., *Kähler Geometry of Loop Spaces*, MCCME Moscow, 2001.
351. Serre, D., *Les invariants du premier ordre de l'équation d'Euler en dimension trois*, C. R. Acad. Sci. Paris Sr. A-B **289** (1979), no. 4, A267–A270; Physica D **13** (1984), no. 1-2, 105–136.
352. Serre, D., *Invariants et dégénérescence symplectique de l'équation d'Euler des fluides parfaits incompressibles*, C. R. Acad. Sci. Paris Sér. I Math. **298** (1984), 349–352.
353. Sheinman, O.K., *Affine Lie algebras on Riemann surfaces*, Funct. Anal. Appl. **27:4** (1993), 266–272.
354. Sheinman, O.K., *Affine Krichever-Novikov algebras, their representations and applications*, in Geometry, topology, and mathematical physics, 297–316, Amer. Math. Soc. Transl. Ser. 2, **212**, AMS, Providence, RI, 2004.
355. Shnirelman, A.I., *The geometry of the group of diffeomorphisms and the dynamics of an ideal incompressible fluid*, Mat. Sb. (N.S.) **128:1** (1985), 82–109. English transl.: Math. USSR-Sb. **56** (1987), 79–105.
356. Shnirelman, A.I., *Attainable diffeomorphisms*, Geom. Funct. Anal. **3:3** (1993), 279–294.
357. Shnirelman, A.I., *Generalized fluid flows, their approximation and applications*, Geom. Funct. Anal. **4:5** (1994), 586–620.
358. Shnirelman, A., *On the nonuniqueness of weak solutions of the Euler equation*, Comm. Pure Appl. Math. **50** (1997), no. 12, 1261–1286; *Weak solutions with decreasing energy of incompressible Euler equations*, Comm. Math. Phys. **210** (2000), no. 3, 541–603.
359. Skryabin, S., *Degree one cohomology for the Lie algebras of derivations*, Lobachevskii J. Math. **14** (2004), 69–107.
360. Spiegel, E.A., *Fluid dynamical form of the linear and nonlinear Schrödinger equations*, Physica D **1** (1980), no. 2, 236–240.
361. Sutherland, B., *Exact results for a quantum many-body system in one dimension. II*, Phys. Rev., **A5**, (1972), 1372–1376.
362. Symes, W.W., *Systems of Toda type, inverse spectral problems, and representation theory*, Invent. Math. **59:1** (1980), 13–51; Invent. Math. **63:3** (1981), 519.
363. Takhtajan, L.A., Teo, L.-P., *Weil-Petersson metric on the universal Teichmüller space*, Mem. Amer. Math. Soc. **183** (2006), no. 861.
364. Taubes, C., *Casson's invariant and gauge theory*, J. Diff. Geom. **31** (1990), 547–599.
365. Terng, C.-L., Uhlenbeck, K., *Poisson actions and scattering theory for integrable systems*, Surveys in differential geometry: integrable systems, Surv. Differ. Geom., **IV**, 315–402, Int. Press, Boston, MA, 1998.
366. Thomas, R.P., *Gauge theory on Calabi-Yau manifolds*, Ph.D. thesis, Oxford (1997).
367. Thomas, R.P., *A holomorphic Casson invariant for Calabi-Yau 3-folds, and bundles on K3 fibrations*, J. Diff. Geom. **54** (2000), 367–438.
368. Thurston, W., *Foliations and groups of diffeomorphisms*, Bull. Amer. Math. Soc. **80** (1974), 304–307.

369. Tsikh, A.K., *Multidimensional Residues and Their Applications*, Translations of Math. Monographs, **103**, AMS, Providence, RI, 1992.

370. Turski, L.A., *Hydrodynamical description of the continuous Heisenberg chain*, Canad. J. Phys. **59:4** (1981), 511–514.

371. Tyurin, A.N., *Symplectic structures on the moduli spaces of vector bundles on algebraic surfaces with $p_g > 0$*, Izv. Akad. Nauk SSSR Ser. Mat. **52:4** (1988), 813–852, 896; English transl.: Math. USSR-Izv. **33:1** (1989), 139–177.

372. Verdier, J.-L. *Les représentations des algèbres de Lie affines: applications à quelques problèmes de physique (d'après E. Date, M. Jimbo, M. Kashiwara, T. Miwa)*, Bourbaki Seminar, Vol. 1981/1982, pp. 365–377, Astérisque, **92-93**, Soc. Math. France, Paris, 1982.

373. Veselov, A.P., *Integrable systems with discrete time, and difference operators*, Funct. Anal. Appl. **22:2** (1988), 83–93.

374. Villani, C., *Topics in Optimal Transportation*, Grad. Studies in Math., **58**, AMS, Providence, RI, 2003.

375. Villani, C., *Optimal Transport, Old and New*, Grundlehren der Mathem. Wissenschaften, Springer, (2008)

376. Virasoro, M.A., *Subsidiary conditions and ghosts in dual resonance models*, Phys. Rev. D **1**, (1970), 1933–1936.

377. Vishik, S.M., Dolzhanskii, F.V., *Analogs of the Euler-Lagrange equations and magnetohydrodynamics equations related to Lie groups*, Sov. Math. Doklady **19** (1978), 149–153.

378. Vizman, C., *Central extensions of semidirect products and geodesic equations*, Phys. Lett. A **330:6** (2004), 460–469.

379. Wagemann, F., *Some remarks on the cohomology of Krichever-Novikov algebras*, Lett. Math. Phys. **47** (1999), 173–177; Lett. Math. Phys. **52** (2000), 349.

380. Wagemann, F., *A two-dimensional analogue of the Virasoro algebra*, J. Geom. Phys. **36** (2000), no. 1-2, 103–116.

381. Wagemann, F., *Density of holomorphic vector fields in meromorphic ones*, preprint, 2001.

382. Wakimoto, M., *Lectures on Infinite-Dimensional Lie Algebra*, World Scientific Publishing Co., Inc., River Edge, NJ, 2001.

383. Warner, F.W., *Foundations of Differentiable Manifolds and Lie Groups*, Graduate Texts in Mathematics, **94**, Springer-Verlag, New York, Berlin, 1983.

384. Weinstein, A., *The local structure of Poisson manifolds*, J. Diff. Geom. **18** (1983), 523–557.

385. Wendt, R., *Weyl's character formula for non-connected Lie groups and orbital theory for twisted affine Lie algebras* J. Funct. Anal. **180** (2001), 31–65.

386. Wendt, R., *A character formula for certain representations of loop groups based on non-simply connected Lie groups*, Math. Z. **247** (2004), 549–580.

387. Witten, E., *Non-abelian bozonization in two dimensions*, Comm. Math. Phys. **92** (1984), 455-472.

388. Witten, E., *Coadjoint orbits of the Virasoro group*, Comm. Math. Phys. **114** (1988), 1–53.

389. Witten, E., *Quantum field theory and the Jones polynomial*, Comm. Math. Phys. **121** (1989), 351–399. The Floer memorial volume, Progr. Math. **133** (1995), Birkhäuser, Basel, 637–678.

390. Witten, E., *Chern-Simons gauge theory as a string theory*, in The Floer memorial volume, Progr. Math. **133** (1995), Birkhäuser, Basel, 637–678.

391. Witten, E., *Perturbative gauge theory as a string theory in twistor space*, Comm. Math. Phys. **252:1-3** (2004), 189–258.

392. Zaharov, V.E., Faddeev, L.D., *The Korteweg-de Vries equation is a completely integrable Hamiltonian system*, Funkt. Anal. i Pril. **5:4** (1971), 18–27.

393. Zuckerman, G.J., *Action principles and global geometry*, Mathematical aspects of string theory (San Diego, Calif., 1986), 259–284, Adv. Ser. Math. Phys., **1**, World Sci. Publishing, Singapore, 1987.

Index